STUDENT'S SOLUTIONS MANUAL

SOLUTIONS PREPARED BY TWIN PRIME EDITORIAL

A SURVEY OF MATHEMATICS WITH APPLICATIONS

EIGHTH EDITION

AND

EXPANDED EIGHTH EDITION

Allen R. Angel
Monroe Community College

Christine D. Abbott
Monroe Community College

Dennis C. Runde
Manatee Community College, Bradenton

PEARSON

Addison
Wesley

Boston San Francisco New York
London Toronto Sydney Tokyo Singapore Madrid
Mexico City Munich Paris Cape Town Hong Kong Montreal

Reproduced by Pearson Addison-Wesley from electronic files supplied by the author.

Copyright © 2009 Pearson Education, Inc.
Publishing as Pearson Addison-Wesley, 75 Arlington Street, Boston, MA 02116.

ISBN-13: 978-0-321-51089-1
ISBN-10: 0-321-51089-5

3 4 5 6 BB 10 09 08

PEARSON

Addison
Wesley

Table of Contents

CHAPTER ONE

CRITICAL THINKING SKILLS

Exercise Set 1.1

1. a) 1, 2, 3, 4, 5, …
    b) Counting numbers

3. **Inductive reasoning** is the process of reasoning to a general conclusion through observations of specific cases.

5. A **counterexample** is a specific case that satisfies the conditions of the conjecture but shows the conjecture is false.

7. Inductive reasoning

9. Inductive reasoning, because a general conclusion was made from observation of specific cases.

11. $5 \times 7 = 35$

13. $1 \ 5(=1+4) \ 10(=4+6) \ 10(=6+4) \ 5(=4+1) \ 1$

15.

17.

19. 25, 30, 35 (Add 5 to previous number.)

21. −1, 1, −1 (Alternate −1 and 1.)

23. $\dfrac{1}{16}, \dfrac{1}{64}, \dfrac{1}{256}$ (Multiply previous number by $\dfrac{1}{4}$.)

25. 36, 49, 64 (The numbers in the sequence are the squares of the counting numbers.)

27. 34, 55, 89 (Each number in the sequence is the sum of the previous two numbers.)

29. There are three letters in the pattern.
     $39 \times 3 = 117$, so the 117th entry is the second R in the pattern. Therefore, the 118th entry is Y.

31. a) 36, 49, 64

     b) Square the numbers 6, 7, 8, 9 and 10.

31. c) $8 \times 8 = 64 \qquad 9 \times 9 = 81$

     72 is not a square number since it falls between the two square numbers 64 and 81.

33. Blue: 1, 5, 7, 10, 12   Purple: 2, 4, 6, 9, 11   Yellow: 3, 8

35. a) $\approx \$6$ billion   b) We are using observation of specific cases to make a prediction.

37.

P	B	P	B
B	P	B	P
P	B	P	B
B	P	B	P

39. a) You should obtain the original number.
     b) You should obtain the original number.
     c) Conjecture: The result is always the original number.
     d)

$$n, 8n, 8n+16, \frac{8n+16}{8} = \frac{8n}{8} + \frac{16}{8} = n+2,$$

$$n+2-2 = n$$

41. a) You should obtain the number 5.

    b) You should obtain the number 5.

    c) Conjecture: No matter what number is chosen, the result is always the number 5.

    d) $n, n+1, n+(n+1) = 2n+1, 2n+1+9 = 2n+10, \dfrac{2n+10}{2} = \dfrac{2n}{2} + \dfrac{10}{2} = n+5, n+5-n = 5$

43. $3+5 = 8$ is one counterexample.

45. Two is a counting number. The sum of 2 and 3 is 5. Five divided by two is $\dfrac{5}{2}$, which is not an even number.

47. One and two are counting numbers. The difference of 1 and 2 is $1-2 = -1$, which is not a counting number.

49. a) The sum of the measures of the interior angles should be $180°$.

    b) Yes, the sum of the measures of the interior angles should be $180°$.

    c) Conjecture: The sum of the measures of the interior angles of a triangle is $180°$.

51. 129, the numbers in positions are found as follows:
$$\begin{array}{cc} a & b \\ c & a+b+c \end{array}$$

53. c

Exercise Set 1.2

(Note: Answers in this section will vary depending on how you round your numbers. The answers may differ from the answers in the back of the textbook. However, your answers should be something near the answers given. All answers are approximate.)

1. $523 + 47.8 + 821.6 + 733 + 92.7 \approx 520 + 50 + 820 + 730 + 90 = 2210$

3. $297,700 \times 4087 \approx 300,000 \times 4000 = 1,200,000,000$

5. $\dfrac{405}{0.049} \approx \dfrac{400}{0.05} = 8000$

7. $0.63 \times 1523 \approx 0.6 \times 1500 = 900$

9. 9% of $2164 \approx 10\%$ of $2000 = 0.10 \times 2000 = 200$

11. $592 \times 2070 \times 992.62$
    $\approx 600 \times 2000 \times 1000 = 1,200,000,000$

13. 27 hours $\times$ \$8.25 per hour
    ≈ 30 hours $\times$ \$8 per hour $= \$240$

15. $\$4.23 + \$2.79 + \$0.79 + \$7.62 + \$12.38 + \$4.99$
    $\approx 4.20 + 2.80 + 0.80 + 7.60 + 12.40 + 5.00 = \$32.80$

17. 57 miles per hour for 3.2 hours
    ≈ 60 miles per hour for 3 hours
    $= 60 \times 3 = 180$ miles

19. $27,453 - 14,292 \approx 27,500 - 14,300 = 13,200$ lb

21. 15% of $\$26.32 \approx 15\%$ of $\$26 = 0.15 \times \$26 = \$3.90$

23. $\dfrac{\$400}{\$23} \approx \dfrac{\$400}{\$25} = 16$

25.

    $(\$29.99 + \$36.99 + \$59.99 + \$49.00) - \$139.99$
    $\approx (\$30 + \$37 + \$60 + \$50) - \$140 = \$177 - \$140$
    $= \$37$

27. $11 \times 8 \times \$1.50 \approx 10 \times 8 \times \$1.50$
    $= 10 \times \$12 = \$120$

29. 100 Mexican pesos $= 100 \times 0.089$ U.S. dollars
    $\approx 100 \times 0.09$ U.S. dollars $= 9$ U.S. dollars
    $\$50 - \$9 = \$41$

31. ≈ 20 miles

33. a) 100

    b) 50

    c) 125

35. a) 5 million

b) 98 million

c) 98 million − 33 million = 65 million

d) 19 million + 79 million + 84 million +
  65 million + 33 million = 280 million

39. 25

43. 150°

47. 9 square units

51.-59. Answers will vary.

37. a) 85%

b) 68% − 53% = 15%

c) 85% of 70 million acres = 59,500,000 acres

d) No, since we are not given the area of each
  state.

41. ≈ 160 bananas

45. 10%

49. 150 feet

61. There are 336 dimples on a regulation golf ball.

Exercise Set 1.3

1. $\dfrac{1 \text{ in.}}{18 \text{ mi}} = \dfrac{4.25 \text{ in.}}{x \text{ mi}}$

$1x = 18(4.25)$

$x = 76.5 \text{ mi}$

3. $\dfrac{3 \text{ ft}}{1.2 \text{ ft}} = \dfrac{48.4 \text{ ft}}{x \text{ ft}}$

$3x = 1.2(48.4)$

$\dfrac{3x}{3} = \dfrac{58.08}{3}$

$x = \dfrac{58.08}{3} = 19.36 \text{ ft}$

5. 5.7% of $27,461 = 0.057 \times \$27,461 = \$1565$
  $\$27,461 + \$1565 = \$29,026$

7. a) 31.2% of $150,000 = 0.312 \times \$150,000$
    $= \$46,800$
  $\$150,000 + \$46,800 = \$196,800$

b) CA: 25.4% of $200,000 = 0.254 \times \$200,000$
    $= \$50,800$

NH: 12.1% of $200,000 = 0.121 \times \$200,000$
    $= \$24,200$

$\$50,800 - \$24,200 = \$26,600$

c) 22.4% of $180,000 = 0.224 \times \$180,000$
    $= \$40,320$
  $\$180,000 + \$40,320 = \$220,320$

9. Denise parks her car for eight hours per day.
  $5[\$2.50 + \$1.00(7 \text{ hours per day})]$
  $= 5[\$2.50 + \$7.00] = 5(\$9.50) = \$47.50$
  Savings: $\$47.50 - \$35.00 = \$12.50$

11. $\$120 + \$80(15) = \$120 + \$1200 = \$1320$
  Savings: $\$1320 - \$1250 = \$70$

13. 15 year mortgage: $\$840.62(12)(15) = \$151,311.60$
  30 year mortgage: $\$620.28(12)(30) = \$223,300.80$
  Savings: $\$223,300.80 - \$151,311.60 = \$71,989.20$

15. a) $\dfrac{86.5}{34} \approx 2.54; \dfrac{91.5}{36} \approx 2.54; \dfrac{96.5}{38} \approx 2.54;$

$\dfrac{101.5}{40} \approx 2.54; \dfrac{106.5}{42} \approx 2.54 \ldots$

So, $48(2.54) \approx 122.$

b) Answers will vary. A close approximation can be
obtained by multiplying the U.S. sizes by 2.54.

17. a) $\dfrac{460}{50} = 9.2 \text{ min}$

b) $\dfrac{1550}{25} = 62 \text{ min}$

c) $\dfrac{1400}{35} = 40 \text{ min}$

d) $\dfrac{1550}{80} + \dfrac{2200}{80} = \dfrac{3750}{80} \approx 47 \text{ min}$

19. a) $1.5 trillion

    b) $\dfrac{\$2.2 \text{ trillion}}{108,819,000}$

      $\approx \dfrac{\$2,200,000,000,000}{100,000,000} = \$22,000$

23. a) $\$620(0.12) = \$74.40$

    b) $\$1200(0.22) = \$264$

    c) The store lost $\$1200 - \$1000 = \$200$ on the purchase.

     Store's profit: $\$264 - \$200 = \$64$

27. a) Short: $\$30 \times 5 = \$150$

      Long: $\$15 \times 5 = \$75$

      $\$150 - \$75 = \$75$; Jeff saves $75.

    b) $6 for first hour, plus $3 \times \$3$ for remaining 3 hours, for a total of $15.

    c) Short: $\$6 + 4 \times \$3 = \$18$

      Long: $15

      Long term is cheaper by $3.

31. Value after first year: $\$1000 + 0.10(\$1000)$

    $= \$1000 + \$100 = \$1100$

    Value after second year: $\$1100 - 0.10(\$1100)$

    $= \$1100 - \$110 = \$990$

    $990 is less than the intial investment of $1000.

21. By mail: $(\$52.80 + \$5.60 + \$8.56) \times 4$

      $= \$66.96 \times 4 = \$267.84$

     Tire store: $\$324 + 0.08 \times \$324$

      $= \$324 + \$25.92 = \$349.92$

     Savings: $\$349.92 - \$267.84 = \$82.08$

25. a) $0.1 \text{ cm}^3 \times 60 \text{ sec} \times 60 \text{ min} \times 24 \text{ hr} \times 365 \text{ days}$

      $= 3,153,600 \text{ cm}^3$

    b) $30 \text{ cm} \times 20 \text{ cm} \times 20 \text{ cm} = 12,000 \text{ cm}^3$

     $0.1 \text{ cm}^3 \times 60 \text{ sec} \times 60 \text{ min} \times 24 \text{ hr} = 8640$

     $\dfrac{12,000}{8640} = 1.38\overline{8} \approx 1.4 \text{ days}$

29. a) Yes, divide the total emissions by the emissions per capita.

    b) $\dfrac{5912.2}{19.8} \approx 298.596 \text{ million} \approx 298.6 \text{ million}$

    c) $\dfrac{4707.3}{3.6} \approx 1307.583 \text{ million} \approx 1.3076 \text{ billion}$

33. a) $\dfrac{\$200}{\$41} \approx 4.87804878$ The maximum number of 10 packs is 4.

   $\$200 - (4 \times \$41) = \$200 - \$164 = \$36$, $\dfrac{\$36}{\$17} = 2.117647059$ Deirdre can also buy two 4 packs.

10 packs	4 packs	Number of rolls	Cost
4	2	$4(10) + 2(4) = 48$	$4(\$41) + 2(\$17) = \$198$
3	4	46	$191
2	6	44	$184
1	9	46	$194
0	11	44	$187

Maximum number of rolls of film is 48.

b) The cost is $198 when she purchases four 10 packs and two 4 packs.

35. a) water/milk: $3(1) = 3$ cups   salt: $3\left(\frac{1}{8}\right) = \frac{3}{8}$ tsp

Cream of wheat: $3(3) = 9$ tbsp $= \frac{9}{16}$ cup (because 16 tbsp = 1 cup)

b) water/milk: $\frac{2+3.75}{2} = \frac{5.75}{2} = 2.875$ cups $= 2\frac{7}{8}$ cups

salt: $\frac{0.25+0.5}{2} = \frac{0.75}{2} = 0.375$ tsp $= \frac{3}{8}$ tsp

cream of wheat: $\frac{0.5+0.75}{2} = \frac{1.25}{2} = 0.625$ cups $= \frac{5}{8}$ cup $= \frac{5}{8}(16 \text{ tbsp}) = 10$ tbsp

c) water/milk: $3\frac{3}{4} - 1 = \frac{15}{4} - \frac{4}{4} = \frac{11}{4} = 2\frac{3}{4}$ cups

salt: $\frac{1}{2} - \frac{1}{8} = \frac{4}{8} - \frac{1}{8} = \frac{3}{8}$ tsp   cream of wheat: $\frac{3}{4} - \frac{3}{16} = \frac{12}{16} - \frac{3}{16} = \frac{9}{16}$ cup = 9 tbsp

d) Differences exist in water/milk because the amount for 4 servings is not twice that for 2 servings. Differences also exist in Cream of Wheat because $\frac{1}{2}$ cup is not twice 3 tbsp.

37. 1 ft^2 would be 12 in. by 12 in.

Thus, 1 ft^2 = 12 in. $\times$ 12 in. = 144 in.^{2}

39. Area of original rectangle = lw

Area of new rectangle = $(2l)(2w) = 4lw$

Thus, if the length and width of a rectangle are doubled, the area is 4 times as large.

41. 11 ft is one-sixth of the pole, so the length is 6×11 ft = 66 ft.

43. Left side: $1(-6) = -6$ Right side: $1(2) = 2$

$2(-2) = -4$ $1(3) = 3$

$-6 + -4 = -10$ $1(6) = 6$

      $2 + 3 + 6 = 11$

Place it at -1 so the left side would total $-10 + -1 = -11$.

45. $30,000 is the difference between one-fourth of the cost and one-fifth of the cost.

$\frac{1}{4} - \frac{1}{5} = \frac{1}{20}$; $20 \times \$3000 = \$60,000.$

The yacht costs $60,000.

47. a) $(4 \times 4) + (3 \times 3) + (2 \times 2) + (1 \times 1)$

$= 16 + 9 + 4 + 1 = 30$

b) $(7 \times 7) + (6 \times 6) + (5 \times 5) + 30$

$= 49 + 36 + 25 + 30 = 140$

49.

51.

8	6	16
18	10	2
4	14	12

53. $6+10+8+4=28; 3+7+5+1=16;$

$10+14+12+8=44$

The sum of the four corner entries is

4 times the number in the center of the middle row.

57. $3\times2\times1=6$ ways

55. $63, 36, 99$

Multiply the number in the center of the middle row by 9.

59.

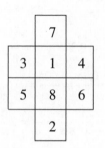

Other answers are possible, but 1 and 8 must appear in the center.

61.

1	2	3	4	5
2	3	4	5	1
3	4	5	1	2
4	5	1	2	3
5	1	2	3	4

Other answers are possible.

63. Mary is the skier.

65. Areas of the colored regions are:

1×1, 1×1, 2×2, 3×3, 5×5, 8×8, 13×13,

21×21; $1+1+4+9+25+64+169+441$

$=714$ square units

67. Let x be the amount Samantha had to start.

After first store: $x - \frac{1}{2}x - 20 = \frac{1}{2}x - 20$

After second store: $\frac{1}{2}\left(\frac{1}{2}x - 20\right) - 20 = \frac{1}{4}x - 30$

This is equal to $0, so the original amount was $120.

Review Exercises

1. 31, 36, 41 (Add 5 to previous number.)

3. -48, 96, -192 (Multiply previous number by -2.)

5. 10, 4, -3 (subtract 1, then 2, then 3, ...)

2. 25, 36, 49 (next three perfect squares)

4. 25, 32, 40 (19 + 6 = 25, 25 + 7 = 32, 32 + 8 = 40)

6. $\frac{3}{8}$, $\frac{3}{16}$, $\frac{3}{32}$ (Multiply previous number by $\frac{1}{2}$.)

7.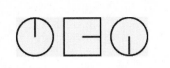

8.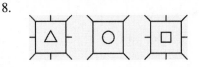

9. c

10. a) The final number is twice the original number.

b) The final number is twice the original number.

c) Conjecture: The final number is twice the original number.

d) $n, 10n, 10n + 5, \dfrac{10n + 5}{5} = \dfrac{10n}{5} + \dfrac{5}{5} = 2n + 1, 2n + 1 - 1 = 2n$

11. This process will always result in an answer of 3. $n, n + 5, 6(n + 5) = 6n + 30, 6n + 30 - 12$

$= 6n + 18, \dfrac{6n + 18}{2} = \dfrac{6n}{2} + \dfrac{18}{2} = 3n + 9, \dfrac{3n + 9}{3} = \dfrac{3n}{3} + \dfrac{9}{3} = n + 3, n + 3 - n = 3$

12. $1^2 + 2^2 = 5, 5$ is an odd number.

(Note: Answers for Ex. 13 - 25 will vary depending on how you round your numbers. The answers may differ from the answers in the back of the textbook. However, your answers should be something near the answers given. All answers are approximate.)

13. $210,302 \times 1992 \approx 210,000 \times 2000 = 420,000,000$

14. $215.9 + 128.752 + 3.6 + 861 + 729$

$\approx 250 + 150 + 0 + 850 + 750 = 2000$

15. 21% of $1012 \approx 20\%$ of 1000

$= 0.20 \times 1000 = 200$

16. Answers will vary.

17. $74 \times \$3.99 \approx 75 \times \$4.00 = \$300$

18. 6% of $\$589 \approx 6\%$ of $600 = 0.06 \times 600 = \$36$

19. $\dfrac{1.1 \text{ mi}}{22 \text{ min}} \approx \dfrac{1 \text{ mi}}{20 \text{ min}} = \dfrac{3 \text{ mi}}{60 \text{ min}} = 3 \text{ mph}$

20. $\$2.49 + \$0.79 + \$1.89 + \$0.10 + \$2.19 + \$6.75$

$\approx \$2 + \$1 + \$2 + \$0 + \$2 + \$7 = \$14.00$

21. $5 \text{ in.} = \dfrac{20}{4} \text{ in.} = 20\left(\dfrac{1}{4}\right) \text{ in.} = 20(0.1) \text{ mi} = 2 \text{ mi}$

22. 0.15 million

23. 0.8 million

24. 13 square units

25. Length $= 1.75$ in., $1.75(12.5) = 21.875 \approx 22$ ft

Height $= 0.625$ in., $0.625(12.5) = 7.8125 \approx 8$ ft

26. $\$2.00 + 7(\$1.50) = \$2.00 + \$10.50 = \$12.50$

Change: $\$20.00 - \$12.50 = \$7.50$

27. $4(\$2.69) = \$10.76$ for four six-packs

Savings: $\$10.76 - \$9.60 = \$1.16$

28. Akala's: 2 hr $= 120$ min, $\dfrac{120}{15} = 8, 8 \times \$15 = \$120$

Berkman's: 2 hr $= 120$ min, $\dfrac{120}{30} = 4,$

$4 \times \$25 = \$100$

Berkman's is the better deal by

$\$120 - \$100 = \$20.00$.

29. To produce the 52 Oscars he found:

$52 \times \$327 = \$17,004$

He was awarded

$\$50,000 - \$17,004 = \$32,996$ more.

30. a) $\dfrac{30}{2500} = \dfrac{x}{24,000}$; $x = \dfrac{30 \times 24,000}{2500} = 288 \text{ lb}$

b) $\dfrac{150}{30} = 5$ bags, and $5 \times 2500 = 12,500 \text{ ft}^2$

31.   10% of \$530 = 0.10 × \$530 = \$53

$53 \times 7 = \$371$

Savings: \$371 − \$60 = \$311

32. $\dfrac{1.5 \text{ mg}}{10 \text{ lb}} = \dfrac{x \text{ mg}}{47 \text{ lb}}$

$10x = 47(1.5)$

$\dfrac{10x}{10} = \dfrac{70.5}{10}$

$x = 7.05 \text{ mg}$

33. $\$4500 - 0.30(\$4500) = \$4500 - \$1350$

$= \$3150 \text{ take-home}$

28% of \$3150 = 0.28 × \$3150 = \$882

34.   9 A.M. Eastern is 6 A.M. Pacific, from 6 A.M. Pacific to 1:35 P.M. Pacific is 7 hr 35 min , 7 hr 35 min − 50 min stop = 6 hr 45 min

35.   3 P.M. − 4 hr = 11 A.M.
July 26, 11:00 A.M.

36.   a) 1 in. × 1 in. = 2.54 cm × 2.54 cm

$= 6.4516 \text{ cm}^2 \approx 6.45 \text{ cm}^2$

b) 1 in. × 1 in. × 1 in.

$= 2.54 \text{ cm} \times 2.54 \text{ cm} \times 2.54 \text{ cm}$

$= 16.387064 \text{ cm}^3 \approx 16.39 \text{ cm}^3$

c) $\dfrac{1 \text{ in.}}{2.54 \text{ cm}} = \dfrac{x \text{ in.}}{1 \text{ cm}}$

$2.54x = 1(1)$

$\dfrac{2.54x}{2.54} = \dfrac{1}{2.54}$

$x = 0.3937... \approx 0.39 \text{ in.}$

37.   Each figure has an additional two dots. To get the hundredth figure, 97 more figures must be drawn, $97(2) = 194$ dots added to the third figure. Thus, $194 + 7 = 201$.

38.

21	7	8	18
10	16	15	13
14	12	11	17
9	19	20	6

39.

23	25	15
13	21	29
27	17	19

40.   59 min 59 sec   Since it doubles every second, the jar was half full 1 second earlier than 1 hour.

41.   6

42.   Nothing. Each friend paid \$9 for a total of \$27; \$25 to the hotel, \$2 to the clerk.
\$25 for the room + \$3 for each friend + \$2 for the clerk = \$30

43.   Let $x =$ the total weight of the four women

$\dfrac{x}{4} = 130, \quad x = 520, \quad \dfrac{520 + 180}{5} = \dfrac{700}{5} = 140 \text{ lb}$

44.   Yes; 3 quarters and 4 dimes, or 1 half dollar, 1 quarter and 4 dimes, or 1 quarter and 9 dimes.
Other answers are possible.

45.   6 cm × 6 cm × 6 cm = 216 cm³

46. Place six coins in each pan with one coin off to the side. If it balances, the heavier coin is the one on the side. If the pan does not balance, take the six coins on the heavier side and split them into two groups of three. Select the three heavier coins and weigh two coins. If the pan balances, it is the third coin. If the pan does not balance, you can identify the heavier coin.

47. $\dfrac{n(n+1)}{2} = \dfrac{500(501)}{2} = \dfrac{250,500}{2} = 125,250$ 48.   16 blue: 4 green $\rightarrow$ 8 blue, 2 yellow $\rightarrow$ 5 blue, 2 white $\rightarrow$ 3 blue

49. 90: 101, 111, 121, 131, 141, 151, 161, 171, 181, 191, ...

50. The fifth figure will be an octagon with sides of equal length. Inside the octagon will be a seven sided figure with each side of equal length. The figure will have one antenna.

51. 61: The sixth figure will have 6 rows of 6 tiles and 5 rows of 5 tiles ($6 \times 6 + 5 \times 5 = 36 + 25 = 61$).

52. Some possible answers are given below. There are other possibilities.

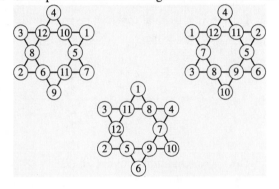

53. a) 2

b) There are 3 choices for the first spot. Once that person is standing, there are 2 choices for the second spot and 1 for the third. Thus, $3 \times 2 \times 1 = 6$.

c) $4 \times 3 \times 2 \times 1 = 24$

d) $5 \times 4 \times 3 \times 2 \times 1 = 120$

e) $n(n-1)(n-2) \cdots 1$, (or $n!$), where $n =$ the number of people in line

Chapter Test

1. 19, 23, 27   (Add 4 to previous number.)

2. $\dfrac{1}{16}, \dfrac{1}{32}, \dfrac{1}{64}$ (Multiply previous number by $\dfrac{1}{2}$.)

3. a) The result is the original number plus 1.

b) The result is the original number plus 1.

c) Conjecture: The result will always be the original number plus 1.

d) $n, 5n, 5n+10, \dfrac{5n+10}{5} = \dfrac{5n}{5} + \dfrac{10}{5} = n+2, n+2-1 = n+1$

(Note: Answers for #4 - #6 will vary depending on how you round your numbers. The answers may differ from the answers in the back of the textbook. However, your answers should be something near the answers given. All answers are approximate.)

4. $0.18 \times 58,000 \approx 0.2 \times 60,000 = 12,000$

5. $\dfrac{210,000}{0.12} \approx \dfrac{210,000}{0.1} \approx 2,100,000$

6.   9 square units

7.   a) $\dfrac{130 \text{ lb}}{63 \text{ in.}} \approx 2.0635$

$\dfrac{2.0635}{63 \text{ in.}} = 0.032754$

$0.032754 \times 703 \approx 23.03$

b) He is in the at risk range.

8. $\$74.39 - \$59.99 = \$14.40$

$\dfrac{\$14.40}{\$0.40} = 36;\ 36$ additional minutes

9. $\dfrac{\$15}{\$2.59} \approx 5.79$

The maximum number of 6 packs is 5.

$\$15.00 - (5 \times \$2.59) = \$15.00 - \$12.95 = \$2.05$

$\dfrac{\$2.05}{\$0.80} = 2.5625$

Thus, two individual cans can be purchased.

6 packs	Indiv. cans	Number of cans
5	2	32
4	5	29
3	9	27
2	12	24
1	15	21
0	18	18

The maximum number of cans is 32.

10. 1 cut yields 2 equal pieces. Cut each of these 2 equal pieces to get 4 equal pieces.

3 cuts $\rightarrow$ 3(2.5 min) = 7.5 min

11. 2.5 in. by 1.875 in.

$\approx 2.5 \times 15.8$ by $1.875 \times 15.8 = 39.5$ in. by 29.625 in.

≈ 39.5 in. by 29.6 in.

(The actual dimensions are 100.5 cm by 76.5 cm.)

12. $\$12.75 \times 40 = \$510$

$\$12.75 \times 1.5 \times 10 = \$191.25$

$\$510 + \$191.25 = \$701.25$

$\$701.25 - \$652.25 = \$49.00$

13.

40	15	20
5	25	45
30	35	10

14. Mary drove the first 15 miles at 60 mph which took $\dfrac{15}{60} = \dfrac{1}{4}$ hr, and the second 15 miles at 30 mph which took $\dfrac{15}{30} = \dfrac{1}{2}$ hr for a total time of $\dfrac{3}{4}$ hr. If she drove the entire 30 miles at 45 mph, the trip would take $\dfrac{30}{45} = \dfrac{2}{3}$ hr (40 min) which is less than $\dfrac{3}{4}$ hr (45 min).

15. $2 \times 6 \times 8 \times 9 \times 13 = 11,232$; 11 does not divide 11,232.

16. 243 jelly beans; $260 - 17 = 243, 234 + 9 = 243, 274 - 31 = 243$

17. a) $3 \times \$3.99 = \$11.97$

b) $9(\$1.75 \times 0.75) = 11.8125 \approx \$11.81$

c) $\$11.97 - \$11.81 = \$0.16$ Using the coupon is least expensive by $0.16.

18. 24 (The first position can hold any of four letters, the second any of the three remaining letters, and so on. $4 \times 3 \times 2 \times 1 = 24$

CHAPTER TWO

SETS

Exercise Set 2.1

1. A **set** is a collection of objects.

3. Description: the set of counting numbers less than 7

   Roster form: $\{1, 2, 3, 4, 5, 6\}$

   Set-builder notation: $\{x | x \in N \text{ and } x < 7\}$

5. A set is **finite** if it either contains no elements or the number of elements in the set is a natural number.

7. Two sets are **equivalent** if they contain the same number of elements.

9. A set that contains no elements is called the **empty set** or **null set**.

11. A **universal set**, symbolized by U, is a set that contains all the elements for any specific discussion.

13. Not well defined, "best" is interpreted differently by different people.

15. Well defined, the contents can be clearly determined.

17. Well defined, the contents can be clearly determined.

19. Infinite, the number of elements in the set is not a natural number.

21. Infinite, the number of elements in the set is not a natural number.

23. Infinite, the number of elements in the set is not a natural number.

25. $\{$ Maine, Maryland, Massachusetts, Michigan, Minnesota, Misssissippi, Missouri, Montana $\}$

27. $\{11, 12, 13, 14, \ldots, 177\}$

29. $B = \{2, 4, 6, 8, \ldots\}$

31. $\{\ \}$ or $\varnothing$

33. $E = \{14, 15, 16, 17, \ldots, 84\}$

35. $\{$ Switzerland, Denmark, Sweden, United Kingdom, Germany, New Zealand $\}$

37. $\{$ Switzerland, Denmark, Sweden, United Kingdom, Germany $\}$

39. $\{2004, 2005\}$

41. $\{1998, 1999, 2000\}$

43. $B = \{x | x \in N \text{ and } 4 < x < 13\}$ or

    $B = \{x | x \in N \text{ and } 5 \leq x \leq 12\}$

45. $C = \{x | x \in N \text{ and } x \text{ is a multiple of } 3\}$

47. $E = \{x | x \in N \text{ and } x \text{ is odd}\}$

49. $C = \{x | x \text{ is February}\}$

51. Set A is the set of natural numbers less than or equal to 7.

53. Set V is the set of vowels in the English alphabet.

55. Set T is the set of species of trees.

57. Set S is the set of seasons.

59. $\{$ Johnson & Johnson, Google, Home Depot $\}$

61. $\{$ United Airlines $\}$

63. $\{1996, 1997, 1998, 1999\}$

65. $\{1998, 1999, 2000, 2001, 2002, 2003, 2004\}$

67. False; $\{e\}$ is a set, and not an element of the set.

69. False; h is not an element of the set.

71. False; 3 is an element of the set.

73. True; *Titanic* is an element of the set.

75. $n(A) = 4$

77. $n(C) = 0$

79. Both; A and B contain exactly the same elements.

81. Neither; the sets have a different number of elements.

83. Equivalent; both sets contain the same number of elements, 3.

85. a) Set A is the set of natural numbers greater than 2. Set B is the set of all numbers greater than 2.

   b) Set A contains only natural numbers. Set B contains other types of numbers, including fractions and decimal numbers.

   c) $A = \{3, 4, 5, 6, \ldots\}$

   d) No

87. Cardinal; 12 tells how many.

89. Ordinal; sixteenth tells Lincoln's relative position.

91. Answers will vary

93. Answers will vary

Exercise Set 2.2

1. Set A is a **subset** of set B, symbolized by $A \subseteq B$, if and only if all the elements of set A are also elements of set B.

3. If $A \subseteq B$, then every element of set A is also an element of set B. If $A \subset B$, then every element of set A is also an element of set B and set $A \neq$ set B.

5. $2^n - 1$, where n is the number of elements in the set.

7. False; Spanish is an element of the set, not a subset.

9. True; the empty set is a subset of every set.

11. True; 5 is not an element of $\{2,4,6\}$.

13. False; the set $\{\varnothing\}$ contains the element $\varnothing$.

15. True; $\{\ \}$ and $\varnothing$ each represent the empty set.

17. False; the set $\{0\}$ contains the element 0.

19. False; $\{\text{swimming}\}$ is a set, not an element.

21. True; the empty set is a subset of every set, including itself.

23. False; no set is a proper subset of itself.

25. $B \subseteq A, B \subset A$

27. $A \subseteq B, A \subset B$

29. $B \subseteq A, B \subset A$

31. $A = B, A \subseteq B, B \subseteq A$

33. $\{\ \}$ is the only subset.

35. $\{\ \}, \{\text{pen}\}, \{\text{pencil}\}, \{\text{pen, pencil}\}$

37. a) $\{\ \}, \{a\}, \{b\}, \{c\}, \{d\}, \{a,b\}, \{a,c\}, \{a,d\},$
   $\{b,c\}, \{b,d\}, \{c,d\}, \{a,b,c\}, \{a,b,d\},$
   $\{a,c,d\}, \{b,c,d\}, \{a,b,c,d\}$

   b) All the sets in part (a) are proper subsets of A except $\{a,b,c,d\}$.

39. False; A could be equal to B.

41. True; every set is a subset of itself.

43. True; ∅ is a proper subset of every set except itself.

45. True; every set is a subset of the universal set.

47. True; ∅ is a proper subset of every set except itself and $U \neq \varnothing$.

49. True; ∅ is a subset of every set.

51. The number of different variations of the house is equal to the number of subsets of {deck, jacuzzi, security system, hardwood flooring}, which is $2^4 = 2 \times 2 \times 2 \times 2 = 16$.

53. The number of different variations is equal to the number of subsets of {call waiting, call forwarding, caller identification, three way calling, voice mail, fax line},

    which is $2^6 = 2 \times 2 \times 2 \times 2 \times 2 \times 2 = 64$.

55. $E = F$ since they are both subsets of each other.

57. a) Yes.

    b) No, c is an element of set D.

    c) Yes, each element of $\{a,b\}$ is an element of set D.

59. A one element set has one proper subset, namely the empty set. A one element set has two subsets, namely itself and the empty set. One is one-half of two. Thus, the set must have one element.

61. Yes

Section 2.3

1.
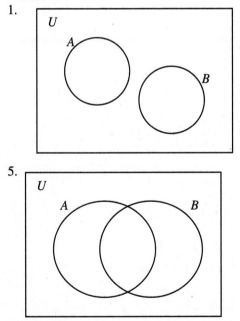

3.
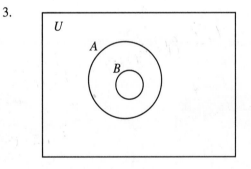

5.

7. Determine the elements that are in the universal set that are not in set A.

9. Select the elements common to both set A and set B.

11. a) *Or* is generally interpreted to mean *union*.

    b) *And* is generally interpreted to mean *intersection*.

13. The difference of two sets A and B is the set of elements that belong to set A but not to set B.

15.   17.

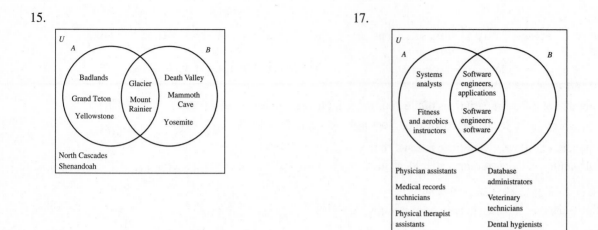

19. The set of animals in U.S. zoos that are not in the San Diego Zoo.

21. The set of insurance companies in the U.S. that do not offer life insurance

23. The set of insurance companies in the U.S. that offer life insurance or car insurance

25. The set of insurance companies in the U.S. that offer life insurance and do not offer car insurance

27. The set of furniture stores in the U.S. that sell mattresses and outdoor furniture

29. The set of furniture stores in the U.S. that do not sell outdoor furniture and sell leather furniture

31. The set of furniture stores in the U.S. that sell mattresses or outdoor furniture or leather furniture

33. $A = \{w, b, c, t, a, h\}$

35. $A \cap B = \{w, b, c, t, a, h\} \cap \{a, h, f, r, d, g\} = \{a, h\}$

37. $A \cup B = \{w, b, c, t, a, h\} \cup \{a, h, f, r, d, g\} = \{w, b, c, t, a, h, f, r, d, g\}$

39. $A' \cap B' = \{w, c, b, t, a, h\}\{a, h, f, r, d, g\} \cap \{w, c, b, t, p, m, z\} = \{p, m, z\}$

41. $A = \{L, \Delta, @, *, \$\}$

43. $U = \{L, \Delta, @, *, \$, R, \square, \alpha, \infty, Z, \Sigma\}$

45. $A' \cup B = \{R, \square, \alpha, \infty, Z, \Sigma\} \cup \{*, \$, R, \square, \alpha\} = \{R, \square, \alpha, \infty, Z, \Sigma, *, \$\}$

47. $A' \cap B = \{L, \Delta, @, *, \$\}' \cap \{*, \$, R, \square, \alpha\} = \{R, \square, \alpha, \infty, Z, \Sigma\} \cap \{*, \$, R, \square, \alpha\} = \{R, \square, \alpha\}$

49. $A \cup B = \{1, 2, 4, 5, 8\} \cup \{2, 3, 4, 6\} = \{1, 2, 3, 4, 5, 6, 8\}$

51. $B' = \{2, 3, 4, 6\}' = \{1, 5, 7, 8\}$

53. $(A \cup B)'$ From #49, $A \cup B = \{1, 2, 3, 4, 5, 6, 8\}$. $(A \cup B)' = \{1, 2, 3, 4, 5, 6, 8\}' = \{7\}$

55. $(A \cup B)' \cap B$: From #53, $(A \cup B)' = \{7\}$. $(A \cup B)' \cap B = \{7\} \cap \{2, 3, 4, 6\} = \{\ \}$

57. $(B \cup A)' \cap (B' \cup A')$: From #53, $(A \cup B)' = (B \cup A)' = \{7\}$.

$(B \cup A)' \cap (B' \cup A') = \{7\} \cap \left(\{2, 3, 4, 6\}' \cup \{1, 2, 4, 5, 8\}'\right) = \{7\} \cap (\{1, 5, 7, 8\} \cup \{3, 6, 7\})$

$= \{7\} \cap \{1, 3, 5, 6, 7, 8\} = \{7\}$

59. $B' = \{b, c, d, f, g\}' = \{a, e, h, i, j, k\}$

61. $A \cap C = \{a, c, d, f, g, i\} \cap \{a, b, f, i, j\} = \{a, f, i\}$

63. $(A \cap C)'$: From #61, $A \cap C = \{a, f, i\}$. $(A \cap C)' = \{a, f, i\}' = \{b, c, d, e, g, h, j, k\}$

65. $A \cup (C \cap B)' = \{a, c, d, f, g, i\} \cup (\{a, b, f, i, j\} \cap \{b, c, d, f, g\})' = \{a, c, d, f, g, i\} \cup \{b, f\}'$

$= \{a, c, d, f, g, i\} \cup \{a, c, d, e, g, h, i, j, k\} = \{a, c, d, e, f, g, h, i, j, k\}$

67. $(A' \cup C) \cup (A \cap B) = \left[\{a,c,d,f,g,i\}' \cup \{a,b,f,i,j\}\right] \cup (\{a,c,d,f,g,i\} \cap \{b,c,d,f,g\})$

$= (\{b,e,h,j,k\} \cup \{a,b,f,i,j\}) \cup \{c,d,f,g\} = \{a,b,e,f,h,i,j,k\} \cup \{c,d,f,g\}$

$= \{a,b,c,d,e,f,g,h,i,j,k\}$, or U

For exercises 69-75: $U = \{1,2,3,4,5,6,7,8,9,10\}$, $A = \{1,2,4,6,9\}$, $B = \{1,3,4,5,8\}$, $C = \{4,5,9\}$

69. $A - B = \{1,2,4,6,9\} - \{1,3,4,5,8\} = \{2,6,9\}$

71. $A - B'$: This leaves only $A \cap B$, which is $\{1,4\}$

73. $(A-B)' = \{2,6,9\}' = \{1,3,4,5,7,8,10\}$

75. $C - A' = \{4,5,9\} - \{3,5,7,8,10\} = \{4,9\}$

For exercises 77-81: $A = \{a,b,c\}$ and $B = \{1,2\}$

77. $\{(a,1),(a,2),(b,1),(b,2),(c,1),(c,2)\}$

79. No; the ordered pairs are not the same.

81. 6

83. $A \cap B = \{1,3,5,7,9\} \cap \{2,4,6,8\} = \{\ \}$

85. $A' \cup B = \{1,3,5,7,9\}' \cup \{2,4,6,8\} = \{2,4,6,8\} \cup \{2,4,6,8\} = \{2,4,6,8\}$, or B

87. $A \cap C' = \{1,3,5,7,9\} \cap \{1,2,3,4,5\}' = \{1,3,5,7,9\} \cap \{6,7,8,9\} = \{7,9\}$

89. $(B \cap C)' = (\{2,4,6,8\} \cap \{1,2,3,4,5\})' = \{2,4\}' = \{1,3,5,6,7,8,9\}$

91. $(C' \cup A) \cap B = \left[\{1,2,3,4,5\}' \cup \{1,3,5,7,9\}\right] \cap \{2,4,6,8\} = (\{6,7,8,9\} \cup \{1,3,5,7,9\}) \cap \{2,4,6,8\}$

$= \{1,3,5,6,7,8,9\} \cap \{2,4,6,8\} = \{6,8\}$

93. $(A \cap B)' \cup C$: From #83, $A \cap B = \{\ \}$.

$(A \cap B)' \cup C = \{\ \}' \cup \{1,2,3,4,5\} = \{1,2,3,4,5,6,7,8,9\} \cup \{1,2,3,4,5\} = \{1,2,3,4,5,6,7,8,9\}$, or U

95. $(A' \cup B') \cap C = \left[\{1,3,5,7,9\}' \cup \{2,4,6,8\}'\right] \cap \{1,2,3,4,5\}$

$= (\{2,4,6,8\} \cup \{1,3,5,7,9\}) \cap \{1,2,3,4,5\} = \{1,2,3,4,5,6,7,8,9\} \cap \{1,2,3,4,5\} = \{1,2,3,4,5\}$, or C

97. A set and its complement will always be disjoint since the complement of a set is all of the elements in the universal set that are not in the set. Therefore, a set and its complement will have no elements in common.

For example, if $U = \{1,2,3\}$, $A = \{1,2\}$, and $A' = \{3\}$, then $A \cap A' = \{\ \}$.

99. Let $A = \{$ customers who owned dogs $\}$ and $B = \{$ customers who owned cats $\}$.

$n(A \cup B) = n(A) + n(B) - n(A \cap B) = 27 + 38 - 16 = 49$

101. a) $A \cup B = \{a,b,c,d\} \cup \{b,d,e,f,g,h\} = \{a,b,c,d,e,f,g,h\}$, $n(A \cup B) = 8$,

$A \cap B = \{a,b,c,d\} \cap \{b,d,e,f,g,h\} = \{b,d\}$, $n(A \cap B) = 2$.

$n(A) + n(B) - n(A \cap B) = 4 + 6 - 2 = 8$

Therefore, $n(A \cup B) = n(A) + n(B) - n(A \cap B)$.

b) Answers will vary.

c) Elements in the intersection of A and B are counted twice in $n(A) + n(B)$.

103. $A \cup B = \{1, 2, 3, 4, \ldots\} \cup \{4, 8, 12, 16, \ldots\} = \{1, 2, 3, 4, \ldots\}$, or A

105. $B \cap C = \{4, 8, 12, 16, \ldots\} \cap \{2, 4, 6, 8, \ldots\} = \{4, 8, 12, 16, \ldots\}$, or B

107. $A \cap C = \{1, 2, 3, 4, \ldots\} \cap \{2, 4, 6, 8, \ldots\} = \{2, 4, 6, 8, \ldots\}$, or C

109. $B' \cap C = \{4, 8, 12, 16, \ldots\}' \cap \{2, 4, 6, 8, \ldots\} = \{0, 1, 2, 3, 5, 6, 7, 9, 10, 11, 13, 14, 15, \ldots\} \cap \{2, 4, 6, 8, \ldots\}$

     $= \{2, 6, 10, 14, 18, \ldots\}$

111. $(A \cap C) \cap B'$: From #107, $A \cap C = C$. $(A \cap C) \cap B' = C \cap B'$.

     From #109, $B' \cap C = C \cap B' = \{2, 6, 10, 14, 18, \ldots\}$

113. $A \cup A' = U$

115. $A \cup \varnothing = A$

117. $A' \cup U = U$

119. $A \cup U = U$

121. If $A \cap B = B$, then $B \subseteq A$.

123. If $A \cap B = \varnothing$, then A and B are disjoint sets.

125. If $A \cap B = A$, then $A \subseteq B$.

Exercise Set 2.4

1. 8

3. Regions II, IV, VI

5. $A \cap B$ is represented by regions II and V. If $A \cap B$ contains 9 elements and region V contains 4 elements, then region II contains $9 - 4 = 5$ elements.

7. a) Yes

     $A \cup B = \{1, 4, 5\} \cup \{1, 4, 5\} = \{1, 4, 5\}$

     $A \cap B = \{1, 4, 5\} \cap \{1, 4, 5\} = \{1, 4, 5\}$

  b) No

  c) No

  c)

$A \cup B$		$A \cap B$	
Set	Regions	Set	Regions
A	I, II	A	I, II
B	II, III	B	II, III
$A \cup B$	I, II, III	$A \cap B$	II

Since the two statements are not represented by the same regions, $A \cup B \neq A \cap B$ for all sets A and B.

9.

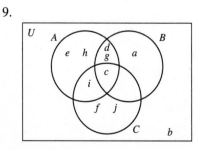

11.

13.

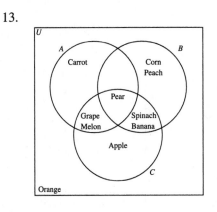

15.

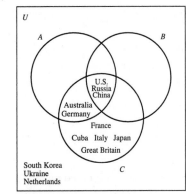

17. Mayo Clinic, V
21. Brigham and Women's Hospital, III
25. Department of Justice, VIII
29. VI
33. III
37. II
41. I
45. VI

49. $B = \{2, 3, 4, 5, 6, 8, 12, 14\}$

53. $(B \cap C)' = \{1, 2, 3, 7, 9, 10, 11, 12, 13, 14\}$

57. $(A \cup C)' = \{2, 11, 12, 13, 14\}$

19. Methodist Hospital, I
23. Department of Energy, II
27. Department of Agriculture, III
31. III
35. V
39. VII
43. VIII

47. $A = \{1, 3, 4, 5, 7, 9\}$

51. $A \cap B = \{3, 4, 5\}$

55. $A \cup B = \{1, 2, 3, 4, 5, 6, 7, 8, 9, 12, 14\}$

59. $A' = \{2, 6, 8, 10, 11, 12, 13, 14\}$

61. $(A \cup B)'$ $A' \cap B'$

Set	Regions	Set	Regions
A	I, II	A	I, II
B	II, III	A'	III, IV
$A \cup B$	I, II, III	B	II, III
$(A \cup B)'$	IV	B'	I, IV
		$A' \cap B'$	IV

Both statements are represented by the same region, IV, of the Venn diagram. Therefore, $(A \cup B)' = A' \cap B'$ for all sets A and B.

63. $A' \cup B'$ $A \cap B$

Set	Regions	Set	Regions
A	I, II	A	I, II
A'	III, IV	B	II, III
B	II, III	$A \cap B$	II
B'	I, IV		
$A' \cup B'$	I, III, IV		

Since the two statements are not represented by the same regions, it is not true that $A' \cup B' = A \cap B$ for all sets A and B.

65. $A' \cap B'$ $(A \cap B)'$

Set	Regions	Set	Regions
A	I, II	A	I, II
A'	III, IV	B	II, III
B	II, III	$A \cap B$	II
B'	I, IV	$(A \cap B)'$	1, III, IV
$A' \cap B'$	IV		

Since the two statements are not represented by the same regions, it is not true that $A' \cap B' = (A \cap B)'$ for all sets A and B.

67. $\left(A' \cap B\right)'$ $A \cup B'$

Set	Regions	Set	Regions
A	I, II	A	I, II
A'	III, IV	B	II, III
B	II, III	B'	I, IV
$A' \cap B$	III	$A \cup B'$	I, II, IV
$\left(A' \cap B\right)'$	I, II, IV		

Both statements are represented by the same regions, I, II, IV, of the Venn diagram. Therefore,

$$\left(A' \cap B\right)' = A \cup B' \text{ for all sets } A \text{ and } B.$$

69. $A \cap (B \cup C)$ $(A \cap B) \cup C$

Set	Regions	Set	Regions
B	II, III, V, VI	A	I, II, IV, V
C	IV , V, VI, VII	B	II, III, V, VI
$B \cup C$	II, III, IV, V, VI, VII	$A \cap B$	II, V
A	I, II, IV, V	C	IV, V, VI, VII
$A \cap (B \cup C)$	II, IV, V	$(A \cap B) \cup C$	II, IV, V, VI, VII

Since the two statements are not represented by the same regions, it is not true that
$A \cap (B \cup C) = (A \cap B) \cup C$ for all sets $A, B,$ and C.

71. $A \cap (B \cup C)$ $(B \cup C) \cap A$

Set	Regions	Set	Regions
B	II, III, V, VI	B	II, III, V, VI
C	IV , V, VI, VII	C	IV, V, VI, VII
$B \cup C$	II, III, IV, V, VI, VII	$B \cup C$	II, III, IV, V, VI, VII
A	I, II, IV, V	A	I, II, IV, V
$A \cap (B \cup C)$	II, IV, V	$(B \cup C) \cap A$	II, IV, V

Both statements are represented by the same regions, II, IV, V, of the Venn diagram.
Therefore, $A \cap (B \cup C) = (B \cup C) \cap A$ for all sets $A, B,$ and C.

73. $A \cap (B \cup C)$ $(A \cap B) \cup (A \cap C)$

Set	Regions	Set	Regions
B	II, III, V, VI	A	I, II, IV, V
C	IV, V, VI, VII	B	II, III, V, VI
$B \cup C$	II, III, IV, V, VI, VII	$A \cap B$	II, V
A	I, II, IV, V	C	IV, V, VI, VII
$A \cap (B \cup C)$	II, IV, V	$A \cap C$	IV, V
		$(A \cap B) \cup (A \cap C)$	II, IV, V

Both statements are represented by the same regions, II, IV, V, of the Venn diagram.
Therefore, $A \cap (B \cup C) = (A \cap B) \cup (A \cap C)$ for all sets $A, B,$ and C.

75. $A \cap (B \cup C)'$ $A \cap (B' \cap C')$

Set	Regions	Set	Regions
B	II, III, V, VI	B	II, III, V, VI
C	IV, V, VI, VII	B'	I, IV, VII, VIII
$B \cup C$	II, III, IV, V, VI, VII	C	IV, V, VI, VII
$(B \cup C)'$	I, VIII	C'	I, II, III, VIII
A	I, II, IV, V	$B' \cap C'$	I, VIII
$A \cap (B \cup C)'$	I	A	I, II, IV, V
		$A \cap (B' \cap C')$	I

Both statements are represented by the same region, I, of the Venn diagram.

Therefore, $A \cap (B \cup C)' = A \cap (B' \cap C')$ for all sets A, B, and C.

77. $(A \cup B)' \cap C$ $(A' \cup C) \cap (B' \cup C)$

Set	Regions	Set	Regions
A	I, II, IV, V	A	I, II, IV, V
B	II, III, V, VI	A'	III, VI, VII, VIII
$A \cup B$	I, II, III, IV, V, VI	C	IV, V, VI, VII
$(A \cup B)'$	VII, VIII	$A' \cup C$	III, IV, V, VI, VII, VIII
C	IV, V, VI, VII	B	II, III, V, VI
$(A \cup B)' \cap C$	VII	B'	I, IV, VII, VIII
		$B' \cup C$	I, IV, V, VI, VII, VIII
		$(A' \cup C) \cap (B' \cup C)$	IV, V, VI, VII, VIII

Since the two statements are not represented by the same regions, it is not true that $(A \cup B)' \cap C = (A' \cup C) \cap (B' \cup C)$ for all sets A, B, and C.

79. $(A \cup B)'$ **81.** $(A \cup B) \cap C'$

83. a) $(A \cup B) \cap C = (\{1, 2, 3, 4\} \cup \{3, 6, 7\}) \cap \{6, 7, 9\} = \{1, 2, 3, 4, 6, 7\} \cap \{6, 7, 9\} = \{6, 7\}$

     $(A \cap C) \cup (B \cap C) = (\{1, 2, 3, 4\} \cap \{6, 7, 9\}) \cup (\{3, 6, 7\} \cap \{6, 7, 9\}) = \varnothing \cup \{6, 7\} = \{6, 7\}$

     Therefore, for the specific sets, $(A \cup B) \cap C = (A \cap C) \cup (B \cap C)$.

b) Answers will vary.

c) $(A \cup B) \cap C$ $(A \cap C) \cup (B \cap C)$

Set	Regions	Set	Regions
A	I, II, IV, V	A	I, II, IV, V
B	II, III, V, VI	C	IV, V, VI, VII
$A \cup B$	I, II, III, IV, V, VI	$A \cap C$	IV, V
C	IV, V, VI, VII	B	II, III, V, VI
$(A \cup B) \cap C$	IV, V, VI	$B \cap C$	V, VI
		$(A \cap C) \cup (B \cap C)$	IV, V, VI

Both statements are represented by the same regions, IV, V, VI, of the Venn diagram.

Therefore, $(A \cup B) \cap C = (A \cap C) \cup (B \cap C)$ for all sets A, B, and C.

85.

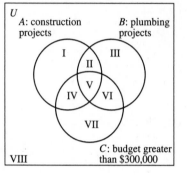

87.   a) A : Office Building Construction Projects, B : Plumbing Projects, C : Budget Greater Than \$300,000

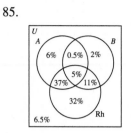

b) Region V; $A \cap B \cap C$

c) Region VI; $A' \cap B \cap C$

d) Region I; $A \cap B' \cap C'$

89.   a)

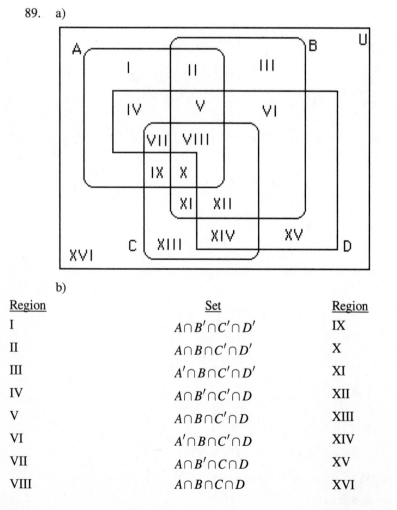

b)

Region	Set	Region	Set
I	$A \cap B' \cap C' \cap D'$	IX	$A \cap B' \cap C \cap D'$
II	$A \cap B \cap C' \cap D'$	X	$A \cap B \cap C \cap D'$
III	$A' \cap B \cap C' \cap D'$	XI	$A' \cap B \cap C \cap D'$
IV	$A \cap B' \cap C' \cap D$	XII	$A' \cap B \cap C \cap D$
V	$A \cap B \cap C' \cap D$	XIII	$A' \cap B' \cap C \cap D'$
VI	$A' \cap B \cap C' \cap D$	XIV	$A' \cap B' \cap C \cap D$
VII	$A \cap B' \cap C \cap D$	XV	$A' \cap B' \cap C' \cap D$
VIII	$A \cap B \cap C \cap D$	XVI	$A' \cap B' \cap C' \cap D'$

Exercise Set 2.5

1.

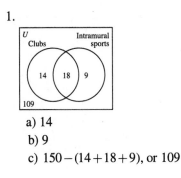

a) 14
b) 9
c) $150 - (14 + 18 + 9)$, or 109

3.

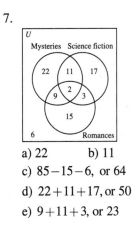

a) 17   b) 12
c) 59, the sum of the numbers in Regions I, II, III

5.

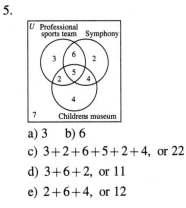

a) 3   b) 6
c) $3 + 2 + 6 + 5 + 2 + 4$, or 22
d) $3 + 6 + 2$, or 11
e) $2 + 6 + 4$, or 12

7.

a) 22   b) 11
c) $85 - 15 - 6$, or 64
d) $22 + 11 + 17$, or 50
e) $9 + 11 + 3$, or 23

9.

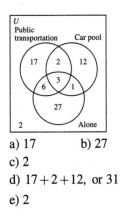

a) 17   b) 27
c) 2
d) $17 + 2 + 12$, or 31
e) 2

11.

a) $30 + 37$, or 67
b) $350 - 25 - 88$, or 237
c) 37   d) 25

13. The Venn diagram shows the number of cars driven by women is 37, the sum of the numbers in Regions II, IV, V. This exceeds the 35 women the agent claims to have surveyed.

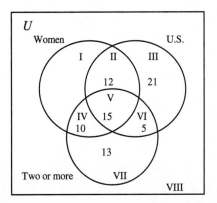

15. First fill in 15, 20 and 35 on the Venn diagram. Referring to the labels in the Venn diagram and the given information, we see that

$a + c = 60$

$b + c = 50$

$a + b + c = 200 - 125 = 75$

Adding the first two equations and subtracting the third from this sum gives $c = 60 + 50 - 75 = 35$.
Then $a = 25$ and $b = 15$. Then $d = 180 - 110 - 25 - 35 = 10$. We now have labeled all the regions except the region outside the three circles, so the number of farmers growing at least one of the crops is $125 + 25 + 110 + 15 + 35 + 10 + 90$, or 410. Thus the number growing none of the crops is $500 - 410$, or 90.

a) 410

b) 35

c) 90

d) $15 + 25 + 10$, or 50

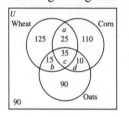

Exercise Set 2.6

1. An **infinite set** is a set that can be placed in a one-to-one correspondence with a proper subset of itself.

3. $\{5, 6, 7, 8, 9, ..., n + 4, ...\}$
   $\quad\downarrow\downarrow\downarrow\downarrow\downarrow\qquad\downarrow$
   $\{8, 9, 10, 11, 12, ..., n + 7, ...\}$

5. $\{3, 5, 7, 9, 11, ..., 2n + 1, ...\}$
   $\quad\downarrow\downarrow\downarrow\downarrow\downarrow\qquad\downarrow$
   $\{5, 7, 9, 11, 13, ..., 2n + 3, ...\}$

7. $\{3, 7, 11, 15, 19, ..., 4n - 1, ...\}$
   $\quad\downarrow\downarrow\ \downarrow\ \downarrow\ \downarrow\qquad\downarrow$
   $\{7, 11, 15, 19, 23, ..., 4n + 3, ...\}$

9. $\{6, 11, 16, 21, 26, ..., 5n+1, ...\}$
   $\quad\downarrow\ \downarrow\ \downarrow\ \downarrow\downarrow\qquad\downarrow$
   $\{11, 16, 21, 26, 31, ..., 5n+6, ...\}$

11. $\left\{\dfrac{1}{2}, \dfrac{1}{4}, \dfrac{1}{6}, \dfrac{1}{8}, ..., \dfrac{1}{2n}, ...\right\}$
    $\qquad\downarrow\downarrow\downarrow\downarrow\qquad\downarrow$
    $\left\{\dfrac{1}{4}, \dfrac{1}{6}, \dfrac{1}{8}, \dfrac{1}{10}, ..., \dfrac{1}{2n+2}, ...\right\}$

13. $\{1, 2, 3, 4, 5, ..., n, ...\}$
    ↓ ↓ ↓ ↓ ↓   ↓
    $\{3, 6, 9, 12, 15, ..., 3n, ...\}$

15. $\{1, 2, 3, 4, 5, ..., n, ...\}$
    ↓ ↓ ↓ ↓ ↓   ↓
    $\{4, 6, 8, 10, 12, ..., 2n + 2, ...\}$

17. $\{1, 2, 3, 4, 5, ..., n, ...\}$
    ↓ ↓ ↓ ↓ ↓   ↓
    $\{2, 5, 8, 11, 14, ..., 3n - 1, ...\}$

19. $\{1, 2, 3, 4, 5, ..., n, ...\}$
    ↓ ↓ ↓ ↓ ↓   ↓
    $\{5, 9, 13, 17, 21, ..., 4n + 1, ...\}$

21. $\{1, 2, 3, 4, 5, ..., n, ...\}$
    ↓ ↓ ↓ ↓ ↓   ↓
    $$\left\{ \frac{1}{3}, \frac{1}{4}, \frac{1}{5}, \frac{1}{6}, \frac{1}{7}, ..., \frac{1}{n+2}, ... \right\}$$

23. $\{1, 2, 3, 4, 5, ..., n, ...\}$
    ↓ ↓ ↓ ↓ ↓   ↓
    $\{1, 4, 9, 16, 25, ..., n^2, ...\}$

25. $\{1, 2, 3, 4, 5, ..., n, ...\}$
    ↓ ↓ ↓ ↓ ↓   ↓
    $\{3, 9, 27, 81, 243, ..., 3^n, ...\}$

27. $=$

29. $=$

31. $=$

Review Exercises

1. True

2. False; the word *best* makes the statement not well defined.

3. True

4. False; no set is a proper subset of itself.

5. False; the elements 6, 12, 18, 24, ... are members of both sets.

6. True

7. False; the two sets do not contain exactly the same elements.

8. True

9. True

10. True

11. True

12. True

13. True

14. True

15. $A = \{7, 9, 11, 13, 15\}$

16. $B = \{\text{Colorado, Nebraska, Missouri, Oklahoma}\}$

17. $C = \{1, 2, 3, 4, ..., 161\}$

18. $D = \{9, 10, 11, 12, ..., 96\}$

19. $A = \{x \mid x \in N \text{ and } 52 < x < 100\}$

20. $B = \{x \mid x \in N \text{ and } x > 42\}$

21. $C = \{x \mid x \in N \text{ and } x < 5\}$

22. $D = \{x \mid x \in N \text{ and } 27 \leq x \leq 51\}$

23. A is the set of capital letters in the English alphabet from E through M, inclusive.

24. B is the set of U.S. coins with a value of less than one dollar.

25. C is the set of the last three lowercase letters in the English alphabet.

26. D is the set of numbers greater than or equal to 3 and less than 9.

27. $A \cap B = \{1, 3, 5, 7\} \cap \{5, 7, 9, 10\} = \{5, 7\}$

28. $A \cup B' = \{1, 3, 5, 7\} \cup \{5, 7, 9, 10\}' = \{1, 3, 5, 7\} \cup \{1, 2, 3, 4, 6, 8\} = \{1, 2, 3, 4, 5, 6, 7, 8\}$

29. $A' \cap B = \{1, 3, 5, 7\}' \cap \{5, 7, 9, 10\} = \{2, 4, 6, 8, 9, 10\} \cap \{5, 7, 9, 10\} = \{9, 10\}$

30. $(A \cup B)' \cup C = (\{1, 3, 5, 7\} \cup \{5, 7, 9, 10\})' \cup \{1, 7, 10\} = \{1, 3, 5, 7, 9, 10\}' \cup \{1, 7, 10\}$
    $= \{2, 4, 6, 8\} \cup \{1, 7, 10\} = \{1, 2, 4, 6, 7, 8, 10\}$

31. $A - B = \{1, 3, 5, 7\} - \{5, 7, 9, 10\} = \{1, 3\}$

32. $A - C' = \{1, 3, 5, 7\} - \{1, 7, 10\}' = \{1, 3, 5, 7\} - \{2, 3, 4, 5, 6, 8, 9\} = \{1, 7\}$

33. $\{(1, 1), (1, 7), (1, 10), (3, 1), (3, 7), (3, 10), (5, 1), (5, 7), (5, 10), (7, 1), (7, 7), (7, 10)\}$

34. $\{(5, 1), (5, 3), (5, 5), (5, 7), (7, 1), (7, 3), (7, 5), (7, 7), (9, 1), (9, 3), (9, 5), (9, 7), (10, 1), (10, 3), (10, 5), (10, 7)\}$

35. $2^4 = 2 \times 2 \times 2 \times 2 = 16$

36. $2^4 - 1 = (2 \times 2 \times 2 \times 2) - 1 = 16 - 1 = 15$

37.

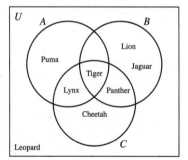

38. $A \cup B = \{a, c, d, f, g, i, k, l\}$

39. $A \cap B' = \{i, k\}$

40. $A \cup B \cup C = \{a, b, c, d, f, g, h, i, k, l\}$

41. $A \cap B \cap C = \{f\}$

42. $(A \cup B) \cap C = \{a, f, i\}$

43. $(A \cap B) \cup C = \{a, b, d, f, h, i, l\}$

44. $(A' \cup B')'$ $A \cap B$

Set	Regions	Set	Regions
A	I, II	A	I, II
A'	III, IV	B	II, III
B	II, III	$A \cap B$	II
B'	I, IV		
$A' \cup B'$	I, III, IV		
$(A' \cup B')'$	II		

Both statements are represented by the same region, II,

of the Venn diagram. Therefore, $(A' \cup B')' = A \cap B$

for all sets A and B.

45. $(A\cup B')\cup(A\cup C')$ $A\cup(B\cap C)'$

Set	Regions	Set	Regions
A	I, II, IV, V	B	II, III, V, VI
B	II, III, V, VI	C	IV, V, VI, VII
B'	I, IV, VII, VIII	$B\cap C$	V, VI
$A\cup B'$	I, II, IV, V, VII, VIII	$(B\cap C)'$	I, II, III, IV, VII, VIII
C	IV, V, VI, VII	A	I, II, IV, V
C'	I, II, III, VIII	$A\cup(B\cap C)'$	I, II, III, IV, V, VII, VIII
$A\cup C'$	I, II, III, IV, V, VIII		
$(A\cup B')\cup(A\cup C')$	I, II, III, IV, V, VII, VIII		

Both statements are represented by the same regions, I, II, III, IV, V, VII, VIII, of the Venn diagram.

Therefore, $(A\cup B')\cup(A\cup C') = A\cup(B\cap C)'$ for all sets $A,B,$ and $C.$

46. II   47. III
48. I   49. IV
50. IV   51. II

52. The company paid $450 since the sum of the numbers in Regions I through IV is 450.

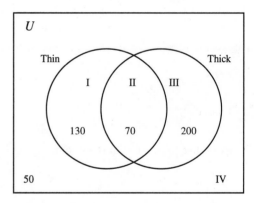

53. a) 315, the sum of the numbers in Regions I through VIII
    b) 10, Region III
    c) 30, Region II
    d) 110, the sum of the numbers in Regions III, VI, VII

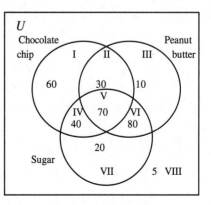

54. a) 38, Region I
 b) 298, the sum of the numbers in Regions I, III, VII
 c) 28, Region VI
 d) 236, the sum of the numbers in Regions I, IV, VII
 e) 106, the sum of the numbers in Regions II, IV, VI

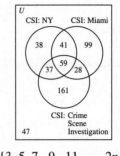

55. $\{2, 4, 6, \quad 8, \quad 10, \ldots, 2n, \ldots\}$
 $\downarrow \downarrow \downarrow \downarrow \ \downarrow \qquad \downarrow$
 $\{4, 6, \ 8, 10, 12, \ldots, 2n + 2, \ldots\}$

56. $\{3, 5, 7, \quad 9, \quad 11, \ldots, 2n + 1, \ldots\}$
 $\downarrow \downarrow \downarrow \downarrow \ \downarrow \qquad \downarrow$
 $\{5, 7, 9, 11, 13, \ldots, 2n + 3, \ldots\}$

57. $\{1, 2, \quad 3, \quad 4, \quad 5, \ldots, \quad n, \ldots\}$
 $\downarrow \downarrow \downarrow \downarrow \downarrow \qquad \downarrow$
 $\{5, 8, \ 11, 14, 17, \ldots, 3n + 2, \ldots\}$

58. $\{1, 2, \ 3, \ 4, \quad 5, \ldots, \quad n, \ldots\}$
 $\downarrow \downarrow \downarrow \downarrow \ \downarrow \qquad \downarrow$
 $\{4, 9, 14, 19, 24, \ldots, 5n - 1, \ldots\}$

Chapter Test

1. True

2. False; the sets do not contain exactly the same elements.

3. True

4. False; the second set has no subset that contains the element 7.

5. False; the empty set is a subset of every set.

6. False; the set has $2^4 = 2 \times 2 \times 2 \times 2 = 16$ subsets.

7. True

8. False; for any set A, $A \cup A' = U$, not $\{\ \}$.

9. True

10. $A = \{1, 2, 3, 4, 5, 6, 7, 8\}$

11. Set A is the set of natural numbers less than 9.

12. $A \cap B = \{3, 5, 7, 9\} \cap \{7, 9, 11, 13\} = \{7, 9\}$

13. $A \cup C' = \{3, 5, 7, 9\} \cup \{3, 11, 15\}' = \{3, 5, 7, 9\} \cup \{5, 7, 9, 13\} = \{3, 5, 7, 9, 13\}$

14. $A \cap (B \cap C)' = \{3, 5, 7, 9\} \cap (\{7, 9, 11, 13\} \cap \{3, 11, 15\})' = \{3, 5, 7, 9\} \cap \{11\}'$

 $= \{3, 5, 7, 9\} \cap \{3, 5, 7, 9, 13, 15\} = \{3, 5, 7, 9\}$, or A.

15. $n(A \cap B') = n(\{3, 5, 7, 9\} \cap \{7, 9, 11, 13\}') = n(\{3, 5, 7, 9\} \cap \{3, 5, 15\}) = n(\{3, 5\}) = 2$

16. $A - B = \{3, 5, 7, 9\} - \{7, 9, 11, 13\} = \{3, 5\}$

17. $A \times C = \{(3, 3), (3, 11), (3, 15), (5, 3), (5, 11), (5, 15), (7, 3), (7, 11), (7, 15), (9, 3), (9, 11), (9, 15)\}$

18.

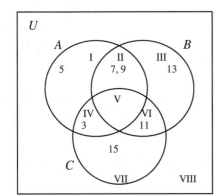

19. $A \cap (B \cup C')$ $(A \cap B) \cup (A \cap C')$

Set	Regions	Set	Regions
B	II, III, V, VI	A	I, II, IV, V
C	IV, V, VI, VII	B	II, III, V, VI
C'	I, II, III, VIII	$A \cap B$	II, V
$B \cup C'$	I, II, III, V, VI, VIII	C	IV, V, VI, VII
A	I, II, IV, V	C'	I, II, III, VIII
$A \cap (B \cup C')$	I, II, V	$A \cap C'$	I, II
		$(A \cap B) \cup (A \cap C')$	I, II, V

Both statements are represented by the same regions, I, II, V, of the Venn diagram.

Therefore, $A \cap (B \cup C') = (A \cap B) \cup (A \cap C')$ for all sets $A, B,$ and C.

20.

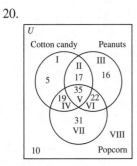

a) 52, the sum of the numbers in Regions I, III, VII
b) 10, Region VIII
c) 93, the sum of the numbers in Regions II, IV, V, VI
d) 17, Region II
e) 38, the sum of the numbers in Regions I, II, III
f) 31, Region VII

21. $\{7, 8, \ 9, \ 10, 11, \ldots, n + 6, \ldots\}$
     $\downarrow \downarrow \downarrow \ \downarrow \ \downarrow \qquad \downarrow$
     $\{8, 9, \ 10, 11, 12, \ldots, n + 7, \ldots\}$

22. $\{1, 2, 3, 4, \ 5, \ldots, \ \ n, \ldots\}$
     $\downarrow \downarrow \downarrow \downarrow \qquad \downarrow$
     $\{1, 3, 5, \ 7, 9, \ldots, 2n - 1, \ldots\}$

CHAPTER THREE

LOGIC

Exercise Set 3.1

1. a) A sentence that can be judged either true or false called a statement.
   b) A simple statement is a sentence that conveys one idea.
   c) Statements consisting of two or more simple statements are called compound statements.

3. *All*, *none* and *some* are quantifiers.

5. a) Some are   b) All are
   c) Some are not   d) None are

7. a) The *exclusive or* means that one or the other event can can occur, but not both. B) Yes; the *inclusive or* means that one or more events can occur simultaneously. c) The *inclusive or* is used in this chapter, unless otherwise stated.

9. compound; disjunction, $\vee$

11. compound; biconditional $\leftrightarrow$

13. compound; conjunction, $\wedge$

15. simple statement

17. compound; negation, $\sim$

19. compound; conjunction, $\wedge$

21. compound; negation, $\sim$

23. Some butterflies are not insects.

25. Some aldermen are running for mayor.

27. All turtles have claws.

29. Some bicycles have three wheels.

31. All pine trees produce pine cones.

33. No pedestrians are in the crosswalk.

35. $\sim p$

37. $\sim q \vee \sim p$

39. $\sim p \rightarrow \sim q$

41. $\sim p \wedge q$

43. $\sim q \leftrightarrow p$

45. $\sim (p \vee q)$

47. Ken Jennings did not win more than \$3 million.

49. Ken Jennings won 74 games of *Jeopardy!* and Ken Jennings won more than \$3 million.

51. If Ken Jennings did not win 74 games of *Jeopardy!* then Ken Jennings won more than \$3 million.

53. Ken Jennings did not win 74 games of *Jeopardy!* or Ken Jennings did not win more than \$3 million.

55. It is false that Ken Jennings won 74 games of *Jeopardy!* and Ken Jennings won more than \$3 million.

57. $(p \wedge \sim q) \vee r$

59. $(p \wedge q) \vee r$

61. $p \rightarrow (q \vee \sim r)$

63. $(r \leftrightarrow q) \wedge p$

65. $q \rightarrow (p \leftrightarrow r)$

67. The water is 70° or the sun is shining, and we do not go swimming.

69. The water is not 70° , and the sun is shining or we go swimming.

71. If we do not go swimming, then the sun is shining and the water is 70°.

73. If the sun is shining then we go swimming, and the water is 70°.

75. The sun is shinning if and only if the water is 70°, and we go swimming.

77. Not permissible. In the list of choices, the connective "or" is the exclusive or, thus one can order either the soup or the salad but not both items.

79. Not permissible. Potatoes and pasta cannot be ordered together.

81. a) w: I bought the watch in Tijuana; p: I paid \$100;
     $w \wedge \sim p$

     b) conjunction

85. a) f: food has fiber; v: food has vitamins;
     h: be healthy; $(f \vee v) \to h$

     b) conditional

89. a) c: classroom is empty; w: is the
     weekend; s: is 7:00 a.m.; $(c \leftrightarrow w) \vee s$

     b) disjunction

83. a) b: below speed limit; p: pulled over;
     $\sim (b \to \sim p)$

     b) negation

87. a) c: may take course; f: fail previous exam;
     p: passed placement test; $c \leftrightarrow (\sim f \vee p)$

     b) biconditional

91. Answers will vary.

Exercise Set 3.2

1. a) $2^2 = 2 \times 2 = 4$ distinct cases

     b)

	p	q
Case 1:	T	T
Case 2:	T	F
Case 3:	F	T
Case 4:	F	F

3. a)

p	q	p	∧	q
T	T	T	T	T
T	F	T	F	F
F	T	F	F	T
F	F	F	F	F
		1	3	2

     b) Only in case 1, when both simple statements are true.

c)

p	q	p	∨	q
T	T	T	T	T
T	F	T	T	F
F	T	F	T	T
F	F	F	F	F
		1	3	2

     d) Only in Case 4, in which both simple statements are false.

5.

p	p	∧	~p
T	T	F	F
F	F	F	T
	1	3	2

7.

p	q	q	∨	~p
T	T	T	T	F
T	F	F	F	T
F	T	T	T	T
F	F	F	T	T
		1	3	2

9.

p	q	~p	∨	~q
T	T	F	F	F
T	F	F	T	T
F	T	T	T	F
F	F	T	T	T
		1	3	2

11.

p	q	~(p ∧ ~ q)
T	T	T T F F
T	F	F T T T
F	T	T F F F
F	F	T F F T
		4 1 3 2

13.

p	q	r	~ q ∨ (p ∧ r)
T	T	T	F T T T T
T	T	F	F F T F F
T	F	T	T T T T T
T	F	F	T T T F F
F	T	T	F F F F T
F	T	F	F F F F F
F	F	T	T T F F T
F	F	F	T T F F F
			1 5 2 4 3

15.

p	q	r	r ∨ (p ∧ ~ q)
T	T	T	T T T F F
T	T	F	F F T F F
T	F	T	T T T T T
T	F	F	F T T T T
F	T	T	T T F F F
F	T	F	F F F F F
F	F	T	T T F F T
F	F	F	F F F F T
			1 5 2 4 3

17.

p	q	r	~ q ∧ (r ∨ ~ p)
T	T	T	F F T T F
T	T	F	F F F F F
T	F	T	T T T T F
T	F	F	T F F F F
F	T	T	F F T T T
F	T	F	F F F T T
F	F	T	T T T T T
F	F	F	T T F T T
			1 5 2 4 3

19.

p	q	r	(~ q ∧ r) ∨ p
T	T	T	F F T T T
T	T	F	F F F T T
T	F	T	T T T T T
T	F	F	T F F T T
F	T	T	F F T F F
F	T	F	F F F F F
F	F	T	T T T T F
F	F	F	T F F F F
			1 3 2 5 4

21. p: The cookies are warm.
    q: The milk is cold.
    In symbolic form the statement is p ∧ q.

p	q	p ∧ q
T	T	T
T	F	F
F	T	F
F	F	F
		1

23. p: I have a new cell phone.
    q: I have a new battery.
    In symbolic form the statement is p ∧ ~ q.

p	q	p	∧	~ q
T	T	T	F	F
T	F	T	T	T
F	T	F	F	F
F	F	F	F	T
		1	3	2

25. p: Jasper Adams is the tutor.
    q: Mark Russo is a secretary.
    In symbolic form the statement is ~ (p ∧ q).

p	q	~ (p ∧ q)
T	T	F T T T
T	F	T T F F
F	T	T F F T
F	F	T F F F
		4 1 3 2

27. p: The copier is out of toner.
    q: The lens is dirty.
    r : The corona wires are broken.
    The statement is p ∨ (q ∨ r).

p	q	r	p ∨ (q ∨ r)
T	T	T	T T T
T	T	F	T T T
T	F	T	T T T
T	F	F	T T F
F	T	T	F T T
F	T	F	F T T
F	F	T	F T T
F	F	F	F F F
			2 3 1

29. p: Congress must act on the bill.
    q: The President signs the bill.
    In symbolic form, the statement is
    p ∧ (q ∨ ~ q).

p	q	p ∧ (q ∨ ~ q)
T	T	T T T T F
T	F	T T F T T
F	T	F F T T F
F	F	F F F T T
		1 5 2 4 3

31. (a) (~ p ∧ r) ∧ q
        (F ∧ T) ∧ F
          F ∧ F
             F
    Therefore the statement is false.
    (b) (~ p ∧ r) ∧ q
        (T ∧ T) ∧ T
           T   ∧ T
              T
    Therefore the statement is true.

33. (a) $(\sim p \vee \sim q) \vee \sim r$

       $(F \vee T) \vee F$

         $T \quad \vee F$

           T

Therefore the statement is true.

(b) $(\sim p \vee \sim q) \vee \sim r$

       $(T \vee F) \vee F$

         $T \quad \vee F$

           T

Therefore the statement is true.

37. (a) $(\sim r \wedge p) \vee q$

       $(F \wedge T) \vee F$

        $F \quad \vee F$

          F

Therefore the statement is false.

(b) $(\sim r \wedge p) \vee q$

       $(F \wedge F) \vee T$

        $F \quad \vee T$

          T

Therefore the statement is true.

41. (a) $(\sim p \vee \sim q) \vee (\sim r \vee q)$

       $(F \vee T) \vee (F \vee F)$

         $T \quad \vee \quad F$

           T

Therefore the statement is true.

(b) $(\sim p \vee \sim q) \vee (\sim r \vee q)$

       $(T \vee F) \vee (F \vee T)$

         $T \quad \vee \quad T$

           T

Therefore the statement is true.

45. E: Virginia borders the Atlantic Ocean.

C: California borders the Indian Ocean.

      $E \quad \vee \quad C$

      $T \quad \vee \quad F$

         T

Therefore the statement is true.

35. (a) $(p \vee \sim q) \wedge \sim (p \wedge \sim r)$

       $(T \vee T) \wedge \sim (T \wedge F)$

         $T \quad \wedge \quad \sim F$

           T

Therefore the statement is true.

(b) $(p \vee \sim q) \wedge \sim (p \wedge \sim r)$

       $(F \vee F) \wedge (F \wedge F)$

         $F \quad \wedge \quad \sim F$

           F

Therefore the statement is false.

39. (a) $(\sim q \vee \sim p) \wedge r$

       $(T \vee F) \wedge T$

         $T \quad \wedge T$

           T

Therefore the statement is true.

(b) $(\sim q \vee \sim p) \wedge r$

       $(F \vee T) \wedge T$

         $T \quad \wedge T$

           T

Therefore the statement is true.

43. $18 \div 3 = 9 \quad \text{or} \quad 56 \div 8 = 7$

       $F \quad \vee \quad T$

          T

Therefore the statement is true.

47. S: Steven Spielberg is a movie director.

H: Tom Hanks is an actor.

M: John Madden is a sports announcer.

      $(S \wedge H) \wedge \sim M$

     $(T \wedge T) \wedge \quad F$

         $T \quad \wedge \quad F$

           F

Therefore the statement is false.

49. IQ: Iraq is in Africa.
    IN: Iran is in South America.
    S: Syria is in the Middle East.
    (IQ ∨ IN) ∧ S
    (F ∨ F) ∧ T
          F   ∧ T
               F
    Therefore the statement is false.

51. p: The United States had the lowest per
       capita CO_2 emissions.
    q: The United States had the lowest per
       capita CO_2 emissions.
    p ∧ q
    F ∧ F
    False

53. p: India had lower per capita CO_2 emissions
       than Japan.
    Q: China had lower per capita CO_2
       emissions than India.
    p ∨ q
    T ∨ F
    True

55. p: 30% of Americans get 6 hours of sleep.
    q: 9% get 5 hours of sleep.
    ~ (p ∧ q)
    ~ (F ∧ T)
      ~F
    True

57. p: 13% of Americans get ≤ 5 hrs. of sleep.
    q: 32% of Americans get ≥ 6 hrs. of sleep.
    r: 30% of Americans get ≥ 8 hrs. of sleep.
    (p ∨ q) ∧ r
    (T ∨ F) ∧ F
       T   ∧ F
    False

59. p ∧ ~q; true when p is true and q is false.

61. p ∨ ~q; true when p is true or when p and
     q are both false.

63. (r ∨ q) ∧ p; true when p is true and when
     either r or q is true.

65. q ∨ (p ∧ ~r); true except when p, q, r have
     truth values TFF, FTF, or FFF.

67. (a) Mr. Duncan qualifies for the loan.
          Mrs. Tuttle qualifies for the loan.
     (b) The Rusineks do not qualify
          because their gross income is too low.

69. (a) Wing Park qualifies for the special
          fare.
     (b) The other 4 do not qualify:
          Gina V. returns after 04/01;
          Kara S. returns on Monday;
          Christos S. does not stay
          at least one Saturday; and
          Alex C. returns on Monday.

71.

p	q	r	[(q ∧ ~r) ∧ (~ p ∨ ~ q)]			∨ (p ∨ ~ r)	
T	T	T	F	F	F	T	T
T	T	F	T	F	F	T	T
T	F	T	F	F	T	T	T
T	F	F	F	F	T	T	T
F	T	T	F	F	T	F	F
F	T	F	T	T	T	T	T
F	F	T	F	F	T	F	F
F	F	F	F	F	T	T	T
			1	3	2	5	4

73. Yes

p	q	r	(p ∧~q) ∨ r	(q ∧ ~r) ∨ p
T	T	T	F T T	F T T
T	T	F	F F F	T T T
T	F	T	T T T	F T T
T	F	F	T T F	F T T
F	T	T	F T T	F F F
F	T	F	F F F	T T F
F	F	T	F T T	F F F
F	F	F	F F F	F F F

Exercise Set 3.3

1. a)

p	q	p	→	q
T	T	T	T	T
T	F	T	F	F
F	T	F	T	T
F	F	F	T	F
		1	3	2

1. c)

p	q	p ↔ q
T	T	T T T
T	F	T F F
F	T	F F T
F	F	F T F
		1 3 2

b) The conditional statement is false only in the case when antecedent is true and the consequent is false, otherwise it is true.

d) The biconditional statement is true when the statements to the left and right of the biconditional symbol match, otherwise, false.

3. a) Substitute the truth values for the simple statement. Then evaluate the compound statement for that specific case.

   b) [(p ↔ q) ∨ (~r → q)] → ~r
       [(T ↔ T) ∨ (~ T → T)] → ~ T
       [T   ∨ (~T → T)] → T
       [T   ∨   T] → T
                   T   → T
                          T

In this specific case the statement is true.

5. A self-contradiction is a compound statement that is false in every case.

7.

p	q	~p → q
T	T	F T T
T	F	F T F
F	T	T T T
F	F	T F F
		1 3 2

9.

p	q	~ (p→~ q)
T	T	T F
T	F	F T
F	T	F T
F	F	F T
		2 1

11.

p	q	~ q	↔	p
T	T	F	F	T
T	F	T	T	T
F	T	F	T	F
F	F	T	F	F
		1	3	2

13.

p	q	p	↔	(q ∨ p)
T	T	T	T	T
T	F	T	T	T
F	T	F	F	T
F	F	F	T	F
		1	3	2

15.

p	q	q → (p → ~ q)
T	T	T F T F F
T	F	F T T T T
F	T	T T F T F
F	F	F T F T T
		4 5 1 3 2

17.

p	q	r	~p	→	(q	∧	r)
T	T	T	F	T	T	T	T
T	T	F	F	T	T	F	F
T	F	T	F	T	F	F	T
T	F	F	F	T	F	F	F
F	T	T	T	T	T	T	T
F	T	F	T	F	T	F	F
F	F	T	T	F	F	F	T
F	F	F	T	F	F	F	F
			4	5	1	3	2

19.

p	q	r	p	↔	(~q	→	r)
T	T	T	T	T			T
T	T	F	T	T			T
T	F	T	T	T			T
T	F	F	T	F			F
F	T	T	F	F			T
F	T	F	F	F			T
F	F	T	F	F			T
F	F	F	F	T			F
			1	3			1

21.

p	q	r	(q	∨	~ r)	↔	~ p
T	T	T	T	T	F	F	F
T	T	F	T	T	T	F	F
T	F	T	F	F	F	T	F
T	F	F	F	T	T	F	F
F	T	T	T	T	F	T	T
F	T	F	T	T	T	T	T
F	F	T	F	F	F	F	T
F	F	F	F	T	T	T	T
			1	3	2	5	4

23.

p	q	r	(~ r	∨	~q)	→	p
T	T	T	F	F	F	T	T
T	T	F	T	T	F	T	T
T	F	T	F	T	T	T	T
T	F	F	T	T	T	T	T
F	T	T	F	F	F	T	F
F	T	F	T	T	F	F	F
F	F	T	F	T	T	F	F
F	F	F	T	T	T	F	F
			1	3	2	5	4

25.

p	q	r	(p → q)	↔	(~q	→	~r)
T	T	T	T	T	F	T	F
T	T	F	T	T	F	T	T
T	F	T	F	T	T	F	F
T	F	F	F	F	T	T	T
F	T	T	T	T	F	T	F
F	T	F	T	T	F	T	T
F	F	T	T	F	T	F	F
F	F	F	T	T	T	T	T
			1	5	2	4	3

27. p: I take niacin; q: I will stay healthy;
r: I will have lower cholesterol

p	q	r	p	→	(q	∧	r)
T	T	T	T	T		T	
T	T	F	T	F		F	
T	F	T	T	F		F	
T	F	F	T	F		F	
F	T	T	F	T		T	
F	T	F	F	T		F	
F	F	T	F	T		F	
F	F	F	F	T		F	
			1	3		2	

29. p: election was fair; q: polling station stayed
open until 8 P.M.; r: we will request a recount

p	q	r	(p	↔	q)	∨	r
T	T	T		T		T	T
T	T	F		T		T	F
T	F	T		F		T	T
T	F	F		F		F	F
F	T	T		F		T	T
F	T	F		F		F	F
F	F	T		T		T	T
F	F	F		T		T	F
				1		3	2

31. p: Mary Andrews sends me an e-mail;
q: we can call her; r: we can write to Mom

p	q	r	(~ p	→	q)	∨	r
T	T	T		T		T	T
T	T	F		T		T	F
T	F	T		T		T	T
T	F	F		T		T	F
F	T	T		T		T	T
F	T	F		T		T	F
F	F	T		F		T	T
F	F	F		F		F	F
				1		3	2

33.

p	p	→	~ p
T	T	F	F
F	F	T	T
	1	3	2

neither

35.

p	q	p	∧	(q ∧ ~ p)
T	T	T	F	F
T	F	T	F	F
F	T	F	F	T
F	F	F	F	F
		1	3	2

self-contradiction

37.

p	q	(~ q	→	p)	∨	~ q
T	T		T		T	F
T	F		T		T	T
F	T		T		T	F
F	F		F		T	T
			1	3		2

tautology

39.

p	q	~p	→	(p ∨ q)
T	T	F	T	T
T	F	F	T	T
F	T	T	T	T
F	F	T	F	F
		1	3	2

not an implication

41.

p	q	(q ∧ p)	→	(p ∧ q)
T	T	T	T	T
T	F	F	T	F
F	T	F	T	F
F	F	F	T	F
		1	3	2

an implication

43.

p	q	[(p → q)	∧	(q → p)]	→	(p ↔ q)
T	T	T	T	T	T	T
T	F	F	F	T	T	F
T	T	T	F	F	T	F
T	F	T	T	T	T	T
		1	3	2	5	4

an implication

45. $\sim p \rightarrow (q \rightarrow r)$
    $F \rightarrow (F \rightarrow T)$
     $T \rightarrow \quad T$
      T

47. $q \leftrightarrow (\sim p \vee r)$
    $F \leftrightarrow \quad T$
     F

49. $(\sim p \wedge \sim q) \vee \sim r$
    $(F \wedge T) \vee F$
     $F \quad \vee F$
      F

51. $(p \wedge r) \leftrightarrow (p \vee \sim q)$
    $(T \wedge T) \leftrightarrow (T \vee T)$
     $T \quad \leftrightarrow \quad T$
      T

53. $(\sim p \leftrightarrow r) \vee (\sim q \leftrightarrow r)$
    $(F \leftrightarrow T) \vee (T \leftrightarrow T)$
     $T \quad \vee \quad T$
      T

55. $\sim [(p \vee q) \leftrightarrow (p \rightarrow \sim r)]$
    $\sim [(T \vee F) \leftrightarrow (T \rightarrow F)]$
     $\sim [T \quad \leftrightarrow \quad F]$
      $\sim F$
      T

57. If $2+7=9$, then $15-3=12$.
$$T \rightarrow T$$
$$T$$

59. A cat has whiskers or a fish can swim, and a chicken lays eggs.
$$(T \vee T) \wedge T$$
$$T \quad \wedge T$$
$$T$$

61. Apple makes computers, if and only if Nike makes sports shoes or Rolex makes watches.
$$T \leftrightarrow (T \vee T)$$
$$T \leftrightarrow \quad T$$
$$T$$

63. Valentine's Day is in February or President's Day is in March, and Thanksgiving Day is in November.
$$(T \vee F) \wedge T$$
$$T \wedge T$$
$$T$$

65. Io has a diameter of 1000–3161 miles, or Thebe may have water, and Io may have atmosphere.
$$(T \vee F) \wedge T$$
$$T \wedge T$$
$$T$$

67. Phoebe has a larger diameter than Rhea if and only if Callisto may have water ice, and Calypso has a diameter of 6–49 miles.
$$(F \leftrightarrow T) \wedge T$$
$$F \quad \wedge T$$
$$F$$

69. The most common cosmetic surgery procedure for females is liposuction or the most common procedure for males is eyelid surger, and 20% of male cosmetic surgery is for nose reshaping.
$$(T \vee F) \wedge F$$
$$T \quad \wedge F$$
$$F$$

For 71–75 *p*: Muhundan spoke at the teachers' conference.
q: Muhundan received the outstanding teacher award
Assume *p* and *q* are true.

71. $q \rightarrow p$
$$T \rightarrow T$$
$$T$$

73. $\sim q \rightarrow p$
$$F \rightarrow T$$
$$T$$

75. $q \rightarrow p$
$$T \rightarrow T$$
$$T$$

77. No, the statement only states what will occur if your sister gets straight A's. If your sister does not get straight A's, your parents may still get her a computer.

79.

p	q	r	[p	∨	(q	→	~ r)]	↔	(p ∧ ~ q)
T	T	T		T		F	F	F	F
T	T	F		T		T	T	F	F
T	F	T		T		T	F	T	T
T	F	F		T		T	T	T	T
F	T	T		F		F	F	T	F
F	T	F		T		T	T	F	F
F	F	T		T		T	F	F	F
F	F	F		T		T	T	F	F
				3		2	1	5	4

81. The statement may be expressed as $(p \rightarrow q) \vee (\sim p \rightarrow q)$, where p: It is a head and q: I win.

p	q	$(p \rightarrow q)$	$\vee$	$(\sim p \rightarrow q)$
T	T	T	T	T
T	F	F	T	T
F	T	T	T	T
F	F	F	T	F
		1	3	2

The statement is a tautology.

83. | Tiger | Boots | Sam | Sue |
|---|---|---|---|
| Blue | Yellow | Red | Green |
| Nine Lives | Whiskas | Friskies | Meow Mix |

Exercise Set 3.4

1. a) Two statements are equivalent if both statements have exactly the same truth values in the answer column of the truth table.

1. b) Construct a truth table for each statement and then compare the columns. If they are identical, then the statements are equivalent. If the answer columns are not identical, then the statements are not equivalent.

3. $\sim (p \wedge q) \Leftrightarrow \sim p \vee \sim q$

   $\sim (p \vee q) \Leftrightarrow \sim p \wedge \sim q$

5. converse $\Leftrightarrow$ inverse;   conditional $\Leftrightarrow$ contrapositive

7. If $p \rightarrow q$ is equivalent to $\sim p \vee q$, then $\sim (p \rightarrow q)$ is equivalent to $\sim (\sim p \vee q)$, which by De Morgan's second law is equivalent to $[\sim (\sim p)] \wedge \sim q$, which is equivalent to $p \wedge \sim q$.

9. $\sim (p \wedge q) \Leftrightarrow \sim p \vee q$ (by law 1) and this is not equivalent to $\sim p \wedge q$.

11. $\sim (p \wedge q) \Leftrightarrow \sim p \vee \sim q$ by law 1 and $\sim (q \vee \sim p) \Leftrightarrow \sim q \wedge \sim (\sim p) \Leftrightarrow \sim q \wedge p$ by law 2. $\sim p \vee \sim q$ is not equivalent to $\sim q \wedge p$, since if p and q are both false, the first statement is true and the second is false.

13. Equivalent by law 2.

15. Yes, equivalent

17. Yes, $\sim (p \rightarrow \sim q) \Leftrightarrow \sim (\sim p \vee \sim q) \Leftrightarrow p \wedge q$

19.

p	q	$p \rightarrow q$	$\sim p \vee q$		
T	T	T	F	T	T
T	F	F	F	F	F
F	T	T	T	T	T
F	F	T	T	T	F
		1	1	3	2

The statements are equivalent.

21.

p	q	$\sim q \rightarrow \sim p$			$p \rightarrow q$
T	T	F	T	F	T
T	F	T	F	F	F
F	T	F	T	T	T
F	F	T	T	T	T
		1	3	2	1

The statements are equivalent.

23.

p	q	r	(p ∨ q) ∨ r	p ∨ (q ∨ r)
T	T	T	T TT	T T T
T	T	F	T TF	T T T
T	F	T	T TT	T T T
T	F	F	T TF	T T F
F	T	T	T TT	F T T
F	T	F	T TF	F T T
F	F	T	F TT	F T T
F	F	F	F FF	F F F
			1 3 2	2 3 1

The statements are equivalent.

25.

p	q	r	p ∧ (q ∨ r)	(p ∧ q) ∨ r
T	T	T	TT TTT	TTTT
T	T	F	TT TTF	TTTT
T	F	T	TT FTT	TFF TT
T	F	F	TF FF F	TFF FF
F	T	T	FF TTT	FFT TT
F	T	F	FF TT F	FFT FF
F	F	T	FF FT T	FFF TT
F	F	F	FF FF F	FFF FF
			1 5 2 4 3	1 3 2 5 4

The statements are not equivalent.

27.

p	q	r	(p → q) ∧ (q → r)	(p → q) → r
T	T	T	T T T	T T T
T	T	F	T F F	T F F
T	F	T	F F T	F T T
T	F	F	F F T	F T F
F	T	T	T T T	T T T
F	T	F	T F F	T F F
F	F	T	T T T	T T T
F	F	F	T T T	T F F
			1 3 2	1 3 2

The statements are not equivalent.

29.

p	q	(p → q) ∧ (q → p)	p ↔ q
T	T	T T T	T
T	F	F F T	F
F	T	T F F	F
F	F	T T T	T
		1 3 2	1

The statements are equivalent.

31. p: The Rocky Mountains are in the East.

q: The Appalachian Mountains are in the West.

In symbolic form, the statement is ~ (p ∧ q).

Applying DeMorgan's Laws we get: ~ p ∨ ~ q.

The Rocky Mountains are not in the East or
the Appalachian Mountains are not in the West.

33. p: The watch was a Swatch watch.

q: The watch was a Swiss Army watch.

In symbolic form, the statement is ~ p ∧ ~ q.

Applying DeMorgan's Laws we get: ~ (p ∨ q).
It is false that the watch was a Swatch watch
or the watch was a Swiss Army watch.

35. p: The hotel has a weight room.

q: The conference center has an auditorium.

In symbolic form, the statement is ~ p ∨ ~ q.

Applying DeMorgan's Laws we get: ~ (p ∧ q).
It is false that the hotel has a weight room and
the conference center has an auditorium.

37. p: Ashely takes the new job.

q: Ashely will move.

r: Ashely will buy a new house in town.

In symbolic form, the statement is
p → (~q ∨ r). Applying DeMorgan's Laws
we get: p → ~ (q ∧ ~r). If Ashely takes the
new job, it is not true that she will move and will
not buy a new house in town.

39. p: Ena selects a new textbook.

q: Ena will have to write a new syllabus.

In symbolic form, the statement is p → q.
Since p → q ⇔ ~ p ∨ q, an equivalent
Statement is: Ena does not select a new textbook
or she will have to write a new syllabus.

41. p: Bob the Tomato visited the nursing home.

q: Bob the Tomato visited the Cub Scout
meeting.

In symbolic form, the statement is p ∨ ~ q.

Since ~p → ~q ⇔ ~ p ∨ q, an equivalent
Statement is: If Bob the Tomato did not visit
the nursing home, then he did not visit the
Cub Scout meeting.

43. p: The plumbers meet in Kansas City.

q: The *Rainmakers* will provide the entertainment.

In symbolic form, the statement is $p \rightarrow q$.

$p \rightarrow q \Leftrightarrow \sim p \vee q$. The plumbers do not meet in KC or the *Rainmakers* will provide the entertainment.

45. p: Chase is hiding.

q: The pitcher is broken.

In symbolic form, the statement is $\sim p \vee q$.

$\sim p \vee q \Leftrightarrow p \rightarrow q$. If Chase is hiding, then the pitcher is broken.

47. p: We go to Cincinnati.

q: We go to the zoo.

In symbolic form, the statement is $\sim(p \rightarrow q)$.

$\sim(p \rightarrow q) \Leftrightarrow p \wedge \sim q$.

We go to Cincinnati and we will not go to the zoo.

49. p: I am cold.

q: The heater is working.

In symbolic form, the statement is $p \wedge \sim q$.

$p \wedge \sim q \Leftrightarrow \sim(p \rightarrow q)$.

It is false that if I am cold then the heater is working.

51. p: Borders has a sale.

q: We will buy $100 worth of books.

In symbolic form, the statement is $\sim(p \rightarrow q)$.

$\sim(p \rightarrow q) \Leftrightarrow p \wedge \sim q$.

Borders has a sale and we will not buy $100 worth of books.

53. p: John Deere will hire new workers.

q: Dubuque will retain the workers.

In symbolic form, the statement is $p \wedge q$.

$p \wedge q \Leftrightarrow \sim(p \rightarrow \sim q)$.

It is false that if John Deere will hire new workers Dubuque will not retain the workers.

55. Converse: If we can finish the quilt in 1 week, then we work every night.

Inverse: If we do not work every night then we cannot finish the quilt in 1 week.

Contrapositive: If we cannot finish the quilt in 1 week then we do not work every night.

57. Converse: If I buy silver jewelry, then I go to Mexico.

Inverse: If I do not go to Mexico, then I do not buy silver jewelry.

Contrapositive: If I do not buy silver jewelry then I do not go to Mexico.

59. Converse: If I scream, then that annoying paper clip (Clippie) shows up on my screen.

Inverse: If Clippie does not show up on my screen, then I will not scream.

Contrapositive: If I do not scream, then Clippie does not show up on my screen.

61. If a natural number is divisible by 10, then it is divisible by 5. True

63. If a natural number is not divisible by 6, then it is not divisible by 3. False

65. If two lines are not parallel, then the two lines intersect in at least one point. True

67. p: Bill Rush is the editor.

    q: Bill Rush is the vice president.

    In symbolic form, the statements are:

    a) ~ p ∨ q, b) q → ~ p, c) p → ~ q

    Statement (c) is the contrapositive of statement. (b). Therefore, statements (b) and (c) are equivalent.

p	q	~p ∨ q	q → ~ p
T	T	F T T	T F F
T	F	F F F	F T F
F	T	T T T	T T T
F	F	T T F	F T T
		1 3 2	1 3 2

Since the truth tables for (a) and (b) are different we conclude that only statements (b) and (c) are equivalent.

71. p: Today is Sunday.

    q: The library is open.

    In symbolic form, the statements are: a) ~ p ∨ q,

    b) p → ~ q, c) q → ~ p. Looking at the truth table for all three statements, we can determine that only statements (b) and (c) are equivalent.

p	q	a) ~ p ∨ q	b) p → ~ q	c) q → ~ p
T	T	F T T	T F F	T F F
T	F	F F F	T T T	F T F
F	T	T T T	F T F	T T T
F	F	T T F	F T T	F T T
		1 3 2	1 3 2	1 3 2

69. p: The office is cool.

    q: The computer is jammed.

    In symbolic form, the statements are: a) ~ p ∧ q, b) ~ p → ~ q, c) ~ (p ∨ ~ q). If we use DeMorgan's Laws on statement (a), we get statement (c). Therefore, statements (a) and (c) are equivalent. If we look at the truth tables for statements (a), (b), and (c), we see that only statements (a) and (c) are equivalent.

p	q	a) ~ p ∧ q	b) ~ p → ~ q	c) ~ (p ∨ ~ q)
T	T	F F T	F T F F	F T T F
T	F	F F F	F T T T	F T T T
F	T	T T T	T F F F	T F F F
F	F	T F F	T T T T	F F T T
		1 3 2	1 3 2	4 1 3 2

73. p: The grass grows.

    q: The trees are blooming.

    In symbolic form, the statements are: a) p ∧ q,

    b) q → ~ p, c) ~ q ∨ ~ p. Using the fact that p → q ⇔ ~ p ∨ q, on statement (b) we get ~ q ∨ ~ p. Therefore, statements (b) and (c) are equivalent. Looking at the truth table for statements (a) and (b) we can conclude that only statements (b) and (c) are equivalent.

p	q	p ∧ q	q → ~ p
T	T	T	T F F
T	F	F	F T F
F	T	F	T T T
F	F	F	F T T
		1	1 3 2

75. p: You drink milk.
    q: Your cholesterol count will be lower.
    In symbolic form, the statements are:
    a) $\sim(\sim p \to q)$, b) $q \leftrightarrow p$, and c) $\sim(p \to \sim q)$.

p	q	$\sim(\sim p \to q)$	$q \leftrightarrow p$	$\sim(p \to \sim q)$
T	T	F FT T	T T T	T T F F
T	F	F FT F	F F T	F T T T
F	T	F TT T	T F F	F F T F
F	F	T TF F	F T F	F F T T
		4 1 3 2	1 3 2	4 1 3 2

Therefore, none of the statements are equivalent.

77. p: The pay is good.
    q: Today is Monday.
    r : I will take the job.
    Looking at the truth tables for statements (a), (b), and (c), we see that none of the statements are equivalent.

p	q	r	a) $(p \wedge q) \to r$	b) $\sim r \to \sim(p \vee q)$	c) $(p \wedge q) \vee r$
T	T	T	T TT	FT F T	T TT
T	T	F	T FF	TF F T	T TF
T	F	T	F TT	FT F T	F TT
T	F	F	F TF	TF F T	F FF
F	T	T	F TT	FT F T	F TT
F	T	F	F TF	TF F T	F FF
F	F	T	F TT	FT T F	F TT
F	F	F	F TF	TT T F	F FF
			1 3 2	1 4 3 2	1 3 2

79. p: The package was sent by Federal Express.
    q: The package was sent by United Parcel Service.
    r : The package arrived on time.
    Using the fact that $p \to q \Leftrightarrow \sim p \vee q$ to rewrite statement (c), we get $p \vee (\sim q \wedge r)$. Therefore, statements (a) and (c) are equivalent. Looking at the truth table for statements (a) and (b), we can conclude that only statements (a) and (c) are equivalent.

p	q	r	a) $p \vee (\sim q \wedge r)$	b) $r \Leftrightarrow (p \vee \sim q)$
T	T	T	T T F F T	T T T T F
T	T	F	T T F F F	F F T T F
T	F	T	T T T T T	T T T T T
T	F	F	T T T F F	F F T T T
F	T	T	F F F F T	T F F F F
F	T	F	F F F F F	F T F F F
F	F	T	F T T T T	T T F T T
F	F	F	F F T F F	F F F T T
			1 5 2 4 3	1 5 2 4 3

81. p: The car needs oil.
    q: The car needs gas.
    r : The car is new.
    In symbolic form, the statements are: a) $p \wedge (q \vee r)$, b) $p \wedge \sim(\sim q \wedge \sim r)$, and c) $p \to (q \vee \sim r)$. If we use DeMorgan's Laws on the disjunction in statement (a), we obtain $p \wedge \sim(\sim q \wedge \sim r)$. Therefore, statements (a) and (b) are equivalent. If we compare the truth tables for (a) and (c) we see that they are not equivalent. Therefore, only statements (a) and (b) are equivalent.

p	q	r	$p \wedge (q \vee r)$	$p \to (q \vee \sim r)$
T	T	T	T T T	T T T T F
T	T	F	T T T	T T T T T
T	F	T	T T T	T F F F F
T	F	F	T F F	T T F T T
F	T	T	F F T	F T T T F
F	T	F	F F T	F T T T T
F	F	T	F F T	F T F F F
F	F	F	F F F	F T F T T
			1 3 2	1 5 2 4 3

83. True. If $p \to q$ is false, it must be of the form $T \to F$. Therefore, the converse must be of the form $F \to T$, which is true.

85. False. A conditional statement and its contrapositive always have the same truth values.

87. If we use DeMorgan's Laws to rewrite $\sim p \vee q$, we get $\sim(p \wedge \sim q)$. Since $\sim p \vee q \Leftrightarrow \sim(p \wedge \sim q)$ and $p \to q \Leftrightarrow \sim p \vee q$, we can conclude that $p \to q \Leftrightarrow \sim(p \wedge \sim q)$. Other answers are possible.

89. Research problem -- Answers will vary.

Exercise Set 3.5

1. a) An argument is valid when its conclusion necessarily follows from the given set of premises.
   b) An argument is invalid or a fallacy when the conclusion does not necessarily follow from premises.

3. Yes. For example,

   $p \rightarrow q$

   $\underline{\quad q \quad}$

   $\therefore p$

   is an invalid argument, but if p happens to be true, then its conclusion is true.

5. Yes. If the conclusion does not follow from the set of premises, then the argument is invalid.

7. a) $p \rightarrow q$ b) If sky is clear, then I'll go to game.

   $\underline{\quad p \quad}$ $\underline{\text{The sky is clear.}}$

   $\therefore q$ I will go to the game.

9. a) $p \rightarrow q$ b) If soil is dry, then grass needs water.

   $\underline{\quad \sim q \quad}$ $\underline{\text{The grass does not need water.}}$

   $\therefore \sim p$ The soil is not dry.

11. a) $p \rightarrow q$ b) If you wash my car, then I pay you $5.

    $\underline{\quad q \quad}$ $\underline{\text{I will give you \$5.}}$

    $\therefore p$ You washed my car.

13. This argument is the fallacy of the inverse, therefore it is invalid.

15. This is the law of detachment, so it is a valid argument.

17. This argument is a disjunctive syllogism and therefore is valid.

19. This argument is the fallacy of the converse. Therefore it is invalid.

21. This argument is the law of contraposition, so it is valid.

23. This argument is the law of syllogism and therefore it is valid.

25.

p	q	r	[(p ↔ q) ∧ (q ∧ r)] → (p ∨ r)				
T	T	T	T	T	T	T	T
T	T	F	T	F	F	T	T
T	F	T	F	F	F	T	T
T	F	F	F	F	F	T	T
F	T	T	F	F	T	T	T
F	T	F	F	F	F	T	F
F	F	T	T	F	F	T	T
F	F	F	T	F	F	T	F
			1	3	2	5	4

The argument is valid.

27.

p	q	r	[(r ↔ p) ∧ (~p ∧ q)] → (p ∧ r)						
T	T	T	T	F	F F T	T	T
T	T	F	F	F	F F T	T	F
T	F	T	T	F	F F F	T	T
T	F	F	F	F	F F F	T	F
F	T	T	F	F	T T T	T	F
F	T	F	T	T	T T T	F	F
F	F	T	F	F	T F F	T	F
F	F	F	T	F	T F F	T	F
			1	5	2 4 3	7	6

The argument is invalid.

29.

p	q	r	[(p → q) ∧ (q ∨ r) ∧ (r ∨ p)] → p						
T	T	T	T	T	T T	T	T T
T	T	F	T	T	T T	T	T T
T	F	T	F	F	T F	T	T T
T	F	F	F	F	F F	T	T T
F	T	T	T	T	T T	T	F F
F	T	F	T	T	T F	F	T F
F	F	T	T	T	T T	T	F F
F	F	F	T	F	F F	F	T F
			1	3	2 5	4	7 6

The argument is invalid.

31.

p	q	r	[(p → q) ∧ (r → ~ p) ∧ (p ∨ r)] → (q ∨ ~ p)
T	T	T	T F TF F F T T TT F
T	T	F	T T FT F T T T TT F
T	F	T	F F TF F F T T FF F
T	F	F	F F FT F F T T FF F
F	T	T	T T TT TT T T TT T
F	T	F	T T FT T F F T TT T
F	F	T	T T TT TT T T FT T
F	F	F	T T FT T F F T FT T
			1 5 24 3 7 6 11 9 10 8

The argument is valid.

33. p: Will Smith wins an Academy Award.
     q: Will Smith retires from acting.

   p → q
   ~p
   ∴ ~q

The argument is invalid (fallacy of the inverse.).

35. p: The baby is a boy.
     q: The baby will be named Alexander Martin.

   p → q
    p
   ∴ q

The argument is valid (law of detachment).

37. p: The guitar is a Les Paul model.
     q: The guitar is made by Gibson.

   p → q
    ~q
   ∴ ~p

This argument is valid by the law of contraposition.

39. p: We planted the garden by the first Friday in April.
     q: We will have potatoes by the Fourth of July.

   p → q
    q
   ∴ p

This is the fallacy of the converse; thus the argument is invalid.

41. p: Sarah Hughes will win an Olympic gold medal in figure skating.
     q: Joey Cheek will win an Olympic gold medal in speed skating.

   p ∨ q
    ~ p
   ∴ q

The argument is a disjunctive syllogism and is therefore valid

43. p: It is cold.
     q: The graduation will be held indoors.
     r: The fireworks will be postponed.
   [(p → q) ∧ (q → r)] → (p → r)

This argument is valid because of the law of syllogism.

45. m: Marie works for the post office
    j: Jim works for target.
    t: Tommy gets an internship.

m	j	t	[(m ∧ j) ∧ (j → t)] → (t → m)				
T	T	T	T	T	T	T	T
T	T	F	T	F	F	T	T
T	F	T	F	F	T	T	T
T	F	F	F	F	T	T	T
F	T	T	F	F	T	T	F
F	T	F	F	F	F	T	T
F	F	T	F	F	T	T	F
F	F	F	F	F	T	T	T
			1	3	2	5	4

The argument is valid.

47. s: It is snowing.
    g: I am going skiing.
    c: I will wear a coat.

s	g	c	[(s ∧ g) ∧ (g → c)] → (s → c)				
T	T	T	T	T	T	T	T
T	T	F	T	F	F	T	F
T	F	T	F	F	T	T	T
T	F	F	F	F	T	T	F
F	T	T	F	F	T	T	T
F	T	F	F	F	F	T	T
F	F	T	F	F	T	T	T
F	F	F	F	F	T	T	T
			1	3	2	5	4

The argument is valid.

49. h: The house has electric heat.
    b: The Flynns will buy the house.
    p: The price is less than $100,000.

h	b	p	[(h → b) ∧ (~p → ~b)] → (h → p)						
T	T	T	T	T	F	T	F	T	T
T	T	F	T	F	T	F	F	T	F
T	F	T	F	F	F	T	T	T	T
T	F	F	F	F	T	T	T	T	F
F	T	T	T	T	F	T	F	T	T
F	T	F	T	F	T	F	F	T	T
F	F	T	T	T	F	T	T	T	T
F	F	F	T	T	T	T	T	T	T
			1	5	2	4	3	7	6

The argument is valid.

51. p: The prescription is called in to Walgreen's.
    q. You pick up the prescription at 4:00 p.m.

    p → q

    ~q

    ∴ ~p

The argument is the law of contraposition and is valid.

53. s: Max is playing Game Boy with the sound off.
    h: Max is wearing headphones.

    s ∨ h

    ~s

    ∴ h

This argument is an example of disjunctive syllogism and is therefore valid.

55. t: The test was easy.
    g: I received a good grade.

t	g	[(t ∧ g) ∧ (~t ∨ ~g)] → ~t						
T	T	T	F	F	F	F	T	F
T	F	F	F	F	T	T	T	F
F	T	F	F	T	T	F	T	T
F	F	F	F	T	T	T	T	T
		1	5	2	4	3	7	6

The argument is valid.

57. c: The baby is crying.
    h: The baby is hungry.

c	h	$[(c \wedge \sim h) \wedge (h \to c)] \to h$
T	T	T F F F T T T
T	F	T T T T T F F
F	T	F F F F F T T
F	F	F F T F T T F
		1 3 2 5 4 7 6

The argument is invalid.

59. f: The football team wins the game. $f \to d$
    d: Dave played quarterback. $d \to \sim s$
    s: The team is in second place. $\therefore f \to s$

Using the law of syllogism $f \to \sim s$, so this
argument is invalid.

61. Your face will break out. (law of detachment)

63. I am stressed out. (disjunctive syllogism)

65. You did not close the deal. (law of contraposition)

67. If you do not pay off your credit card bill, then
    the bank makes money. (law of syllogism)

69. No. An argument is <u>invalid</u> only when the
    conjunction of the premises is true and the
    conclusion is false.

Exercise Set 3.6

1. a) It is a valid argument.
   b) It is an invalid argument.

3. a)   b)   c)

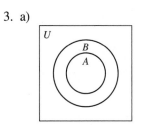
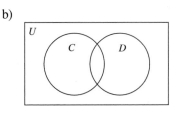
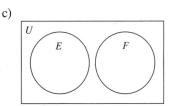

5. Yes. If the conjunction of the premises is false in all
   cases, then the argument is valid regardless of the
   truth value of the conclusion.

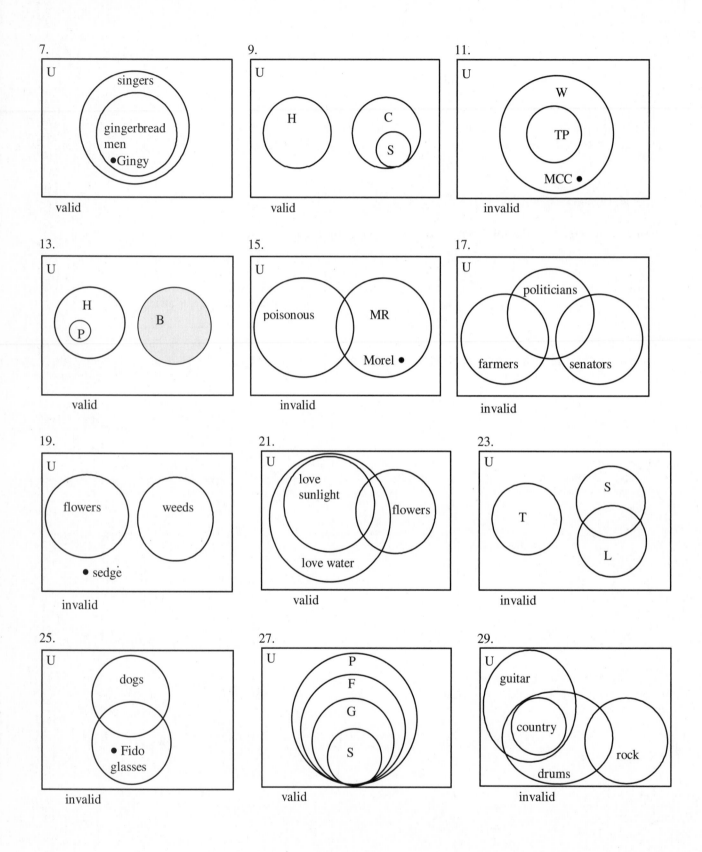

31. $[(p \to q) \wedge (p \vee q)] \to \sim p$ can be expressed as a set statement by $[(P' \cup Q) \cap (P \cup Q)] \subseteq P'$. If this statement is true, then the argument is valid; otherwise, the argument is invalid.

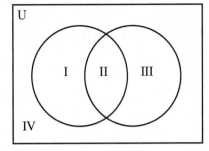

Set	Regions
$P' \cup Q$	II, III, IV
$P \cup Q$	I, II, III
$(P' \cup Q) \cap (P \cup Q)$	II, III
P'	III, IV

Since $(P' \cup Q) \cap (P \cup Q)$ is not a subset of P', the argument is invalid.

Exercise Set 3.7

1. a) In a series circuit, the current can take only one path.

   b) $\wedge$ (and)

3. One of the two switches will always be open.

5. $p \wedge q$

	Light	
p	q	$p \wedge q$
T	T	T
T	F	F
F	T	F
F	F	F

7. $(p \vee q) \wedge \overline{q}$

		Light		
p	q	$(p \vee q)$	$\wedge$	$\overline{q}$
T	T	T	F	F
T	F	T	T	T
F	T	T	F	F
F	F	F	F	T

9. $(p \wedge q) \wedge [(p \wedge \overline{q}) \vee r]$

   It is clear from the $(p \wedge q)$ condition that both p and q must be closed for the bulb to light. In this case the upper branch of the parallel portion of the circuit is open since it includes $\overline{q}$, so for the bulb to light, r must be closed. Thus the bulb lights only if p, q and r are all T.

11. $p \vee q \vee (r \wedge \overline{p})$

   Reading from the circuit, we can see that the only case in which the bulb will *not* light is: p is F, q is F, and r is F.

13.

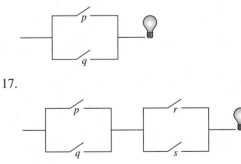

15.

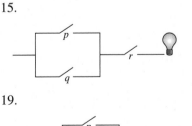

17.

19.

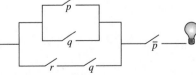

21. $p \vee q$; $\overline{p} \wedge \overline{q}$

Not equivalent; in fact, by De Morgan's laws,
$(p \vee q) \Leftrightarrow \overline{(\overline{p} \wedge \overline{q})}$, so the first circuit will light
the bulb exactly when the second one does not.

23. $\big[(p \wedge q) \vee r\big] \wedge p$; $(q \vee r) \wedge p$

Clearly both circuits light the bulb only if p is T.
In this case $p \wedge q$ has the same value as q alone,
so $(p \wedge q) \vee r$ will have the same value as $q \vee r$.
Thus the two circuits are equivalent.

25. $(p \vee \overline{p}) \wedge q \wedge r$; $p \wedge q \wedge r$

Not equivalent, since the first circuit will light the bulb when p is F, q is T and r is T, while
p being F breaks the second circuit.

27. a)

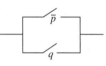

b)

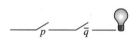

Review Exercises

1. Some gift cards are exchangeable.

2. Some bears are not mammals.

3. No women are presidents.

4. All pine trees are green.

5. The coffee is Maxwell House or the coffee is hot.

6. The coffee is not hot and the coffee is strong.

7. If the coffee is hot, then the coffee is strong and it is not Maxwell House.

8. The coffee is Maxwell House if and only if the coffee is not strong.

9. The coffee is not Maxwell House, if and only if the coffee is strong and the coffee is not hot.

10. The coffee is Maxwell House or the coffee is not hot, and the coffee is not strong.

11. $r \wedge q$

12. $p \rightarrow r$

13. $(r \rightarrow q) \vee \sim p$

14. $(q \leftrightarrow p) \wedge \sim r$

15. $(r \wedge q) \vee \sim p$

16. $\sim (r \wedge q)$

17.

p	q	(p	∨	q)	∧	~	p
T	T		T		F	F	
T	F		T		F	F	
F	T		T		T	T	
F	F		F		F	T	
			1		3	2	

18.

p	q	q	↔	(p	∨	~ q)
T	T	T	T	T	T	F
T	F	F	F	T	T	T
F	T	T	F	F	F	F
F	F	F	F	F	T	T
		1	5	2	4	3

19.

p	q	r	(p ∨ q)	↔	(p ∨ r)
T	T	T	T	T	T
T	T	F	T	T	T
T	F	T	T	T	T
T	F	F	T	T	T
F	T	T	T	T	T
F	T	F	T	F	F
F	F	T	F	F	T
F	F	F	F	T	F
			1	3	2

20.

p	q	r	p	∧	(~ q	∨	r)
T	T	T	T	T	F	T	T
T	T	F	T	F	F	F	F
T	F	T	T	T	T	T	T
T	F	F	T	T	T	T	F
F	T	T	F	F	F	T	T
F	T	F	F	F	F	F	F
F	F	T	F	F	T	T	T
F	F	F	F	F	T	T	F
			4	5	1	3	2

21.

p	q	r	P	→	(q	∧	~ r)
T	T	T	T	F	T	F	F
T	T	F	T	T	T	T	T
T	F	T	T	F	F	F	F
T	F	F	T	F	F	F	T
F	T	T	F	T	T	F	F
F	T	F	F	T	T	T	T
F	F	T	F	T	F	F	F
F	F	F	F	T	F	F	T
			4	5	1	3	2

22.

p	q	r	(p ∧ q)	→	~ r
T	T	T	T	F	F
T	T	F	T	T	T
T	F	T	F	T	F
T	F	F	F	T	T
F	T	T	F	T	F
F	T	F	F	T	T
F	F	T	F	T	F
F	F	F	F	T	T
			1	3	2

23. The premises are true but the conclusion is false and $T \rightarrow F$ is false, so the statement is false.

24. Since $17 + 4 = 21$ and $3 - 9 = -6$, both of the statements joined by *and* are true, so the compound statement is true.

25. p: Oregon borders the Pacific Ocean.
    q: California borders the Atlantic Ocean.
    r: Minnesota is south of Texas.
    $(p \vee q) \rightarrow r$
    $(T \vee F) \rightarrow F$
    $\quad T \;\; \rightarrow F$
    $\qquad\quad F$

26. p: $15 - 7 = 22$ $(p \vee q) \wedge r$
    q: $4 + 9 = 13$ $(F \vee T) \wedge T$
    r : $9 - 8 = 1$ $T \;\; \wedge T$
                               T

27. $(p \rightarrow \sim r) \vee (p \wedge q)$
    $(T \rightarrow T\,) \vee (T \wedge F)$
    $\quad\; T \quad \vee \quad F$
    $\qquad\qquad T$

28. $(p \vee q) \leftrightarrow (\sim r \wedge p)$
    $(T \vee F) \leftrightarrow (T \wedge T)$
    $\quad T \quad \leftrightarrow \quad T$
    $\qquad\qquad T$

29. $\sim r \leftrightarrow [(p \vee q) \leftrightarrow \sim p]$
    $\; T \leftrightarrow [(T \vee F) \leftrightarrow F]$
    $\; T \leftrightarrow [\quad T \quad \leftrightarrow F]$
    $\; T \leftrightarrow \qquad\quad F$
    $\qquad F$

30. $\sim [(q \wedge r) \rightarrow (\sim p \vee r)]$
    $\sim [(F \wedge F) \rightarrow (\; F \vee F)]$
    $\sim [\quad F \quad \rightarrow \qquad F]$
    $\qquad\quad \sim T$
    $\qquad\qquad F$

31.

p	q	$\sim p \vee \sim q$	$\sim p \leftrightarrow q$
T	T	F F F	F F T
T	F	F T T	F T F
F	T	T T F	T T T
F	F	T T T	T F F
		1 3 2	1 3 2

The statements are not equivalent.

32. Using the fact that $(p \rightarrow q) \Leftrightarrow (\sim p \vee q)$, we can conclude that $\sim p \rightarrow \sim q \Leftrightarrow p \vee \sim q$.

33.

p	q	r	$\sim p \vee (q \wedge r)$	$(\sim p \vee q) \wedge (\sim p \vee r)$
T	T	T	F T T	F T T T F T T
T	T	F	F F F	F T T F F F F
T	F	T	F F F	F F F F F T T
T	F	F	F F F	F F F F F F F
F	T	T	T T T	T T T T T T T
F	T	F	T T F	T T T T T T F
F	F	T	T T F	T T F T T T T
F	F	F	T T F	T T F T ·T F
			2 3 1	1 3 2 7 4 6 5

The statements are equivalent.

34.

p	q	(~ q	→	p)	∧	p		~	(~ p	↔	q)	∨	p
T	T	F	T	T	T	T		T	F	F	T	T	T
T	F	T	T	T	T	T		F	F	T	F	T	T
F	T	F	T	F	F	F		F	T	T	T	F	F
F	F	T	F	F	F	F		T	T	F	F	T	F
		1	3	2	5	4		4	1	3	2	6	5

The statements are not equivalent.

35. p: Bobby Darin sang *Mack the Knife*.

   q: Elvis wrote *Memphis*.

In symbolic form, the statement is p ∧ ~q. We are given that p ∧ ~q ⇔ ~(p → q). So an equivalent statement is: It is false that if Bobby Darin sang *Mack the Knife* then Elvis wrote *Memphis*.

36. p: Lynn Swann played for the Steelers.

   q: Jack Tatum played for the Raiders.

In symbolic form, the statement is p ∨ q. We are given that ~p ∨ q ⇔ (p → q), so p ∨ q ⇔ (~p → q). Thus an equivalent statement is: If Lynn Swann did not play for the Steelers then Jack Tatum played for the Raiders.

37. p: Altec Lansing only produces speakers.

   q: Harmon Kardon only produces stereo receivers.

The symbolic form is ~ (p ∨ q).

Using De Morgan's Laws, we get

~ (p ∨ q) ⇔ ~ p ∧ ~ q.

Altec Lansing does not produce only speakers and Harmon Kardon does not produce only stereo receivers.

38. p: Travis Tritt won an Academy Award.

   q: Randy Jackson does commercials for Milk Bone dog biscuits.

The symbolic form is ~ p ∧ ~ q.

Using De Morgan's Laws, we get

~ p ∧ ~ q ⇔ ~ (p ∨ q). It is false that Travis Tritt won an Academy Award or Randy Jackson does commercials for Milk Bone dog biscuits.

39. p: The temperature is above 32 degrees Fahrenheit.

   q: We will go ice fishing at O'Leary's Lake.

The symbolic form is ~ p → q.

Using DeMorgan's Laws, we get ~ p → q ⇔ p ∨ q.

The temperature is above 32 degrees Fahrenheit or we will go ice fishing at O'Leary's Lake.

40. Converse: If you soften your opinion, then you hear a new voice today.

Inverse: If you don't hear a new voice today then you don't soften your opinion

Contrapositive: If you don't soften your opinion, then you don't hear a new voice today.

41. Converse: If we are going to learn the table's value then we take the table to *Antiques Roadshow*.

Inverse: If we do not take the table to *Antiques Roadshow*, then we will not learn the table's value.

Contrapositive: If we are not going to learn the table's value then we do not take the table to *Antiques Roadshow*.

42. Converse: If Maureen Gerald is helping at school, then she is not in attendance.

Inverse: If Maureen Gerald is in attendance, then she is not helping at school.

Contrapositive: If Maureen Gerald is not helping at school, then she is in attendance.

43. Converse: If we do not buy a desk at Miller's
    Furniture, then the desk is made by Winner's Only
    and is in the Rose catalog.
    Inverse: If we will buy a desk at Miller's furniture,
    then the desk is not made by Winner's Only or it
    is not in the Rose catalog.
    Contrapositive: If the desk is not made by
    Winner's Only or is not in the Rose catalog,
    then we will buy a desk at Miller's Furniture.

44. Converse: If I let you attend the prom,
    then you get straight A's on your report
    card.
    Inverse: If you do not get straight A's on
    your report card, then I will not let you
    attend the prom.
    Contrapositive: If I do not let you attend
    the prom, then you do not get straight
    A's on your report card.

45. p: You read 10 books in the summer.
    q: You reach your goal.
    In symbolic form, the statements are: a) $p \rightarrow q$,
    b) $\sim p \vee q$, and c) $\sim (p \wedge \sim q)$. Using the fact that
    $p \rightarrow q$ is equivalent to $\sim p \vee q$, statements (a) and (b)
    are equivalent. Using DeMorgan's Laws on
    statement (b) we get $\sim (p \wedge \sim q)$.

    Therefore all 3 statements are equivalent.

46. p: The screwdriver is on the workbench.
    q: The screwdriver is on the counter.
    In symbolic form, the statements are: a) $p \leftrightarrow \sim q$,
    b) $\sim q \rightarrow \sim p$, and c) $\sim (q \wedge \sim p)$. Looking at the truth
    tables for statements (a), (b), and (c) we can conclude
    that none of the statements are equivalent.

		a)			b)			c)		
p	q	$p \leftrightarrow \sim q$			$\sim q \rightarrow \sim p$			$\sim (q \wedge \sim p)$		
T	T	T	F	F	F	T	F	T	TF	F
T	F	T	T	T	T	F	F	T	FF	F
F	T	F	T	F	F	T	T	F	TT	T
F	F	F	F	T	T	T	T	T	FF	T
		1	3	2	1	3	2	4	1 3	2

47. p: $2 + 3 = 6$.
    q: $3 + 1 = 5$.
    In symbolic form, the statements are: a) $p \rightarrow q$,
    b) $p \leftrightarrow \sim q$, and c) $\sim q \rightarrow \sim p$.
    Statement (c) is the contrapositive of statement
    (a). Therefore statements (a) and (c) are equivalent.
    For p and q false, (a) and (c) are true but (b) is false,
    so (b) is not equivalent to (a) and (c).

48. p: The sale is on Tuesday.

q: I have money.

r : I will go to the sale.

In symbolic form the statements are: a) $(p \wedge q) \to r$, b) $r \to (p \wedge q)$, and c) $r \vee (p \wedge q)$. The truth table for statements (a), (b), and (c) shows that none of the statements are equivalent.

p	q	r	$(p \wedge q) \to r$			$r \to (p \wedge q)$			$r \vee (p \wedge q)$		
T	T	T	T	T T		T T	T		T T	T	
T	T	F	T	F F		F T	T		F T	T	
T	F	T	F	T T		T F	F		T T	F	
T	F	F	F	T F		F T	F		F F	F	
F	T	T	F	T T		T F	F		T T	F	
F	T	F	F	T F		F T	F		F F	F	
F	F	T	F	T T		T F	F		T T	F	
F	F	F	F	T F		F T	F		F F	F	
			1	3 2		1 3	2		1 3	2	

49.

p	q	$[(p \to q) \wedge \sim p] \to q$				
T	T	T	F F	T T		
T	F	F	F F	T F		
F	T	T	T T	T T		
F	F	T	T T	F F		
		1	3 2	5 4		

The argument is invalid.

50.

p	q	r	$[(p \wedge q) \wedge (q \to r)] \to (p \to r)$				
T	T	T	T	T T	T	T
T	T	F	T	F F	T	F
T	F	T	F	F T	T	T
T	F	F	F	F T	T	F
F	T	T	F	F T	T	T
F	T	F	F	F F	T	T
F	F	T	F	F T	T	T
F	F	F	F	F T	T	T
			1	3 2	5	4

The argument is valid.

51. p: Jose Macias is the manager.

q: Kevin Geis is the coach.

r: Tim Weisman is the umpire

$p \to q$

$q \to r$

$\therefore p \to r$

This argument is in the form of the law of syllogism so it is valid.

52. p: We eat at Joe's. $p \to q$

p: We get heartburn. $q \vee r$

r : We take Tums. $\therefore \sim q$

If p is F, q is T, and r is either T or F, the premises are both true but the conclusion is false, so the argument is invalid.

53.

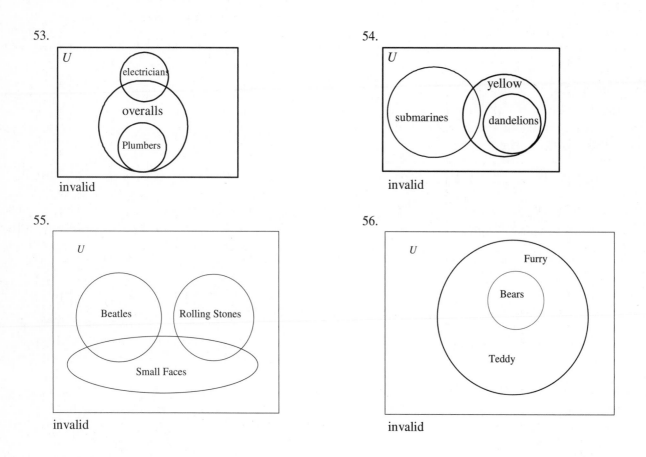

invalid

54.

invalid

55.

invalid

56.

invalid

57. a) $p \wedge [(q \wedge r) \vee \overline{p}]$

b) For the bulb to be on, the first switch on the left must be closed, so p is T. The eliminates the bottom branch of the parallel portion, so both switches on the top branch must be closed. Thus the bulb lights exactly when p, q, and r are all T

58.

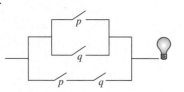

59. Symbolically the two circuits are $(p \vee q) \wedge (\overline{q} \vee \overline{p})$ and $(p \wedge \overline{q}) \vee (q \wedge \overline{p})$.

p	q	$\overline{p}$	$\overline{q}$	$(p \vee q)$	$\wedge$	$(\overline{q} \vee \overline{p})$	p	q	$\overline{p}$	$\overline{q}$	$(p \wedge \overline{q})$	$\vee$	$(q \wedge \overline{p})$
T	T	F	F	T	F	F	T	T	F	F	F	F	F
T	F	F	T	T	T	T	T	F	F	T	T	T	F
F	T	T	F	T	T	T	F	T	T	F	F	T	T
F	F	T	T	F	F	T	F	F	T	T	F	F	F

The next-to last columns of these truth tables are identical, so the circuits are equivalent.

Chapter Test

1. $(p \land r) \lor \sim q$

2. $(r \to q) \lor \sim p$

3. $\sim (r \leftrightarrow \sim q)$

4. Ann is not the secretary and Elaine is the president, if and only if Dick is not the vice president.

5. If Ann is the secretary or Dick is not the vice president, then Elaine is the president.

6.

p	q	r	[~ (p → r)] ∧ q			
T	T	T	F	T	F	T
T	T	F	T	F	T	T
T	F	T	F	T	F	F
T	F	F	T	F	F	F
F	T	T	F	T	F	T
F	T	F	F	T	F	T
F	F	T	F	T	F	F
F	F	F	F	T	F	F
			2	1	4	3

7.

p	q	r	(q ↔ ~ r) ∨ p				
T	T	T	T	F	F	T	T
T	T	F	T	T	T	T	T
T	F	T	F	T	F	T	T
T	F	F	F	F	T	T	T
F	T	T	T	F	F	F	F
F	T	F	T	T	T	T	F
F	F	T	F	T	F	T	F
F	F	F	F	F	T	F	F
			1	3	2	5	4

8. p: $2 + 6 = 8$
   q: $7 - 12 = 5$
   $p \lor q$
   $T \lor F$
   T

9. p: A scissors can cut paper.
   q: A dime equals 2 nickels.
   r : Louisville is a city in Kentucky.
   $\quad (p \lor q) \leftrightarrow r$
   $\quad (T \lor T) \leftrightarrow T$
   $\quad\quad T \quad \leftrightarrow T$
   $\quad\quad\quad T$

10. $(r \lor q) \leftrightarrow (p \land \sim q)$
    $(T \lor F) \leftrightarrow (T \land T)$
    $\quad T \quad \leftrightarrow \quad T$
    $\quad\quad\quad T$

11. $[\sim(r \to \sim p)] \land (q \to p)$
    $[\sim(T \to F\ )] \land (F \to T)$
    $[\ \sim (F)\quad] \land \quad T$
    $\quad\quad T \quad\quad \land \quad T$
    $\quad\quad\quad\quad T$

12. By DeMorgan's Laws ,
    $\sim p \lor q \Leftrightarrow \sim (\sim (\sim p) \land \sim q).$
    and this is equivalent to
    $\sim(p \land \sim q).$

13. p: The bird is red.
    q: It is a cardinal.
    In symbolic form the statements
    are: a) $p \to q$, b) $\sim p \lor q$,
    and c) $\sim p \to \sim q$.
    Statement (c) is the inverse of
    statement (a) and thus they cannot
    be equivalent. Using the fact that
    $p \to q \Leftrightarrow \sim p \lor q$, to rewrite
    statement (a) we get $\sim p \lor q$.
    Therefore statements (a) and (b)
    are equivalent.

14. p: The test is today.   q: The concert is tonight. In symbolic form the
     statements are: a) ~ (p ∨ q), b) ~ p ∧ ~ q, and ~ p → ~ q.
     Applying DeMorgan's Law to statement (a) we get: ~ p ∧ ~ q.
     Therefore statements (a) and (b) are equivalent. When we compare
     the truth tables for statements (a), (b), and (c) we see that only
     statements (a) and (b) are equivalent.

p	q	~ (p ∨ q)		~ p ∧ ~ q			~ p → ~ q		
T	T	F	T	F	F	F	F	T	F
T	F	F	T	F	F	T	F	T	T
F	T	F	T	T	F	F	T	F	F
F	F	T	F	T	T	T	T	T	T
		2	1	1	3	2	1	3	2

15. s: The soccer team won the game.
     f: Sue played fullback.
     p: The team is in second place.
     This argument is the law of
     syllogism and therefore it is
     valid.

$$s \rightarrow f$$
$$\underline{f \rightarrow p}$$
$$s \rightarrow p$$

     This argument is the law of
     syllogism and therefore it is valid.

16.

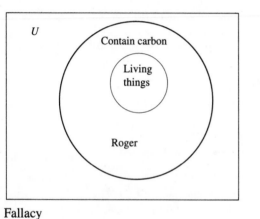

Fallacy

17. Some highways are not roads

18. Nick did not play football or Max
     did not play baseball.

19. Converse: If today is Saturday, then the
     garbage truck comes.
     Inverse: If the garbage truck does not
     come today, then today is not Saturday.
     Contrapositive: If today is not Saturday,
     then the garbage truck does not come.

20.

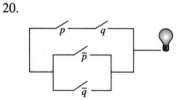

CHAPTER FOUR

SYSTEMS OF NUMERATION

Exercise Set 4.1

1. A **number** is a quantity, and it answers the question, "How many?" A **numeral** is a symbol used to represent the number.

3. A **system of numeration** consists of a set of numerals and a scheme or rule for combining the numerals to represent numbers.

5. The Hindu-Arabic numeration system

7. In a **multiplicative system**, there are numerals for each number less than the base and for powers of the base. Each numeral less than the base is multiplied by a numeral for the power of the base, and these products are added to obtain the number.

9. $100+100+100+10+10+10+10$
$+1+1+1+1+1 = 345$

11. $1000+1000+100+100+100+100+10+10$
$+1+1+1 = 2423$

13. $100,000+100,000+100,000+10,000$
$+10,000+10,000+1000+1000+1000+1000$
$+100+100+10+1+1+1+1 = 334,214$

15. ∩∩∩∩|||

17. ⸗⸗∩∩∩|||||

19. ◁||||||⸗999999999∩∩∩|||||

21. $5+1+1+1 = 8$

23. $(50-10)+1+1+1 = 43$

25. $1000+100+100+10+10+10+5+1 = 1236$

27. $1000+1000+(1000-100)+(50-10)+5+1$
$= 2946$

29. $10(1000)+1000+1000+500+100+50+10$
$+5+1 = 12,666$

31. $9(1000)+(500-100)+50+10+(5-1) = 9464$

33. XXVII

35. CCCXLI

37. MMV

39. $\overline{\text{IV}}$DCCXCIII

41. $\overline{\text{IX}}$CMXCIX

43. $\overline{\text{XX}}$DCXLIV

45. $7(10)+4 = 74$

47. $4(1000)+8(10)+1 = 4081$

49. $8(1000)+5(100)+5(10) = 8550$

51. $4(1000)+3 = 4003$

53. 五十三

55. 三百七十八

61

57.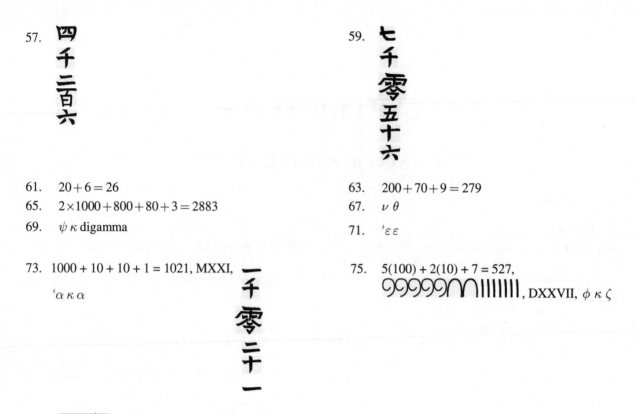

59.

61. $20 + 6 = 26$

63. $200 + 70 + 9 = 279$

65. $2 \times 1000 + 800 + 80 + 3 = 2883$

67. $\nu\,\theta$

69. $\psi\,\kappa$ digamma

71. $'\varepsilon\varepsilon$

73. $1000 + 10 + 10 + 1 = 1021$, MXXI,
$'\alpha\,\kappa\,\alpha$

75. $5(100) + 2(10) + 7 = 527$,
, DXXVII, $\phi\,\kappa\,\zeta$

77. $\overline{\text{CMXCIX}}$CMXCIX

79. Advantage: You can write some numbers more compactly.
Disadvantage: There are more numerals to memorize.

81. Advantage: You can write some numbers more compactly.
Disadvantage: There are more numerals to memorize.
The Hindu-Arabic system has fewer symbols, more compact notation, the inclusion of zero, and the capability of expressing decimal numbers and fractions.

83. MM

85. a) , b) Answers will vary.

Exercise Set 4.2

1. Positional value system

3. $40 \rightarrow$ four tens, $400 \rightarrow$ four hundreds

5. a) 10
   b) 0, 1, 2, 3, 4, 5, 6, 7, 8, 9

7. Write each digit times its corresponding positional value.

9. a) There may be confusion because numbers could be interpreted in different ways. For example, ❙ could be interpreted to be either 1 or 60.
   b) ❙❙ ＜❙❙❙ for both numbers; $133 = 2(60) + 13(1)$ and $7980 = 133(60)$

11. $1, 20, 18 \times 20, 18 \times (20)^2, 18 \times (20)^3$

13. $(2 \times 10) + (3 \times 1)$

15. $(3 \times 100) + (5 \times 10) + (9 \times 1)$

17. $(8 \times 100) + (9 \times 10) + (7 \times 1)$

19. $(4 \times 1000) + (3 \times 100) + (8 \times 10) + (7 \times 1)$

21. $(1 \times 10,000) + (6 \times 1000) + (4 \times 100) + (0 \times 10) + (2 \times 1)$

23. $(3 \times 100,000) + (4 \times 10,000) + (6 \times 1000) + (8 \times 100) + (6 \times 10) + (1 \times 1)$

25. $(10 + 1 + 1 + 1 + 1)(1) = 14$

27. $(10 + 1 + 1 + 1)(60) + (1 + 1 + 1 + 1)(1) = 13(60) + 4(1) = 780 + 4 = 784$

29. $1(60^2) + (10 + 10 + 1)(60) + (10 - (1 + 1))(1) = 3600 + 21(60) + (10 - 2)(1) = 3600 + 1260 + 8 = 4868$

31. 35 is 35 units.

33. 471 is 7 groups of 60 and 51 units remaining.

35. 3685 is 1 group of 3600, 1 group of 60, and 25 units remaining.

37. $2(20) + 17(1) = 40 + 17 = 57$

39. $13(18 \times 20) + 0(20) + 2(1) = 4680 + 0 + 2 = 4682$

41. $11(18 \times 20) + 2(20) + 0(1) = 3960 + 40 + 0 = 4000$

43.

45.

$$\begin{array}{r} 14 \\ 20\overline{\smash{\big)}\,297} \\ \underline{280} \\ 17 \end{array}$$

$297 = 14(20) + 17(1)$

47.

$$\begin{array}{r} 6 \\ 360\overline{\smash{\big)}\,2163} \\ \underline{2160} \\ 3 \end{array}$$

$2163 = 6(360) + 0(20) + 3(1)$

49. Hindu-Arabic:

$5(18 \times 20) + 7(20) + 4(1) = 1800 + 140 + 4 = 1944$

Babylonian: $1944 = 32(60) + 24(1)$

51.

$(\triangle \times \bigcirc^2) + (\square \times \bigcirc) + (\lozenge \times 1)$

53. a) No largest number; The positional values above 18×20 are $18 \times 20^2, 18 \times 20^3, \ldots$

b) $999,999 = 6(18 \times 20^3) + 18(18 \times 20^2) + 17(18 \times 20) + 13(20) + 19(1)$

55. $2(60) + 23(1) = 120 + 23 = 143$

   23

   $143 + 23 = 166$

   $166 = 2(60) + 46(1)$

57. $7(18 \times 20) + 6(20) + 15(1) = 2520 + 120 + 15 = 2655$

   $6(18 \times 20) + 7(20) + 13(1) = 2160 + 140 + 13 = 2313$

   $2655 + 2313 = 4968$

   $4968 = 13(18 \times 20) + 14(20) + 8(1)$

59. Advantages: In general, a place-value system is more compact; large and small numbers can be written more easily; there are fewer symbols to memorize.

   Disadvantage: If many of the symbols in the numeral represent zero, then a place-value system may be less compact.

61. Answers will vary.

Exercise Set 4.3

1. a) Multiply each digit by its positional value and find the sum of these products.

   b) Divide the base 10 numeral by the highest power of the base that is smaller than the base 10 numeral. Record the quotient and repeat the procedure with the remainder, using the next lower power of the base. Continue until the remainder is smaller than the base. The list of quotients followed by the remainder is the numeral in the new base.

3. a) The last digit is equal to the base; all digits should represent numbers smaller than the base.

   b) The only digits used in base 2 are 0 and 1, so 3 cannot be part of a base 2 numeral.

   c) There is no digit L in base 12.

   d) There is no digit G in base 16.

5. $1_2 = 1 \times 1 = 1$

7. $23_5 = (2 \times 5) + (3 \times 1) = 13$

9. $270_8 = (2 \times 64) + (7 \times 8) + (0 \times 1) = 184$

11. $309_{12} = (3 \times 144) + (0 \times 12) + (9 \times 1) = 441$

13. $573_{16} = (5 \times 256) + (7 \times 16) + (3 \times 1) = 1395$

15. $110101_2 = 32 + 16 + 4 + 1 = 53$

17. $7654_8 = (7 \times 512) + (6 \times 64) + (5 \times 8) + (4 \times 1) = 4012$

19. $A91_{12} = (10 \times 144) + (9 \times 12) + 1 = 1549$

21. $C679_{16} = (12 \times 4096) + (6 \times 256) + (7 \times 16) + 9 = 50,809$

23. To convert 6 to base 2   ... 16 8 4 2 1

$$
\begin{array}{ccc}
\underline{1} & \underline{1} & \underline{0} \\
4\,\overline{)\,6} & 2\,\overline{)\,2} & 0\,\overline{)\,0} \\
\underline{4} & \underline{2} & \underline{0} \\
2 & 0 & 0
\end{array}
$$

$6 = 110_2$

25. To convert 347 to base 3   ... 243 81 27 9 3 1

$$
\begin{array}{cccccc}
\underline{1} & \underline{1} & \underline{0} & \underline{2} & \underline{1} & \underline{2} \\
243\,\overline{)\,347} & 81\,\overline{)\,104} & 27\,\overline{)\,26} & 9\,\overline{)\,23} & 3\,\overline{)\,5} & 1\,\overline{)\,2} \\
\underline{243} & \underline{81} & \underline{0} & \underline{18} & \underline{3} & \underline{2} \\
104 & 23 & 23 & 5 & 2 & 0
\end{array}
$$

$347 = 110212_3$

27. To convert 53 to base 4   ... 256 64 16 4 1

$$
\begin{array}{ccc}
\underline{3} & \underline{1} & \underline{1} \\
16\,\overline{)\,53} & 4\,\overline{)\,5} & 1\,\overline{)\,1} \\
\underline{48} & \underline{4} & \underline{1} \\
5 & 1 & 0
\end{array}
$$

$53 = 311_4$

29. To convert 102 to base 5   ... 625 125 25 5 1

$$
\begin{array}{ccc}
\underline{4} & \underline{0} & \underline{2} \\
25\,\overline{)\,102} & 5\,\overline{)\,2} & 1\,\overline{)\,2} \\
\underline{100} & \underline{0} & \underline{2} \\
2 & 2 & 0
\end{array}
$$

$102 = 402_5$

31. To convert 1098 to base 8   ... 4096 512 64 8 1

$$
\begin{array}{cccc}
\underline{2} & \underline{1} & \underline{1} & \underline{2} \\
512\,\overline{)\,1098} & 64\,\overline{)\,74} & 8\,\overline{)\,10} & 1\,\overline{)\,2} \\
\underline{1024} & \underline{64} & \underline{8} & \underline{2} \\
74 & 10 & 2 & 0
\end{array}
$$

$1098 = 2112_8$

33. To convert 1432 to base 9   ... 6561 729 81 9 1

$$
\begin{array}{cccc}
\underline{1} & \underline{8} & \underline{6} & \underline{1} \\
729\,\overline{)\,1432} & 81\,\overline{)\,703} & 9\,\overline{)\,55} & 1\,\overline{)\,1} \\
\underline{729} & \underline{648} & \underline{54} & \underline{1} \\
703 & 55 & 1 & 0
\end{array}
$$

$1432 = 1861_9$

35. To convert 9004 to base 12   ... 20,736 1728 144 12 1

$$
\begin{array}{cccc}
\underline{5} & \underline{2} & \underline{6} & \underline{4} \\
1728\,\overline{)\,9004} & 144\,\overline{)\,364} & 12\,\overline{)\,76} & 1\,\overline{)\,4} \\
\underline{8640} & \underline{288} & \underline{72} & \underline{4} \\
364 & 76 & 4 & 0
\end{array}
$$

$9004 = 5264_{12}$

37. To convert 493 to base 16 ... 65,536   4096   256   16   1

$$\begin{array}{r} 1 \\ 256\overline{\big)\,493} \\ \underline{256} \\ 237 \end{array}\qquad \begin{array}{r} 14 \quad = \text{E} \\ 16\overline{\big)\,237} \\ \underline{224} \\ 13 \end{array}\qquad \begin{array}{r} 13 \quad = \text{D} \\ 1\overline{\big)\,13} \\ \underline{13} \\ 0 \end{array}$$

$$493 = 1\text{ED}_{16}$$

39. To convert 9455 to base 16 ... 65,536   4096   256   16   1

$$\begin{array}{r} 2 \\ 4096\overline{\big)\,9455} \\ \underline{8192} \\ 1263 \end{array}\qquad \begin{array}{r} 4 \\ 256\overline{\big)\,1263} \\ \underline{1024} \\ 239 \end{array}\qquad \begin{array}{r} \text{E} \quad = 14 \\ 16\overline{\big)\,239} \\ \underline{224} \\ 15 \end{array}\qquad \begin{array}{r} 15 \quad = \text{F} \\ 1\overline{\big)\,15} \\ \underline{15} \\ 0 \end{array}$$

$$9455 = 24\text{EF}_{16}$$

41. To convert 2009 to base 3 ... 2187   729   243   81   27   9   3   1

$$\begin{array}{r} 2 \\ 729\overline{\big)\,2009} \\ \underline{1458} \\ 551 \end{array}\quad \begin{array}{r} 2 \\ 243\overline{\big)\,551} \\ \underline{486} \\ 65 \end{array}\quad \begin{array}{r} 0 \\ 81\overline{\big)\,65} \\ \underline{0} \\ 65 \end{array}\quad \begin{array}{r} 2 \\ 27\overline{\big)\,65} \\ \underline{54} \\ 11 \end{array}\quad \begin{array}{r} 1 \\ 9\overline{\big)\,11} \\ \underline{9} \\ 2 \end{array}\quad \begin{array}{r} 0 \\ 3\overline{\big)\,2} \\ \underline{0} \\ 2 \end{array}\quad \begin{array}{r} 2 \\ 1\overline{\big)\,2} \\ \underline{2} \\ 0 \end{array}$$

$$2009 = 2202102_3$$

43. To convert 2009 to base 5 ... 3125   625   125   25   5   1

$$\begin{array}{r} 3 \\ 625\overline{\big)\,2009} \\ \underline{1875} \\ 134 \end{array}\quad \begin{array}{r} 1 \\ 125\overline{\big)\,134} \\ \underline{125} \\ 9 \end{array}\quad \begin{array}{r} 0 \\ 25\overline{\big)\,9} \\ \underline{0} \\ 9 \end{array}\quad \begin{array}{r} 1 \\ 5\overline{\big)\,9} \\ \underline{9} \\ 4 \end{array}\quad \begin{array}{r} 4 \\ 1\overline{\big)\,4} \\ \underline{4} \\ 0 \end{array}$$

$$2009 = 31014_5$$

45. To convert 2009 to base 7 ... 2401   343   49   7   1

$$\begin{array}{r} 5 \\ 343\overline{\big)\,2009} \\ \underline{1715} \\ 294 \end{array}\quad \begin{array}{r} 6 \\ 49\overline{\big)\,294} \\ \underline{294} \\ 0 \end{array}\quad \begin{array}{r} 0 \\ 7\overline{\big)\,0} \\ \underline{0} \\ 3 \end{array}\quad \begin{array}{r} 0 \\ 1\overline{\big)\,0} \\ \underline{0} \\ 0 \end{array}$$

$$2009 = 5600_7$$

47. To convert 2009 to base 11 ... 14,641   1331   121   11   1

$$\begin{array}{r} 1 \\ 1331\overline{\big)\,2009} \\ \underline{1331} \\ 678 \end{array}\quad \begin{array}{r} 5 \\ 121\overline{\big)\,678} \\ \underline{605} \\ 73 \end{array}\quad \begin{array}{r} 6 \\ 11\overline{\big)\,73} \\ \underline{66} \\ 7 \end{array}\quad \begin{array}{r} 7 \\ 1\overline{\big)\,7} \\ \underline{7} \\ 7 \end{array}$$

$$2009 = 1567_{11}$$

49. To convert 2009 to base 15 ... 3375   225   15   1

$$\begin{array}{r} 8 \\ 225\overline{\big)\,2009} \\ \underline{1800} \\ 209 \end{array}\quad \begin{array}{r} 13 \quad = \text{D} \\ 15\overline{\big)\,209} \\ \underline{195} \\ 14 \end{array}\quad \begin{array}{r} 14 \quad = \text{E} \\ 1\overline{\big)\,14} \\ \underline{14} \\ 0 \end{array}$$

$$2009 = 8\text{DE}_{15}$$

51. $4(5) + 3(1) = 20 + 3 = 23$

53. $3(5^2) + 0(5) + 3(1) = 3(25) + 0 + 3$
$$= 75 + 0 + 3 = 78$$

55. To convert ... 25   5   1

$$\begin{array}{r} 3 = \ominus \\ 5\overline{\big)\,19} \\ \underline{15} \\ 4 \end{array}\qquad \begin{array}{r} 4 = \oslash \\ 1\overline{\big)\,4} \\ \underline{4} \\ 0 \end{array}$$

$$19 = \ominus\,\oslash_{\,5}$$

57. To convert   ... 125 25 5 1

$$\frac{2}{25\overline{)74}}=\bigcirc \qquad \frac{4}{5\overline{)24}}=\bigcirc \qquad \frac{4}{1\overline{)4}}=\bigcirc$$
$$\frac{50}{24} \qquad\qquad \frac{20}{4} \qquad\qquad \frac{4}{0}$$

$$74=\bigcirc\bigcirc\bigcirc{}_5$$

59. $3(4)+2(1)=12+2=14$

61.

$$2\left(4^2\right)+1(4)+0(1)=2(16)+4+0=32+4+0=36$$

For #63-65, blue = 0 = b, red = 1 = r, gold = 2 = go, green = 3 = gr

63. To convert   ... 16 4 1

$$\frac{2}{4\overline{)10}}=\boxed{go} \qquad \frac{2}{1\overline{)2}}=\boxed{go}$$
$$\frac{8}{2} \qquad\qquad \frac{2}{0}$$

$$10=\boxed{go}\,\boxed{go}{}_4$$

65. To convert   ... 64 16 4 1

$$\frac{3}{16\overline{)60}}=\boxed{gr} \quad \frac{3}{4\overline{)12}}=\boxed{gr} \quad \frac{0}{1\overline{)0}}=\boxed{b}$$
$$\frac{48}{12} \qquad\qquad \frac{12}{0} \qquad\qquad \frac{0}{0}$$

$$60=\boxed{gr}\,\boxed{gr}\,\boxed{b}{}_4$$

67. a) Each remainder is multiplied by the proper power of 5.

b)

5	683		
5	136	3	↑
5	27	1	↑
5	5	2	↑
5	1	0	↑
	0	1	↑

$$683=10213_5$$

c)

8	763		
8	95	3	↑
8	11	7	↑
8	1	3	↑
	0	1	↑

$$763=1373_8$$

69. Answers will vary.

71. $1\left(b^2\right)+1(b)+1=43$

$b^2+b+1=43$

$b^2+b-42=0$

$(b+7)(b-6)=0$

$b+7=0$ or $b-6=0$

$b=-7$ or $b=6$

Since the base cannot be negative, $b=6$.

73. a) $3(4^4)+1(4^3)+2(4^2)+3(4)+0(1)=3(256)+64+2(16)+12+0=768+64+32+12+0=876$

b) To convert   ... 256 64 16 4 1

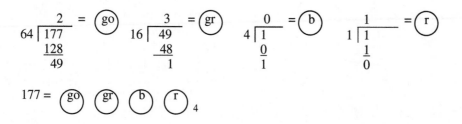

$$177 = \text{(go)} \text{(gr)} \text{(b)} \text{(r)}_4$$

75. Answers will vary.

Exercise Set 4.4

1. a) $b^0 = 1, b^1 = b, b^2, b^3, b^4$

b) $2^0 = 1, 2^1 = 2, 2^2, 2^3, 2^4$

3. No; there is no 3 in base 3.

5. Answers will vary.

7.
$$\begin{array}{r} 21_3 \\ \underline{20_3} \\ 111_3 \end{array}$$

9.
$$\begin{array}{r} 1234_5 \\ \underline{341_5} \\ 2130_5 \end{array}$$

11.
$$\begin{array}{r} 799_{12} \\ \underline{218_{12}} \\ 9B5_{12} \end{array}$$

13.
$$\begin{array}{r} 1112_3 \\ \underline{1011_3} \\ 2200_3 \end{array}$$

15.
$$\begin{array}{r} 14631_7 \\ \underline{6040_7} \\ 24001_7 \end{array}$$

17.
$$\begin{array}{r} 1110_2 \\ \underline{110_2} \\ 10100_2 \end{array}$$

19.
$$\begin{array}{r} 201_3 \\ \underline{-120_3} \\ 11_3 \end{array}$$

21.
$$\begin{array}{r} 2138_9 \\ \underline{-1207_9} \\ 831_9 \end{array}$$

23.
$$\begin{array}{r} 1101_2 \\ \underline{-111_2} \\ 110_2 \end{array}$$

25.
$$\begin{array}{r} 1001_2 \\ \underline{-110_2} \\ 11_2 \end{array}$$

27.
$$\begin{array}{r} 4223_7 \\ \underline{-304_7} \\ 3616_7 \end{array}$$

29.
$$\begin{array}{r} 2100_3 \\ \underline{-1012_3} \\ 1011_3 \end{array}$$

31.
$$\begin{array}{r} 22_3 \\ \underline{\times\ 2_3} \\ 121_3 \end{array}$$

33.
$$\begin{array}{r} 647_8 \\ \underline{\times\ 5_8} \\ 4103_8 \end{array}$$

35.
$$\begin{array}{r} 512_6 \\ \times\ 23_6 \\ \hline 2340 \\ 1424 \\ \hline 21020_6 \end{array}$$

37.
$$\begin{array}{r} B12_{12} \\ \times\ 83_{12} \\ \hline 2936 \\ 7494 \\ \hline 77676_{12} \end{array}$$

39.
$$\begin{array}{r} 111_2 \\ \times\ 101_2 \\ \hline 111 \\ 000 \\ 111 \\ \hline 100011_2 \end{array}$$

41.
$$\begin{array}{r} 316_7 \\ \times\ 16_7 \\ \hline 2541 \\ 316 \\ \hline 6031_7 \end{array}$$

43. $1_2 \times 1_2 = 1_2$

$$
\begin{array}{r}
101_2 \\
1_2\ \overline{)\ 101_2} \\
\underline{1} \\
00 \\
\underline{0} \\
01 \\
\underline{01} \\
0
\end{array}
$$

45. $2_6 \times 1_6 = 2_6$ $\quad 223_6 \quad R1_6$

$2_6 \times 2_6 = 4_6$

$2_6 \times 3_6 = 10_6$

$2_6 \times 4_6 = 12_6$

$2_6 \times 5_6 = 14_6$

$$
\begin{array}{r}
223_6 \quad R1_6 \\
2_6\ \overline{)\ 451_6} \\
\underline{4} \\
5 \\
\underline{4} \\
11 \\
\underline{10} \\
1
\end{array}
$$

47. $2_4 \times 1_4 = 2_4$

$2_4 \times 2_4 = 10_4$

$2_4 \times 3_4 = 12_4$

$$
\begin{array}{r}
123_4 \\
2_4\ \overline{)\ 312_4} \\
\underline{2} \\
11 \\
\underline{10} \\
12 \\
\underline{12} \\
0
\end{array}
$$

49. $2_4 \times 1_4 = 2_4$

$2_4 \times 2_4 = 10_4$

$2_4 \times 3_4 = 12_4$

$$
\begin{array}{r}
103_4 \quad R1_4 \\
2_4\ \overline{)\ 213_4} \\
\underline{2} \\
01 \\
\underline{00} \\
13 \\
\underline{12} \\
1
\end{array}
$$

51. $3_5 \times 1_5 = 3_5$

$3_5 \times 2_5 = 11_5$

$3_5 \times 3_5 = 14_5$

$3_5 \times 4_5 = 22_5$

$$3_5 \overline{\smash)224_5} \quad 41_5 \quad R1_5$$

$$\underline{22}$$

$$04$$

$$\underline{\ 3}$$

$$1$$

53. $6_7 \times 1_7 = 6_7$

$6_7 \times 2_7 = 15_7$

$6_7 \times 3_7 = 24_7$

$6_7 \times 4_7 = 33_7$

$6_7 \times 5_7 = 42_7$

$6_7 \times 6_7 = 51_7$

$$6_7 \overline{\smash)404_7} \quad 45_7 \quad R2_7$$

$$\underline{33}$$

$$44$$

$$\underline{42}$$

$$2$$

55. 3_5

    $+\ 3_5$

    $11_5 = $ ⊖⊖ $_5$

57. 23_5

    $+\ 13_5$

    $41_5 = $ ◯⊖ $_5$

For #59-65, blue = 0 = b, red = 1 = r, gold = 2 = go, green = 3 = gr

59. 3_4

    $+\ 2_4$

    $11_4 = $ (r)(r) $_4$

61. 12_4

    $+\ 30_4$

    $102_4 = $ (r)(b)(go) $_4$

63. 31_4

    $-\ 13_4$

    $12_4 = $ (r)(go) $_4$

65. 231_4

    $-\ 103_4$

    $122_4 = $ (r)(go)(go) $_4$

67. $2302_5 = 2(5^3) + 3(5^2) + 0(5) + 2(1) = 2(125) + 3(25) + 0 + 2 = 250 + 75 + 0 + 2 = 327$

69. FAB_{16}

    $\times\ \ 4_{16}$

    $2C$

    28

    $\underline{3C}$

    $3EAC_{16}$

71. $20_4 \times 1_4 = 20_4$

$20_4 \times 2_4 = 100_4$

$20_4 \times 3_4 = 120_4$

$$20_4 \overline{\smash)223_4} \quad 11_4 \quad R3_4$$

$$\underline{20}$$

$$23$$

$$\underline{20}$$

$$3$$

73. a) 462_8

    $\times 35_8$

    2772

    1626

    21252_8

b) $462_8 = 4(8^2) + 6(8) + 2(1) = 4(64) + 48 + 2 = 256 + 48 + 2 = 306$

$35_8 = 3(8) + 5(1) = 24 + 5 = 29$

c) $306 \times 29 = 8874$

d) $21252_8 = 2(8^4) + 1(8^3) + 2(8^2) + 5(8) + 2(1)$

$= 2(4096) + 512 + 2(64) + 40 + 2$

$= 8192 + 512 + 128 + 40 + 2 = 8874$

e) Yes, in part a), the numbers were multiplied in base 8 and then converted to base 10 in part d). In part b), the numbers were converted to base 10 first, then multiplied in part c).

75. Orange = 0; purple = 1; turquoise = 2; red = 3
77. Answers will vary.

Exercise Set 4.5

1. Duplation and mediation, lattice multiplication and Napier rods

3. a) Answers will vary.

   b)

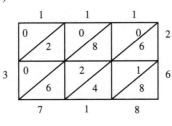

143 × 26 = 3718

5. 9 – 171
   4 – 342
   2 – 684
   1 – 1368
       1539

7. 27 – 53
  13 – 106
   6 – 212
   3 – 424
   1 – 848
      1431

9. 35 – 236
  17 – 472
   8 – 944
   4 – 1888
   2 – 3776
   1 – 7552
      8260

11. 93 –   93
  46 – 186
  23 – 372
  11 – 744
   5 – 1488
   2 – 2976
   1 – 5952
      8649

13.

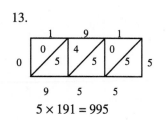

$$5 \times 191 = 995$$

15.

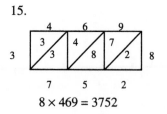

$$8 \times 469 = 3752$$

17.

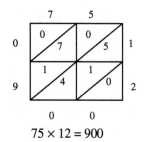

$$75 \times 12 = 900$$

19.

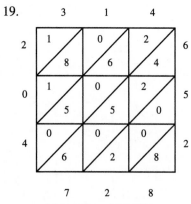

$$314 \times 652 = 204{,}728$$

21.

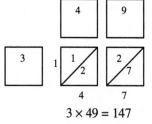

$$3 \times 49 = 147$$

23.

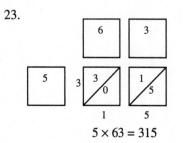

$$5 \times 63 = 315$$

25.

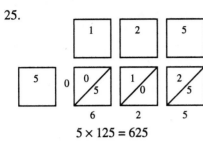

$5 \times 125 = 625$

27.

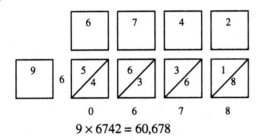

$9 \times 6742 = 60{,}678$

29.   a) 253×46; Place the factors of 8 until the
correct factors and placements are found
so the rest of the rectangle can be completed.

b)

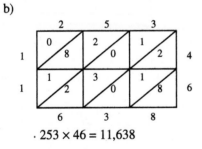

$\cdot\ 253 \times 46 = 11{,}638$

31. a) 4×382; Place the factors of 12 until the correct
factors and placements are found so the rest
can be completed.

b)

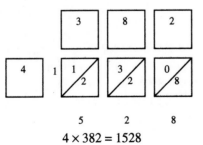

$4 \times 382 = 1528$

33. 13 – 22
     ~~6 – 44~~
     3 – 88
     1 – <u>176</u>
     286 =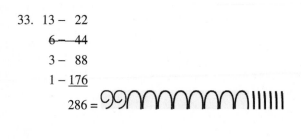

35.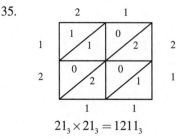

$21_3 \times 21_3 = 1211_3$

37. a) $1000 + 500 + 100 + 100 + 50 + 10 + 10 + 5 + 1 = 1776$
     b) Answers will vary.
39. Answers will vary.

Review Exercises

1. $1000 + 1000 + 100 + 10 + 10 + 1 = 2121$
2. $1000 + 100 + 100 + 10 + 1 + 1 + 1 + 1 = 1214$
3. $10 + 100 + 100 + 100 + 1 + 1000 = 1311$
4. $100 + 10 + 1000 + 1 + 1000 + 1 + 1 + 1 = 2114$
5. $1000 + 1000 + 100 + 100 + 100 + 10 + 1 + 1 + 1 + 1 = 2314$
6. $100 + 100 + 10 + 1 + 1000 + 1000 + 1 + 100 = 2312$
7. *bbba*
8. *cbbaaaaa*
9. *ccbbbbbbbbbaaa*
10. *ddaaaaaaaa*
11. *dddddcccccccccbbbbba*
12. *ddcccbaaaa*
13. $3(10) + 2 = 30 + 2 = 32$
14. $8(10) + 5 = 80 + 5 = 85$
15. $7(100) + 4(10) + 9 = 700 + 40 + 9 = 749$
16. $4(1000) + 6(10) + 8 = 4000 + 60 + 8 = 4068$
17. $5(1000) + 6(100) + 4(10) + 8 = 5000 + 600 + 40 + 8 = 5648$
18. $6(1000) + 9(100) + 5 = 6000 + 900 + 5 = 6905$
19. *cxg*
20. *ayexg*
21. *hyfxb*
22. *czixd*
23. *fzd*
24. *bza*
25. $9(10) + 3(1) = 90 + 3 = 93$
26. $2(100) + 3(1) = 200 + 3 = 203$
27. $5(100) + 6(10) + 8(1) = 500 + 60 + 8 = 568$
28. $4(10,000) + 6(1000) + 8(100) + 8(10) + 3(1) = 40,000 + 6000 + 800 + 80 + 3 = 46,883$
29. $6(10,000) + 4(1000) + 4(100) + 8(10) + 1 = 60,000 + 4000 + 400 + 80 + 1 = 64,481$
30. $6(10,000) + 5(100) + 2(10) + 9(1) = 60,000 + 500 + 20 + 9 = 60,529$
31. la
32. tpd
33. vrc
34. BArg
35. ODvog
36. QFvrf
37.
38. MDCCLXXVI

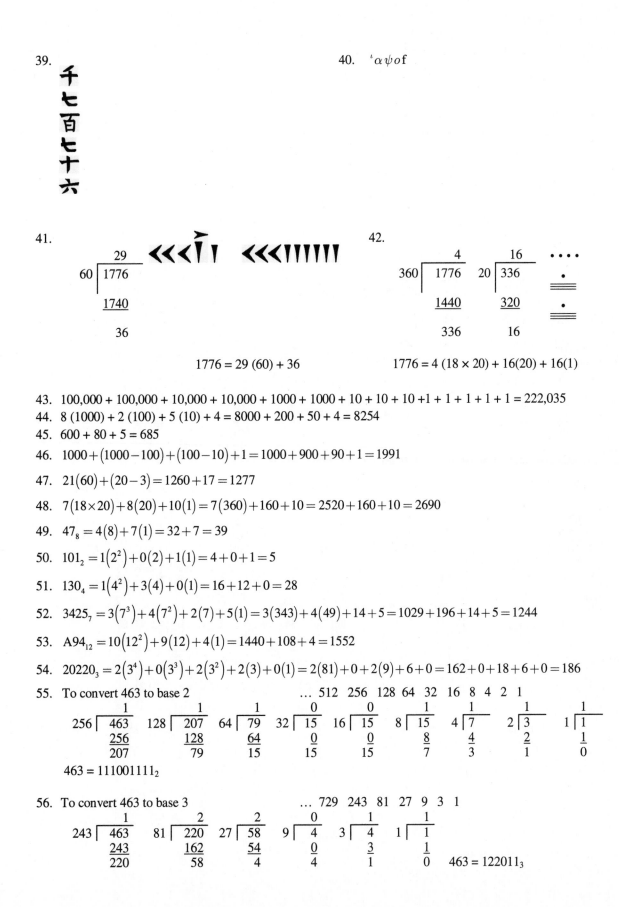

39.

40. $'\alpha\psi o f$

41.

$$\begin{array}{r} 29 \\ 60 \overline{\smash{\big)}\ 1776} \\ \underline{1740} \\ 36 \end{array}$$

1776 = 29 (60) + 36

42.

$$\begin{array}{r} 4 \\ 360 \overline{\smash{\big)}\ 1776} \\ \underline{1440} \\ 336 \end{array} \qquad \begin{array}{r} 16 \\ 20 \overline{\smash{\big)}\ 336} \\ \underline{320} \\ 16 \end{array}$$

1776 = 4 (18 × 20) + 16(20) + 16(1)

43. 100,000 + 100,000 + 10,000 + 10,000 + 1000 + 1000 + 10 + 10 + 10 + 1 + 1 + 1 + 1 + 1 = 222,035

44. 8 (1000) + 2 (100) + 5 (10) + 4 = 8000 + 200 + 50 + 4 = 8254

45. 600 + 80 + 5 = 685

46. $1000 + (1000 - 100) + (100 - 10) + 1 = 1000 + 900 + 90 + 1 = 1991$

47. $21(60) + (20 - 3) = 1260 + 17 = 1277$

48. $7(18 \times 20) + 8(20) + 10(1) = 7(360) + 160 + 10 = 2520 + 160 + 10 = 2690$

49. $47_8 = 4(8) + 7(1) = 32 + 7 = 39$

50. $101_2 = 1(2^2) + 0(2) + 1(1) = 4 + 0 + 1 = 5$

51. $130_4 = 1(4^2) + 3(4) + 0(1) = 16 + 12 + 0 = 28$

52. $3425_7 = 3(7^3) + 4(7^2) + 2(7) + 5(1) = 3(343) + 4(49) + 14 + 5 = 1029 + 196 + 14 + 5 = 1244$

53. $A94_{12} = 10(12^2) + 9(12) + 4(1) = 1440 + 108 + 4 = 1552$

54. $20220_3 = 2(3^4) + 0(3^3) + 2(3^2) + 2(3) + 0(1) = 2(81) + 0 + 2(9) + 6 + 0 = 162 + 0 + 18 + 6 + 0 = 186$

55. To convert 463 to base 2   ... 512 256 128 64 32 16 8 4 2 1

$$\begin{array}{c} 1 \\ 256 \overline{\smash{\big)}\ 463} \\ \underline{256} \\ 207 \end{array} \quad \begin{array}{c} 1 \\ 128 \overline{\smash{\big)}\ 207} \\ \underline{128} \\ 79 \end{array} \quad \begin{array}{c} 1 \\ 64 \overline{\smash{\big)}\ 79} \\ \underline{64} \\ 15 \end{array} \quad \begin{array}{c} 0 \\ 32 \overline{\smash{\big)}\ 15} \\ \underline{0} \\ 15 \end{array} \quad \begin{array}{c} 0 \\ 16 \overline{\smash{\big)}\ 15} \\ \underline{0} \\ 15 \end{array} \quad \begin{array}{c} 1 \\ 8 \overline{\smash{\big)}\ 15} \\ \underline{8} \\ 7 \end{array} \quad \begin{array}{c} 1 \\ 4 \overline{\smash{\big)}\ 7} \\ \underline{4} \\ 3 \end{array} \quad \begin{array}{c} 1 \\ 2 \overline{\smash{\big)}\ 3} \\ \underline{2} \\ 1 \end{array} \quad \begin{array}{c} 1 \\ 1 \overline{\smash{\big)}\ 1} \\ \underline{1} \\ 0 \end{array}$$

$463 = 111001111_2$

56. To convert 463 to base 3   ... 729 243 81 27 9 3 1

$$\begin{array}{c} 1 \\ 243 \overline{\smash{\big)}\ 463} \\ \underline{243} \\ 220 \end{array} \quad \begin{array}{c} 2 \\ 81 \overline{\smash{\big)}\ 220} \\ \underline{162} \\ 58 \end{array} \quad \begin{array}{c} 2 \\ 27 \overline{\smash{\big)}\ 58} \\ \underline{54} \\ 4 \end{array} \quad \begin{array}{c} 0 \\ 9 \overline{\smash{\big)}\ 4} \\ \underline{0} \\ 4 \end{array} \quad \begin{array}{c} 1 \\ 3 \overline{\smash{\big)}\ 4} \\ \underline{3} \\ 1 \end{array} \quad \begin{array}{c} 1 \\ 1 \overline{\smash{\big)}\ 1} \\ \underline{1} \\ 0 \end{array} \quad 463 = 122011_3$$

57. To convert 463 to base 4 ... 1024 256 64 16 4 1

$$256\overline{)463} \quad 64\overline{)207} \quad 16\overline{)15} \quad 4\overline{)15} \quad 1\overline{)3}$$

	1		3		0		3		3

$256\overline{)463}$ $64\overline{)207}$ $16\overline{)15}$ $4\overline{)15}$ $1\overline{)3}$

256   192   0   12   3

207   15   15   3   0 $463 = 13033_4$

58. To convert 463 to base 8 ... 512 64 8 1

    7   1   7

$64\overline{)463}$ $8\overline{)15}$ $1\overline{)7}$

448   8   7

15   7   0 $463 = 717_8$

59. To convert 463 to base 12 ... 1728 144 12 1

    3   2   7

$144\overline{)463}$ $12\overline{)31}$ $1\overline{)7}$

432   24   7

31   7   0 $463 = 327_{12}$

60. To convert 463 to base 16 ... 65,536 4096 256 16 1

    1   12 $= C$ 15 $= F$

$256\overline{)463}$ $16\overline{)207}$ $1\overline{)15}$

256   192   15

207   15   0 $493 = 1CF_{16}$

61.
$$\begin{array}{r} 52_7 \\ \underline{55_7} \\ 140_7 \end{array}$$

62.
$$\begin{array}{r} 10110_2 \\ \underline{11001_2} \\ 101111_2 \end{array}$$

63.
$$\begin{array}{r} 9B_{12} \\ \underline{87_{12}} \\ 166_{12} \end{array}$$

64.
$$\begin{array}{r} 2B9_{16} \\ \underline{456_{16}} \\ 70F_{16} \end{array}$$

65.
$$\begin{array}{r} 3024_5 \\ \underline{4023_5} \\ 12102_5 \end{array}$$

66.
$$\begin{array}{r} 3407_8 \\ \underline{7014_8} \\ 12423_8 \end{array}$$

67.
$$\begin{array}{r} 4032_7 \\ \underline{-321_7} \\ 3411_7 \end{array}$$

68.
$$\begin{array}{r} 1001_2 \\ \underline{-101_2} \\ 100_2 \end{array}$$

69.
$$\begin{array}{r} A7B_{12} \\ \underline{-95_{12}} \\ 9A6_{12} \end{array}$$

70.
$$\begin{array}{r} 4321_5 \\ \underline{-442_5} \\ 3324_5 \end{array}$$

71.
$$\begin{array}{r} 1713_8 \\ \underline{-1243_8} \\ 450_8 \end{array}$$

72.
$$\begin{array}{r} F64_{16} \\ \underline{-2A3_{16}} \\ CC1_{16} \end{array}$$

73.
$$\begin{array}{r} 1011_2 \\ \underline{\times\ 101_2} \\ 1011 \\ 0000 \\ \underline{1011} \\ 110111_2 \end{array}$$

74.
$$\begin{array}{r} 221_3 \\ \underline{\times\ 22_3} \\ 1212 \\ \underline{1212} \\ 21102_3 \end{array}$$

75.
$$\begin{array}{r} 34_5 \\ \underline{\times 21_5} \\ 34 \\ \underline{123} \\ 1314_5 \end{array}$$

76.
$$\begin{array}{r} 476_8 \\ \underline{\times\ 23_8} \\ 1672 \\ \underline{1174} \\ 13632_8 \end{array}$$

77.
$$\begin{array}{r} 126_{12} \\ \underline{\times\ 47_{12}} \\ 856 \\ \underline{4A0} \\ 5656_{12} \end{array}$$

78.
$$\begin{array}{r} 1A3_{16} \\ \underline{\times\ 12_{16}} \\ 346 \\ \underline{1A3} \\ 1D76_{16} \end{array}$$

79.
$2_3 \times 1_3 = 2_3$
$2_3 \times 2_3 = 11_3$
$2_3 \times 3_3 = 20_3$
$2_3 \times 4_3 = 22_3$

$$2_3 \overline{)120_3} = 21_3 \ \text{R} \ 1_3$$
$$\underline{11}$$
$$10$$
$$\underline{\ 2}$$
$$1$$

80.
$2_4 \times 1_4 = 2_4$
$2_4 \times 2_4 = 10_4$
$2_4 \times 3_4 = 12_4$

$$2_4 \overline{)320_4} = 130_4$$
$$\underline{2}$$
$$12$$
$$\underline{12}$$
$$0$$
$$\underline{0}$$
$$0$$

81.
$3_5 \times 1_5 = 3_5$
$3_5 \times 2_5 = 11_5$
$3_5 \times 3_5 = 14_5$
$3_5 \times 4_5 = 22_5$

$$3_5 \overline{)130_5} = 23_5 \ \text{R} 1_5$$
$$\underline{11}$$
$$20$$
$$\underline{14}$$
$$1$$

82.
$4_6 \times 1_6 = 4_6$
$4_6 \times 2_6 = 12_6$
$4_6 \times 3_6 = 20_6$
$4_6 \times 4_6 = 24_6$
$4_6 \times 5_6 = 32_6$

$$4_6 \overline{)3020_6} = 433_6$$
$$\underline{24}$$
$$22$$
$$\underline{20}$$
$$20$$
$$\underline{20}$$
$$0$$

83.
$3_6 \times 1_6 = 3_6$
$3_6 \times 2_6 = 10_6$
$3_6 \times 3_6 = 13_6$
$3_6 \times 4_6 = 20_6$
$3_6 \times 5_6 = 23_6$

$$3_6 \overline{)2034_6} = 411_6 \ \text{R} 1_6$$
$$\underline{20}$$
$$03$$
$$\underline{\ 3}$$
$$04$$
$$\underline{\ 3}$$
$$1$$

84.
$6_8 \times 1_8 = 6_8$
$6_8 \times 2_8 = 14_8$
$6_8 \times 3_8 = 22_8$
$6_8 \times 4_8 = 30_8$
$6_8 \times 5_8 = 36_8$
$6_8 \times 6_8 = 44_8$
$6_8 \times 7_8 = 52_8$

$$6_8 \overline{)5072_8} = 664_8 \ \text{R} 2_8$$
$$\underline{44}$$
$$47$$
$$\underline{44}$$
$$32$$
$$\underline{30}$$
$$2$$

85.
~~142~~   ~~24~~
71 - 48
35 - 96
17 - 192
~~8~~   ~~384~~
~~4~~   ~~768~~
~~2~~   ~~1536~~
1 - 3072
    3408

86.

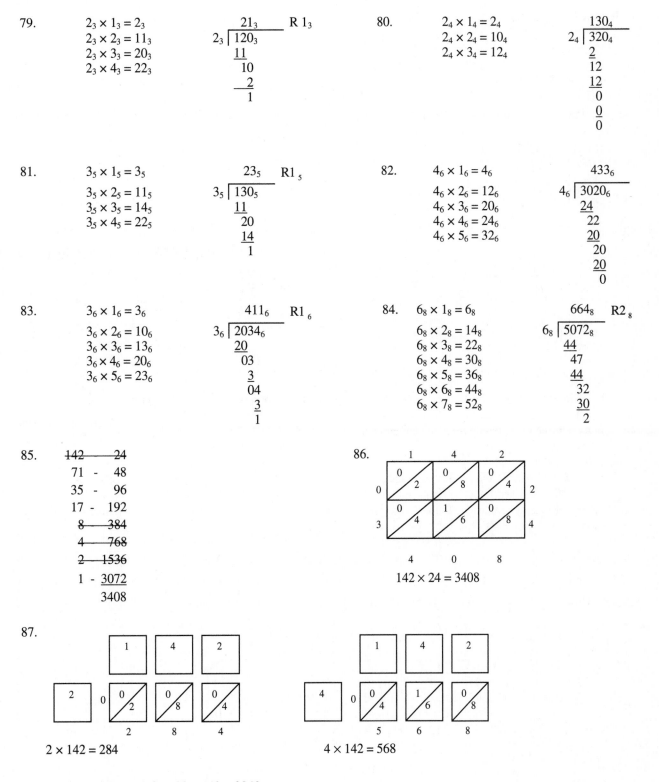

$142 \times 24 = 3408$

87.

$2 \times 142 = 284$

$4 \times 142 = 568$

$2 \times 142 = 284$, therefore $20 \times 142 = 2840$

Therefore, $142 \times 24 = 2840 + 568 = 3408$.

Chapter Test

1. A **number** is a quantity and answers the question "How many?" A **numeral** is a symbol used to represent the number.

2. $1000 + 1000 + (500 - 100) + 50 + 10 + 10 + 10 + (5 - 1) = 2484$

3. $21(60) + 15(1) = 1260 + 15 = 1275$

4. $8(1000) + 0 + 9(10) = 8000 + 0 + 90 = 8090$

5. $2(18 \times 20) + 12(20) + 9(1) = 2(360) + 240 + 9 = 720 + 240 + 9 = 969$

6. $100{,}000 + 10{,}000 + 10{,}000 + 1000 + 1000 + 100 + 10 + 10 + 10 + 10 + 1 + 1 = 122{,}142$

7. $2(1000) + 700 + 40 + 5 = 2000 + 700 + 40 + 5 = 2745$

8.

9. $^{\iota}\beta\upsilon o f$

10.

$$
\begin{array}{r} 3 \\ \hline 360\,|\,1434 \\ \underline{1080} \\ 354 \end{array}
\qquad
\begin{array}{r} 17 \\ \hline 20\,|\,354 \\ \underline{340} \\ 14 \end{array}
$$

$$1434 = 3(18 \times 20) + 17(20) + 14(1)$$

11.

$$
\begin{array}{r} 26 \\ \hline 60\,|\,1596 \\ \underline{1560} \\ 36 \end{array}
$$

$$1596 = 26(60) + 36(1)$$

12. MMMDCCVI

13. In an additive system, the number represented by a particular set of numerals is the sum of the values of the numerals.

14. In a multiplicative system, there are numerals for each number less than the base and for powers of the base. Each numeral less than the base is multiplied by a numeral for the power of the base, and these products are added to obtain the number.

15. In a ciphered system, the number represented by a particular set of numerals is the sum of the values of the numerals. There are numerals for each number up to and including the base and multiples of the base.

16. In a place-value system, each number is multiplied by a power of the base. The position of the numeral indicates the power of the base by which it is multiplied.

17. $23_4 = 2(4) + 3(1) = 8 + 3 = 11$

18. $403_5 = 4(5^2) + 0(5) + 3(1) = 4(25) + 0 + 3 = 100 + 0 + 3 = 103$

19. $101101_2 = 1(2^5) + 0(2^4) + 1(2^3) + 1(2^2) + 0(2) + 1(1) = 32 + 0 + 8 + 4 + 0 + 1 = 45$

20. $3A7_{12} = 3(12^2) + 10(12) + 7(1) = 3(144) + 120 + 7 = 432 + 120 + 7 = 559$

21. To convert 36 to base 2   ... 64 32 16 8 4 2 1

$$
\begin{array}{r} 1 \\ \hline 32\,|\,36 \\ \underline{32} \\ 4 \end{array}
\begin{array}{r} 0 \\ \hline 16\,|\,4 \\ \underline{0} \\ 4 \end{array}
\begin{array}{r} 0 \\ \hline 8\,|\,4 \\ \underline{0} \\ 4 \end{array}
\begin{array}{r} 1 \\ \hline 4\,|\,4 \\ \underline{4} \\ 0 \end{array}
\begin{array}{r} 0 \\ \hline 2\,|\,0 \\ \underline{0} \\ 0 \end{array}
\begin{array}{r} 0 \\ \hline 1\,|\,0 \\ \underline{0} \\ 0 \end{array}
$$

$$36 = 100100_2$$

22. To convert 93 to base 8   ... 512 64 8 1

$$
\begin{array}{r} 1 \\ \hline 64\,|\,93 \\ \underline{64} \\ 29 \end{array}
\begin{array}{r} 3 \\ \hline 8\,|\,29 \\ \underline{24} \\ 5 \end{array}
\begin{array}{r} 5 \\ \hline 1\,|\,5 \\ \underline{5} \\ 0 \end{array}
$$

$$93 = 135_8$$

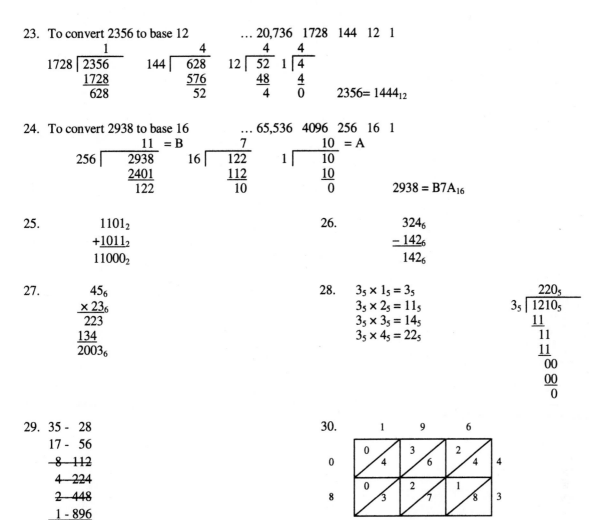

23. To convert 2356 to base 12   ... 20,736   1728   144   12   1

$$
\begin{array}{r}
1 \\
1728\,\overline{)2356} \\
\underline{1728} \\
628
\end{array}
\quad
\begin{array}{r}
4 \\
144\,\overline{)628} \\
\underline{576} \\
52
\end{array}
\quad
\begin{array}{r}
4 \\
12\,\overline{)52} \\
\underline{48} \\
4
\end{array}
\quad
\begin{array}{r}
4 \\
1\,\overline{)4} \\
\underline{4} \\
0
\end{array}
\quad 2356 = 1444_{12}
$$

24. To convert 2938 to base 16   ... 65,536   4096   256   16   1

$$
\begin{array}{r}
11 \;=B \\
256\,\overline{)2938} \\
\underline{2401} \\
122
\end{array}
\quad
\begin{array}{r}
7 \\
16\,\overline{)122} \\
\underline{112} \\
10
\end{array}
\quad
\begin{array}{r}
10 \;=A \\
1\,\overline{)10} \\
\underline{10} \\
0
\end{array}
\quad 2938 = B7A_{16}
$$

25.
$$
\begin{array}{r}
1101_2 \\
+\,1011_2 \\
\hline
11000_2
\end{array}
$$

26.
$$
\begin{array}{r}
324_6 \\
-\,142_6 \\
\hline
142_6
\end{array}
$$

27.
$$
\begin{array}{r}
45_6 \\
\times\,23_6 \\
\hline
223 \\
134 \\
\hline
2003_6
\end{array}
$$

28. $3_5 \times 1_5 = 3_5$

      $3_5 \times 2_5 = 11_5$

      $3_5 \times 3_5 = 14_5$

      $3_5 \times 4_5 = 22_5$

$$
\begin{array}{r}
220_5 \\
3_5\,\overline{)1210_5} \\
\underline{11} \\
11 \\
\underline{11} \\
00 \\
\underline{00} \\
0
\end{array}
$$

29.   35 - 28

     17 - 56

     ~~8 - 112~~

     ~~4 - 224~~

     ~~2 - 448~~

     1 - 896

       980

30.

$43 \times 196 = 8428$

CHAPTER FIVE

NUMBER THEORY AND THE REAL NUMBER SYSTEM

Exercise Set 5.1

1. Number theory is the study of numbers and their properties.

3. a) *a* divides *b* means that *b* divided by *a* has a remainder of zero.
   b) *a* is divisible by *b* means that *a* divided by *b* has a remainder of zero.

5. A composite number is a natural number that is divisible by a number other than itself and 1. Any natural number that is not prime is composite.

7. a) The least common multiple (LCM) of a set of natural numbers is the smallest natural number that is divisible (without remainder) by each element of the set.
   b) Determine the prime factorization of each number. Then find the product of the prime factors with the largest exponent that appear in any of the prime factorizations.
   c) The prime factors with the largest exponent that appear in any of the factorizations are 2^4 and 5. The LCM of 16 and 40 is $2^4 \cdot 5$, or 80.

$$\begin{array}{c|c} 2 & 16 \\ 2 & 8 \\ 2 & 4 \\ & 2 \end{array}$$
$$16 = 2^4$$

9. Mersenne Primes are prime numbers of the form $2^n - 1$ where n is a prime number.

11. Goldbach's conjecture states that every even number greater than or equal to 4 can be represented as the sum of two (not necessarily distinct) prime numbers.

13. The prime numbers between 1 and 100 are: 2, 3, 5, 7, 11, 13, 17, 19, 23, 29, 31, 37, 41, 43, 47, 53, 59, 61, 67, 71, 73, 79, 83, 89, 97.

15. True; since $27 \div 7 = 4$

17. False; 26 is a multiple of 13.

19. False.

21. True; if a number is divisible by 10, then it is also divisible by 5.

23. False; if a number is divisible by 3, then the sum of the number's digits is divisible by 3.

25. True; since $2 \cdot 3 = 6$.

27. Divisible by 3, 5 and 9.

29. Divisible by 2, 3, 6 and 9.

31. Divisible by 2, 3, 4, 5, 6, 8, and 10.

33. $2 \cdot 3 \cdot 4 \cdot 5 \cdot 6 = 720$. (other answers are possible)

35.
$$\begin{array}{c|c} 2 & 48 \\ 2 & 24 \\ 2 & 12 \\ 2 & 6 \\ & 3 \end{array}$$
$$48 = 2^4 \cdot 3$$

37.
$$\begin{array}{c|c} 2 & 168 \\ 2 & 84 \\ 2 & 42 \\ 3 & 21 \\ & 7 \end{array}$$
$$168 = 2^3 \cdot 3 \cdot 7$$

39.
$$\begin{array}{c|c} 2 & 332 \\ 2 & 166 \\ & 83 \end{array}$$
$$332 = 2^2 \cdot 83$$

41.

3	513
3	171
3	57
	19

$513 = 3^3 \cdot 19$

43.

2	1336
2	668
2	334
	167

$1336 = 2^3 \cdot 167$

45.

3	2001
23	667
	29

$2001 = 3 \cdot 23 \cdot 29$

47. The prime factors of 21 and 18 are: $6 = 3 \cdot 2$, $15 = 3 \cdot 7$

a) The common factor is 3, thus, the GCD = 3.

b) The factors with the greatest exponent that appear in either are 2, 3, 7. Thus, the LCM = $2 \cdot 3 \cdot 7 = 42$.

49. The prime factors of 20 and 35 are: $20 = 2^2 \cdot 5$,

$35 = 5 \cdot 7$

a) The common factor is 5; thus the GCD = 5.

b) The factors with the greatest exponent that appear in either are: 2^2, 5 and 7; thus, the LCM = $2^2 \cdot 5 \cdot 7 = 140$

51. The prime factors of 40 and 900 are: $40 = 2^3 \cdot 5$,

$900 = 2^2 \cdot 3^2 \cdot 5^2$

a) The common factors are: 2^2, 5; thus, the GCD = $2^2 \cdot 5 = 20$.

b) The factors with the greatest exponent that appear in either are: $2^3, 3^2, 5^2$; thus, the LCM = $2^2 \cdot 3^2 \cdot 5^2 = 1800$

53. The prime factors of 96 and 212 are: $96 = 2^5 \cdot 3$,

$212 = 2^2 \cdot 53$

a) The common factors are: 2^2; thus, the GCD = $2^2 = 4$.

b) The factors with the greatest exponent that appear in either are: 2^5, 3, 53; thus, the LCM = $2^5 \cdot 3 \cdot 53 = 5088$

55. The prime factors of 24, 48, and 128 are: $24 = 2^3 \cdot 3$,

$48 = 2^4 \cdot 3$, $128 = 2^7$

a) The common factors are: 2^3; thus, the GCD = $2^3 = 8$.

b) The factors with the greatest exponent that appear in any are: 2^7, 3; thus, LCM = $2^7 \cdot 3 = 384$

57. Use the list of primes generated in exercise 13. The next two sets of twin primes are: 17, 19, 29, 31.

59. (a) 10, 21 Yes; (b) 22, 26 No; (c) 27, 28 Yes; (d) 85, 119 Yes

61. Fermat number = $2^{2^n} + 1$, where n is a natural number. $2^{2^1} + 1 = 5$, $2^{2^2} + 1 = 2^4 + 1 = 17$, $2^{2^3} + 1 = 2^8 + 1 = 257$. These numbers are prime.

63. $2 \times 60, 3 \times 40, 4 \times 30, 5 \times 24, 6 \times 20, 8 \times 15, 10 \times 12,$
$12 \times 10, 15 \times 8, 20 \times 6, 24 \times 5, 30 \times 4, 40 \times 3, 60 \times 2$

65. The gcd of 30 and 14 is 210 days.

67. The gcd of 70 and 175 is 35 cars.

69 The gcd of 150 and 180 is 30 trees.

71. The lcm of 5 and 6 is 30 days.

73. A number is divisible by 15 if both 3 and 5 divide the number.

75. $40 \div 15 = 2$ with rem. $= 10$
$15 \div 10 = 1$ with rem. $= 5$
$10 \div 5 = 2$ with rem. $= 0$
Thus, gcd of 15 and 40 is 5.

77. $105 \div 35 = 3$ with rem. $= 0$.
Thus, gcd of 105 and 35 is 35.

79. $180 \div 150 = 1$ with rem. $= 30$.
$150 \div 30 = 5$ with rem. $= 0$.
Thus, the gcd of 150 and 180 is 30.

81. The proper factors of 12 are: 1, 2, 3, 4, and 6.
$1 + 2 + 3 + 4 + 6 = 16 \neq 12$
Thus, 12 is not a perfect #.

83. The proper factors of 48 are: 1, 2, 3, 4, 6, 8, 12, 16, and 24.
$1 + 2 + 3 + 4 + 6 + 8 + 12 + 16 + 24 = 76$
Thus, 48 is not a perfect #.

85. a) $60 = 2^2 \cdot 3^1 \cdot 5^1$ Adding 1 to each exponent and then multiplying these numbers, we get $(2+1)(1+1)(1+1) = 3 \cdot 2 \cdot 2 = 12$ divisors of 60 b) They are 1, 2, 3, 4, 5, 6, 10, 12, 15, 20, 30, and 60.

87. One of the numbers must be divisible by 3 and at least one must be even, so their product will be divisible by 2 and 3 and thus by 6.

89. $54036 = (54,000 + 36)$; $54,000 \div 18 = 3,000$ and $36 \div 18 = 2$
Thus, since $18 \,|\, 54000$ and $18 \,|\, 36$, $18 \,|\, 54036$.

91. $8 = 2+3+3$, $9 = 3+3+3$, $10 = 2+3+5$, $11 = 2+2+7$, $12 = 2+5+5$, $13 = 3+3+7$, $14 = 2+5+7$, $15 = 3+5+7$, $16 = 2+7+7$, $17 = 5+5+7$, $18 = 2+5+11$, $19 = 3+5+11$, $20 = 2+7+11$.

93. Denmark, Kangaroo, Orange. The sum of the digits minus 5 will always be 4, so the first letter for the country will always be D.

Exercise Set 5.2

1. Begin at zero, draw an arrow to the value of the first number. From the tip of that arrow draw another arrow by moving a number of spaces equal to the value of the second number. Be sure to move left if the number is negative and move right if the number is positive. The sum of the two numbers is at the tip of the second arrow.

3. To rewrite a subtraction problem as an addition problem, rewrite the subtraction sign as an addition sign and change the second number to its additive inverse.

5. The product of two numbers with like signs is a positive number, and the product of two numbers with unlike signs is a negative number.

7. $-4 + 7 = 3$

9. $-9 + 5 = -4$

11. $[6 + (-11)] + 0 = -5 + 0 = -5$

13. $[(-3) + (-4)] + 9 = -7 + 9 = 2$

15. $[(-23) + (-9)] + 11 = [-32] + 11 = -21$

17. $1 - 7 = -6$

19. $-5 - 4 = -9$

21. $-5 - (-3) = -5 + 3 = -2$

23. $14 - 20 = 14 + (-20) = -6$

25. $[5 + (-3)] - 4 = 2 - 4 = 2 + (-4) = -2$

27. $-5 \cdot 6 = -30$

29. $(-8)(-8) = 64$

31. $[(-8)(-2)] \cdot 6 = 16 \cdot 6 = 96$

33. $(5 \cdot 6)(-2) = (30)(-2) = -60$

35. $[(-3)(-6)] \cdot [(-5)(8)] = (18)(-40) = -720$

37. $-28 \div (-4) = 7$

39. $11 \div (-11) = -1$

41. $\dfrac{56}{-8} = -7$

43. $\dfrac{-210}{14} = -15$

45. $144 \div (-3) = -48$

47. True; every whole number is an integer.

49. False; the difference of two negative integers may be positive, negative, or zero.

51. True; the product of two integers with like signs is a positive integer.

53. True; the quotient of two integers with unlike signs is a negative number.

55. False; the sum of a positive integer and a negative integer could be pos., neg., or zero.

57. $(7 + 11) \div 3 = 18 \div 3 = 6$

59. $[(-7)(-6)] - 22 =$
$42 - 22 = 20$

61. $(4 - 8)(3) = (-4)(3) = -12$

63. $[2 + (-17)] \div 3 = [-15] \div 3 = -5$

65. $[(-22)(-3)] \div (2 - 13) =$
$= 66 \div (-11) = -6$

67. $-9, -6, -3, 0, 3, 6$

69. $-6, -5, -4, -3, -2, -1$

71. $11,250 - 95 + 61 - 65 - 42 =$
$11,250 + 61 - (95 + 65 + 42) =$
$11,311 - 202 = 11,109$

73. $14,495 - (-282) =$
$14,495 + 282 = 14,777$ feet

75. $-6 + 4 - 1 + 12 =$
$-7 + 16 = 9$; 9 yards; no

77. a) $+ 1 - (-8) = + 1 + 8 = 9$.
There is a 9 hr. time diff.
b) $- 5 - (-7) = -5 + 7 = 2$.
There is a 2 hr. time diff.

79.

$$\frac{-1+2-3+4-5+\ldots 99+100}{1-2+3-4+5\ldots+99-100} =$$

$$\frac{50}{-50} = -1$$

81. $0 + 1 - 2 + 3 + 4 - 5 + 6 - 7 - 8 + 9 = 1$ (other answers are possible)

Exercise Set 5.3

1. Rational numbers is the set of all numbers of the form p/q, where p and q are integers, and $q \neq 0$.

3. a) Divide both the numerator and the denominator by their greatest common divisor.

   b) $\dfrac{28}{35} = \dfrac{28 \div 7}{35 \div 7} = \dfrac{4}{5}$

5. For positive mixed numbers, multiply the denominator of the fraction by the integer preceding it. Add this product to the numerator. This sum is the numerator of the improper fraction; the denominator is the same as the denominator of the mixed number. For negative mixed numbers, you can temporarily ignore the negative sign, perform the conversion described above, and then reattach the negative sign.

7. a) The reciprocal of a number is 1 divided by the number.

   b) The reciprocal of -2 is $\dfrac{1}{-2} = -\dfrac{1}{2}$

9. a) To add or subtract two fractions with a common denominator, we add or subtract their numerators and keep the common denominator.

   b) $\dfrac{11}{24} + \dfrac{3}{24} = \dfrac{14}{24} = \dfrac{14 \div 2}{24 \div 2} = \dfrac{7}{12}$ c) $\dfrac{37}{48} - \dfrac{13}{48} = \dfrac{24}{48} = \dfrac{24 \div 24}{48 \div 24} = \dfrac{1}{2}$

11. We can multiply a fraction by the number one in the form of c/c (where c is a nonzero integer) and the number will maintain the same value.

13. GCD of 5 and 10 is 5.

$$\frac{5}{10} = \frac{5 \div 5}{10 \div 5} = \frac{1}{2}$$

15. GCD of 24 and 54 is 6.

$$\frac{24}{54} = \frac{24 \div 6}{54 \div 6} = \frac{4}{9}$$

17. GCD of 95 and 125 is 5.

$$\frac{95}{125} = \frac{95 \div 5}{125 \div 5} = \frac{19}{25}$$

19. GCD of 112 and 176 is 16.

$$\frac{112}{176} = \frac{112 \div 16}{176 \div 16} = \frac{7}{11}$$

21. GCD of 45 and 495 is 45.

$$\frac{45}{495} = \frac{45 \div 45}{495 \div 45} = \frac{1}{11}$$

23.

$$3\frac{5}{8} = \frac{(8)(3)+5}{8} = \frac{24+5}{8} = \frac{29}{8}$$

25. $-1\frac{15}{16} = -\frac{-((1)(16)+15)}{16}$

$$= -\frac{16+15}{16} = -\frac{31}{16}$$

27. $-4\frac{15}{16} = -\frac{(4)(16)+15}{16}$

$$= -\frac{64+15}{16} = -\frac{79}{16}$$

29. $1\frac{1}{8} = \frac{(1)(8)+1}{8} = \frac{8+1}{8} = \frac{9}{8}$

31. $1\frac{7}{8} = \frac{(1)(8)+7}{8} = \frac{8+7}{8} = \frac{15}{8}$

33. $\frac{13}{5} = \frac{10+3}{5} = \frac{(2)(5)+3}{5} = 2\frac{3}{5}$

35. $-\frac{73}{6} = \frac{-(72+1)}{6}$

$$= \frac{-(12 \cdot 6 + 1)}{6} = -12\frac{1}{6}$$

37. $-\frac{878}{15} = -\frac{870+8}{15}$

$$= -\frac{(58)(15)+8}{15} = -58\frac{8}{15}$$

39. $\frac{7}{10} = 0.7$

41. $\frac{2}{9} = .\overline{2}$

43. $\frac{3}{8} = 0.375$

45. $\frac{13}{6} = 2.1\overline{6}$

47. $\frac{85}{15} = 5.\overline{6}$

49. $\frac{75}{100} = \frac{75 \div 25}{100 \div 25} = \frac{3}{4}$

51.

$$0.045 = \frac{45}{1000} = \frac{45 \div 5}{1000 \div 5} = \frac{9}{200}$$

53. $0.2 = \frac{2}{10} = \frac{1}{5}$

55. $0.0131 = \frac{131}{10,000}$

57. $.0001 = \frac{1}{10000}$

59. Let $n = 0.\overline{1}$, $10n = 1.\overline{1}$

$$10n = 1.\overline{1}$$
$$-n = 0.\overline{1}$$
$$\overline{9n = 1.0}$$

$$\frac{9n}{9} = \frac{1}{9} = n$$

61. Let $n = 1.\overline{9}$, $10n = 19.\overline{9}$

$$10n = 19.\overline{9}$$
$$-n = 1.\overline{9}$$
$$\overline{9n = 18.0}$$

$$\frac{9n}{9} = \frac{18}{9} = 2 = n$$

63. Let

$$n = 1.\overline{36}, \quad 100n = 136.\overline{36}$$

$$100n = 136.\overline{36}$$
$$-n = 1.\overline{36}$$
$$\overline{99n = 135.0}$$

$$\frac{99n}{99} = \frac{135}{99} = \frac{15}{11} = n$$

65. Let $n = 2.0\overline{5}$,

$\quad 10n = 20.\overline{5}, \quad 100n = 205.\overline{5}$

$100n = 205.\overline{5}$

$\underline{-10n = 20.\overline{5}}$

$90n = 185.0$

$\dfrac{90n}{90} = \dfrac{185}{90} = \dfrac{37}{18} = n$

67. Let $n = 3.4\overline{78}$,

$\quad 1000n = 3478.\overline{78}$

$1000n = 3478.\overline{78}$

$\underline{-10n = 34.\overline{78}}$

$990n = 3444.0$

$\dfrac{990n}{990} = \dfrac{3444}{990} = \dfrac{574}{165} = n$

69. $\dfrac{1}{2} \bullet \dfrac{4}{5} = \dfrac{1 \bullet 4}{2 \bullet 5} = \dfrac{4}{10} = \dfrac{4 \div 2}{10 \div 2} = \dfrac{2}{5}$

71. $\dfrac{-3}{8} \bullet \dfrac{-16}{15} = \dfrac{48}{120} = \dfrac{2}{5}$

73. $\dfrac{7}{8} \div \dfrac{8}{7} = \dfrac{7}{8} \bullet \dfrac{7}{8} = \dfrac{49}{64}$

75. $\left(\dfrac{3}{5} \bullet \dfrac{4}{7} \right) \div \dfrac{1}{3} = \dfrac{12}{35} \div \dfrac{1}{3} = \dfrac{12}{35} \bullet \dfrac{3}{1} = \dfrac{36}{35}$

77. $\left[\left(-\dfrac{2}{3} \right)\left(\dfrac{5}{8} \right) \right] \div \left(-\dfrac{7}{16} \right) = \left(-\dfrac{10}{24} \right) \bullet \left(-\dfrac{16}{7} \right) = \dfrac{160}{168} = \dfrac{20}{21}$

79. The lcm of 3 and 4 is 12.

$\dfrac{1}{4} + \dfrac{2}{3} = \left(\dfrac{1}{4} \bullet \dfrac{3}{3} \right) + \left(\dfrac{2}{3} \bullet \dfrac{4}{4} \right) = \dfrac{3}{12} + \dfrac{8}{12} = \dfrac{11}{12}$

81. The lcm of 11 and 22 is 22.

$\dfrac{2}{11} + \dfrac{5}{22} = \left(\dfrac{2}{11} \bullet \dfrac{2}{2} \right) + \dfrac{5}{22} = \dfrac{4}{22} + \dfrac{5}{29} = \dfrac{9}{22}$

83. The lcm of 9 and 54 is 54.

$\dfrac{5}{9} - \dfrac{7}{54} = \left(\dfrac{5}{9} \bullet \dfrac{6}{6} \right) - \dfrac{7}{54} = \dfrac{30}{54} - \dfrac{7}{54} = \dfrac{23}{54}$

85. The lcm of 12, 48, and 72 is 144.

$\dfrac{1}{12} + \dfrac{1}{48} + \dfrac{1}{72} = \left(\dfrac{1}{12} \bullet \dfrac{12}{12} \right) + \left(\dfrac{1}{48} \bullet \dfrac{3}{3} \right) + \left(\dfrac{1}{72} \bullet \dfrac{2}{2} \right)$

$= \dfrac{12}{144} + \dfrac{3}{144} + \dfrac{2}{144} = \dfrac{17}{144}$

87. The lcm of 30, 40, and 50 is 600.

$\dfrac{1}{30} - \dfrac{3}{40} - \dfrac{7}{50} = \left(\dfrac{1}{30} \bullet \dfrac{20}{20} \right) \left(\dfrac{3}{40} \bullet \dfrac{15}{15} \right) \left(\dfrac{7}{50} \bullet \dfrac{12}{12} \right)$

$= \dfrac{20}{600} - \dfrac{45}{600} - \dfrac{84}{600} = -\dfrac{109}{600}$

89. $\dfrac{2}{3} + \dfrac{1}{8} = \dfrac{2 \bullet 8 + 3 \bullet 1}{3 \bullet 8} = \dfrac{16 + 3}{23} = \dfrac{19}{24}$

91. $\dfrac{5}{6} - \dfrac{7}{8} = \dfrac{5 \bullet 4 - 7 \bullet 3}{24} = \dfrac{20 - 21}{24} = \dfrac{-1}{24}$

93. $\dfrac{3}{8} + \dfrac{5}{12} = \dfrac{3 \bullet 12 + 8 \bullet 5}{8 \bullet 12} = \dfrac{36 + 40}{96} = \dfrac{76}{96} = \dfrac{19}{24}$

95. $\left(\dfrac{2}{3} \bullet \dfrac{9}{10} \right) + \dfrac{2}{5} = \dfrac{18}{30} + \dfrac{2}{5} = \dfrac{18}{30} + \left(\dfrac{2}{5} \bullet \dfrac{6}{6} \right) =$

$= \dfrac{18}{30} + \dfrac{12}{30} = \dfrac{30}{30} = 1$

97. $\left(\dfrac{3}{4} + \dfrac{1}{6} \right) \div \left(2 - \dfrac{7}{6} \right) = \left(\dfrac{3}{4} \bullet \dfrac{3}{3} + \dfrac{1}{6} \bullet \dfrac{2}{2} \right) \div \left(\dfrac{2}{1} \bullet \dfrac{6}{6} - \dfrac{7}{6} \right) = \left(\dfrac{9}{12} + \dfrac{2}{12} \right) \div \left(\dfrac{12}{6} - \dfrac{7}{6} \right) = \dfrac{11}{12} \div \dfrac{5}{6} = \dfrac{11}{12} \bullet \dfrac{6}{5} = \dfrac{66}{60} = \dfrac{11}{10}$

99. $\left(3\dfrac{4}{9} \right) \div \left(4 + \dfrac{2}{3} \right) = \left(\dfrac{3}{1} \bullet \dfrac{9}{9} - \dfrac{4}{9} \right) \div \left(\dfrac{4}{1} \bullet \dfrac{3}{3} + \dfrac{2}{3} \right) = \left(\dfrac{27}{9} - \dfrac{4}{9} \right) \div \left(\dfrac{12}{3} + \dfrac{2}{3} \right) = \dfrac{23}{9} \div \dfrac{14}{3} = \dfrac{23}{9} \bullet \dfrac{3}{14} = \dfrac{69}{126} = \dfrac{23}{42}$

101. $\quad 71\dfrac{5}{8} \quad \rightarrow \quad 70\dfrac{13}{8}$

$\quad \underline{-69\dfrac{7}{8}} \quad \rightarrow \quad \underline{-69\dfrac{7}{8}}$

$\quad\quad\quad\quad\quad 1\dfrac{6}{8} \rightarrow 1\dfrac{3}{4} \text{ inches}$

103.

$$14\left(8\frac{5}{8}\right)=14\left(\frac{69}{8}\right)=\frac{966}{8}=\frac{966\div2}{8\div2}=\frac{483}{4}=120.75"$$

105.

$$1-\left(\frac{1}{4}+\frac{1}{5}+\frac{1}{2}\right)2\frac{1}{4}+3\frac{7}{8}+4\frac{1}{4}=2\frac{4}{16}+3\frac{14}{16}+4\frac{4}{16}$$

$$=9\frac{22}{16}=10\frac{6}{16}$$

$$20\frac{5}{16}-10\frac{6}{16}=19\frac{21}{16}-10\frac{6}{16}=9\frac{15}{16}"$$

107.

$$1-\left(\frac{1}{2}+\frac{2}{5}\right)=1-\left(\frac{5}{10}+\frac{4}{10}\right)=1-\frac{9}{10}=\frac{10}{10}-\frac{9}{10}=\frac{1}{10}$$

Student tutors represent 0.1 of the budget.

109. $4\frac{1}{2}+30\frac{1}{4}+24\frac{1}{8}=4\frac{4}{8}+30\frac{2}{8}+24\frac{1}{8}=58\frac{7}{8}$ inches

111. a) $1\frac{49}{60},2\frac{48}{60},9\frac{6}{60},6\frac{3}{60},2\frac{9}{60},\frac{22}{60}$

b) $1\frac{49}{60}+2\frac{48}{60}+9\frac{6}{60}+6\frac{3}{60}+2\frac{9}{60}+\frac{22}{60}$

$$=(1+2+9+6+2+0)+\frac{49+48+6+3+9+22}{60}$$

$$=20+\frac{137}{60}=22\frac{17}{60} \text{ or } 22 \text{ hours, 17 minutes}$$

113. $8\frac{3}{4}$ ft $=\left(\frac{35}{4}\cdot\frac{12}{1}\right)$ in. $=105$ in.

$$\left[105-(3)\left(\frac{1}{8}\right)\right]\div 4=\left[\frac{840}{8}-\frac{3}{8}\right]\div 4=\frac{837}{8}\cdot\frac{1}{4}=\frac{837}{32}=26\frac{5}{32}. \quad \text{The length of each piece is } 26\frac{5}{32} \text{ in.}$$

115. a) $20+18\frac{3}{8}\div 2=20+9\frac{3}{16}=29\frac{3}{16}$ in.

b) $26\frac{1}{4}+6\frac{3}{4}=33$ in.

c) $26\frac{1}{4}+\left(6\frac{3}{4}-\frac{1}{4}\right)=26\frac{1}{4}+6\frac{2}{4}=32\frac{3}{4}$ in.

117. $\dfrac{0.10+0.11}{2}=\dfrac{0.21}{2}=0.105$

119. $\dfrac{-2.176+(-2.175)}{2}=\dfrac{-4.351}{2}=-2.1755$

121. $\dfrac{4.872+4.873}{2}=\dfrac{9.745}{2}=4.8725$

123. $\left(\dfrac{1}{3}+\dfrac{2}{3}\right)\div 2=\dfrac{3}{3}\cdot\dfrac{1}{2}=\dfrac{3}{6}=\dfrac{1}{2}$

125. $\left(\dfrac{1}{100}+\dfrac{1}{10}\right)\div 2=\dfrac{11}{100}\cdot\dfrac{1}{2}=\dfrac{11}{200}$

127.

$$\left(\frac{1}{10}+\frac{1}{100}\right)\div 2=\left(\frac{10}{100}+\frac{1}{100}\right)\cdot\frac{1}{2}=\frac{11}{100}\cdot\frac{1}{2}=\frac{11}{200}$$

129. a) Water (or milk): $\left(1+1\frac{3}{4}\right)\div 2=\left(\frac{4}{4}+\frac{7}{4}\right)\cdot\frac{1}{2}=\frac{11}{4}\cdot\frac{1}{2}=\frac{11}{8}=1\frac{3}{8}$ cup;

      Oats: $\left(\frac{1}{2}+1\right)\div 2=\frac{3}{2}\cdot\frac{1}{2}=\frac{3}{4}$ cup

   b) Water (or milk): $1+\frac{1}{2}=1\frac{1}{2}$ cup;

      Oats: $\frac{1}{2}+\frac{1}{2}\cdot\frac{1}{2}=\frac{3}{4}$ cup

131. a) $\frac{1}{8}$ b) $\frac{1}{16}$ c) 5 times   d) 6 times

Exercise Set 5.4

1. A rational number can be written as a ratio of two integers, p/q, with q not equal to zero. Numbers that cannot be written as the ratio of two integers are called irrational numbers.

3. A perfect square number is any number that is the square of a natural number.

5. a) To add or subtract two or more square roots with the same radicand, add or subtract their coefficients and then multiply by the common radical.

   b) $3\sqrt{5}+8\sqrt{5}-6\sqrt{5}=11\sqrt{5}-6\sqrt{5}=5\sqrt{5}$

7. a) Multiply both the numerator and denominator by the same number that will result in the radicand in the denominator becoming a perfect square.

   b) $\dfrac{2}{\sqrt{3}}=\dfrac{2}{\sqrt{3}}\cdot\dfrac{\sqrt{3}}{\sqrt{3}}=\dfrac{2\sqrt{3}}{\sqrt{9}}=\dfrac{2\sqrt{3}}{3}$

9. $\sqrt{25}=5$ rational

11. $\frac{3}{5}$ rational

13. Irrational; non-terminating, non-repeating decimal

15. Rational; quotient of two integers

17. Irrational; non-terminating, non-repeating decimal

19. $\sqrt{16}=4$

21. $\sqrt{100}=10$

23. $-\sqrt{169}=-13$

25. $-\sqrt{81}=-9$

27. $-\sqrt{100}=-10$

29. 2, rational, integer, natural

31. $\sqrt{25}=5$, rat'l, integer., nat'l

33. irrational

35. rational

37. rational

39. $\sqrt{12}=\sqrt{4}\sqrt{3}=2\sqrt{3}$

41. $\sqrt{48}=\sqrt{3}\sqrt{16}=4\sqrt{3}$

43. $\sqrt{63}=\sqrt{9}\sqrt{7}=3\sqrt{7}$

45. $\sqrt{84}=\sqrt{4}\sqrt{21}=2\sqrt{21}$

47. $\sqrt{162}=\sqrt{81}\sqrt{2}=9\sqrt{2}$

49. $3\sqrt{5}+2\sqrt{5}=(3+2)\sqrt{5}=5\sqrt{5}$

51. $5\sqrt{18}-7\sqrt{8}=5\left(\sqrt{9}\sqrt{2}\right)-7\left(\sqrt{4}\sqrt{2}\right)$

    $=15\sqrt{2}-14\sqrt{2}=(15-14)\sqrt{2}=\sqrt{2}$

53. $4\sqrt{12}-7\sqrt{27}=4\sqrt{4}\sqrt{3}-7\sqrt{9}\sqrt{3}$

    $=4\cdot 2\sqrt{3}-7\cdot 3\sqrt{3}=8\sqrt{3}-21\sqrt{3}$

    $=-13\sqrt{3}$

55.
$$5\sqrt{3}+7\sqrt{12}-3\sqrt{75}$$
$$=5\sqrt{3}+7\cdot2\sqrt{3}-3\cdot5\sqrt{3}$$
$$=5\sqrt{3}+14\sqrt{3}-15\sqrt{3}$$
$$=(5+14-15)\sqrt{3}=4\sqrt{3}$$

57.
$$\sqrt{8}-3\sqrt{50}+9\sqrt{32}$$
$$=2\sqrt{2}-3\cdot5\sqrt{2}+9\cdot4\sqrt{2}$$
$$=2\sqrt{2}-15\sqrt{2}+36\sqrt{2}$$
$$=(2-15+36)\sqrt{2}=23\sqrt{2}$$

59.
$$\sqrt{3}\sqrt{27}=\sqrt{3\cdot27}=\sqrt{81}=9$$

61. $\sqrt{6}\cdot\sqrt{10}=\sqrt{2}\sqrt{3}\sqrt{2}\sqrt{5}$
$$=\sqrt{4}\sqrt{15}=2\sqrt{15}$$

63. $\sqrt{10}\cdot\sqrt{20}=\sqrt{200}$
$$=\sqrt{100}\cdot\sqrt{2}=10\sqrt{2}$$

65. $\dfrac{\sqrt{20}}{\sqrt{5}}=\sqrt{\dfrac{20}{5}}=\sqrt{4}=2$

67. $\dfrac{\sqrt{72}}{\sqrt{8}}=\sqrt{9}=3$

69. $\dfrac{1}{\sqrt{5}}=\dfrac{1}{\sqrt{5}}\dfrac{\sqrt{5}}{\sqrt{5}}=\dfrac{\sqrt{5}}{5}$

71. $\dfrac{\sqrt{3}}{\sqrt{7}}\cdot\dfrac{\sqrt{7}}{\sqrt{7}}=\dfrac{\sqrt{21}}{7}$

73.
$$\dfrac{\sqrt{20}}{\sqrt{3}}=\dfrac{\sqrt{20}}{\sqrt{3}}\dfrac{\sqrt{3}}{\sqrt{3}}=\dfrac{\sqrt{60}}{\sqrt{9}}$$
$$=\dfrac{\sqrt{4}\sqrt{15}}{3}=\dfrac{2\sqrt{15}}{3}$$

75. $\dfrac{\sqrt{5}}{\sqrt{3}}=\dfrac{\sqrt{5}}{\sqrt{3}}\cdot\dfrac{\sqrt{3}}{\sqrt{3}}=\dfrac{\sqrt{15}}{3}$

77.
$$\dfrac{\sqrt{10}}{\sqrt{6}}\cdot\dfrac{\sqrt{6}}{\sqrt{6}}=\dfrac{\sqrt{60}}{6}$$
$$=\dfrac{2\sqrt{15}}{6}=\dfrac{\sqrt{15}}{3}$$

79. $\sqrt{5}$ is between 2 and 3 since $\sqrt{5}$ is between $\sqrt{4}=2$ and $\sqrt{9}=3$. $\sqrt{5}$ is between 2 and 2.5 since 5 is closer to 4 than to 9. Using a calculator $\sqrt{5}\approx2.2$.

81. $\sqrt{107}$ is between 10 and 11 since $\sqrt{107}$ is between $\sqrt{100}=10$ and $\sqrt{121}=11$. $\sqrt{107}$ is between 10 and 10.5 since 107 is closer to 100 than to 121. Using a calculator $\sqrt{107}\approx10.3$.

83. $\sqrt{170}$ is between 13 and 14 since $\sqrt{170}$ is between $\sqrt{169}=13$ and $\sqrt{196}=14$. $\sqrt{170}$ is between 13 and 13.5 since 170 is closer to 169 than to 196. Using a calculator $\sqrt{170}\approx13.04$.

85. False. The result may be a rational number or an irrational number.

87. True

89. False. The result may be a rational number or an irrational number.

91. $\sqrt{3}+5\sqrt{3}=6\sqrt{3}$

93. $\sqrt{2}\cdot\sqrt{3}=\sqrt{6}$

95. No. $2\neq1.414$ since $\sqrt{2}$ is an irrational number and 1.414 is a rational number.

97. No. 3.14 and $\dfrac{22}{7}$ are rational numbers, π is an irrational number.

99.
$$\sqrt{4\cdot16}=\sqrt{4}\sqrt{16}$$
$$\sqrt{64}\ =\ 2\cdot4$$
$$8\ =\ 8$$

101. a) $s = \sqrt{\dfrac{4}{0.04}} = \sqrt{100} = 10$ mph

     b) $s = \sqrt{\dfrac{16}{0.04}} = \sqrt{400} = 20$ mph

     c) $s = \sqrt{\dfrac{64}{0.04}} = \sqrt{1600} = 40$ mph

     d) $s = \sqrt{\dfrac{256}{0.04}} = \sqrt{6400} = 80$ mph

103. a) $\sqrt{0.04} = 0.2$ a terminating decimal and thus it is rational.

     b) $\sqrt{0.7} = \sqrt{\dfrac{7}{10}} = \dfrac{\sqrt{70}}{10}$; $\sqrt{70}$ is irrational since the only integers with rational square roots are the perfect squares and 70 is not a perfect square.

     Thus $\dfrac{\sqrt{70}}{10} = \sqrt{0.7}$ is irrational.

105. a) $\left(44 \div \sqrt{4}\right) \div \sqrt{4} = \left(44 \div 2\right) \div 2 = 22 \div 2 = 11$

     b) $\left(44 \div 4\right) + \sqrt{4} = 11 + 2 = 13$

     c) $4 + 4 + 4 + \sqrt{4} = 12 + 2 = 14$

     d) $\sqrt{4}\left(4 + 4\right) + \sqrt{4} = 2(8) + 2 = 16 + 2 = 18$

Exercise Set 5.5

1. The set of real numbers is the union of the rational numbers and the irrational numbers.

3. If the given operation is preformed on any two elements of the set and the result is an element of the set, then the set is <u>closed</u> under the given operation.

5. The order in which two numbers are multiplied does not make a difference in the result.   Ex. $2 \cdot 3 = 3 \cdot 2$

7. The associative property of addition states that when adding three real numbers, parentheses may be placed around any two adjacent numbers. $(a + b) + c = a + (b + c)$

9. Not closed. (e.g., $3 - 5 = -2$ is not a natural number).

11. Closed. The product of two natural numbers is a natural number.

13. Closed. The sum of two integers is an integer.

15. Closed. The product of two integers is an integer.

17. Closed   19. Not closed   21. Not closed   23. Not closed

25. Closed   27. Closed

29. Commutative property of addition. The order $5 + x$ is changed to $x + 5$. .

31. $(-2) + (-3) = -5 = (-3) + (-2)$

33. No. $5 - 3 = 2$, but $3 - 5 = -2$

35. $[(-3) + (-5)] + (-7) = (-8) + (-7) = -15$

     $(-3) + [(-5)] + (-7)] = (-3) + (-12) = -15$

37. No.

     $(8 \div 4) \div 2 = 2 \div 2 = 1$, but $8 \div (4 \div 2) = 8 \div 2 = 4$

39. No. $(8 \div 4) \div 2 = 2 \div 2 = 1$,

     but $8 \div (4 \div 2) = 8 \div 2 = 4$

41. $3(y + 5) = 3 \cdot y + 3 \cdot 5$

     Distributive property

43. $(7 \cdot 8) \cdot 9 = 7 \cdot (8 \cdot 9)$

    Associative property of multiplication

45. $(24 + 7) + 3 = 24 + (7 + 3)$

    Associative property of addition

47. $\sqrt{3} \cdot 7 = 7 \cdot \sqrt{3}$

    Commutative property of multiplication

49. $-1(x + 4) = (-1) \cdot x + (-1) \cdot 4$

    Distributive property

51. $\sqrt{5} \cdot 2 = 2 \cdot \sqrt{5}$

    Commutative property of multiplication

53. $(r + s) \cdot t = (r \cdot t) + (s \cdot t)$

    Distributive property

55. $(f \cdot g) + (j \cdot h) = (g \cdot f) + (j \cdot h)$

    Commutative property of multiplication

57. $4(z + 1) = 4z + 4$

59. $-\dfrac{3}{4}(x - 12) = -\dfrac{3}{4}x + \dfrac{3}{4} \cdot 12 = -\dfrac{3}{4}x + 9$

61. $6\left(\dfrac{x}{2} + \dfrac{2}{3}\right) = \dfrac{6x}{2} + \dfrac{12}{3} = 3x + 4$

63. $32\left(\dfrac{1}{16}x - \dfrac{1}{32}\right) = \dfrac{32x}{16} - \dfrac{32}{32} = 2x - 1$

65. $\sqrt{2}\left(\sqrt{8} - \sqrt{2}\right) = \sqrt{16} - \sqrt{4} - 4 - 2 = 2$

67. $5\left(\sqrt{2} + \sqrt{3}\right) = 5\sqrt{2} + 5\sqrt{3}$

69. a) Distributive property

    b) Associative property of addition

71. a) Distributive property

    b) Associative property of addition;

    c) Commutative property of addition

    d) Associative property of addition

73. a) Distributive property

    b) Commutative property of addition;

    c) Associative property of addition

    d) Commutative property of addition

75. Yes. You can either feed your dog first or give your dog water first.

77. No. The clothes must be washed first before being dried.

79. Yes. Can be done in either order; either fill the car with gas or wash the windshield.

81. No. Pressing the keys will have no effect if there are no batteries in place.

83. Yes. The order does not matter.

85. Yes. The order does not matter.

87. Yes. The final result will be the same regardless of the order of the events.

89. Baking pizzelles: mixing eggs into the batter, or mixing sugar into the batter.; Yard work: mowing the lawn, or trimming the bushes

91. No. $0 \div a = 0$ but $a \div 0$ is undefined.

Exercise Set 5.6

1. 2 is the base and 3 is the exponent.

3. a) If m and n are natural numbers and a is any real number, then $a^m a^n = a^{m+n}$

   b) $2^3 \cdot 2^4 = 2^{3+4} = 2^7 = 128$

5. a) If a is any real number except 0, then $a^0 = 1$.

   b) $7^0 = 1$

7. a) If m and n are natural numbers and a is any real number, then $\left(a^m\right)^n = a^{m \cdot n}$

   b) $\left(3^2\right)^4 = 3^{2 \cdot 4} = 3^8 = 6561$

9. a) Since 1 raised to any exponent equals +1, then $-1^{500} = (-1)\left(1^{500}\right) = (-1)(1) = -1$

   b) Since -1 raised to an even exponent equals 1, then number $(-1)^{500} = \left((-1)^2\right)^{250} = (1)^{250} = 1$

   c) In -1^{501}, -1 is not raised to the 501st power, but +1 is; so $-1^{501} = (-1)\left(1^{501}\right) = (-1)(1) = -1$

   d) Since -1 is raised to an odd exponent is -1, then $(-1)^{501} = -1$

11. a) If the exponent is positive, move the decimal point in the number to the right the same number of places as the exponent adding zeros where necessary. If the exponent is negative, move the decimal point in the number to the left the same number of places as the exponent adding zeros where necessary.

    b) $2.91 \times 10^{-5} = 0.0000291$

    c) $7.02 \times 10^6 = 7,020,000$

13. a) $3^2 = 3 \cdot 3 = 9$

    b) $2^3 = 2 \cdot 2 \cdot 2 = 8$

15. a) $(-5)^2 - (-5)(-5) = 25$

    b) $-5^2 = -(5)(5) = -25$

17. a) $-2^4 = -(2)(2)(2)(2) = -16$

    b) $(-2)^4 = (-2)(-2)(-2)(-2) = 16$

19. a) $-4^3 = -(4)(4)(4) = -64$

    b) $(-4)^3 = (-4)(-4)(-4) = -64$

21. a) $\left(\frac{1}{8}\right)^2 = \left(\frac{1}{8}\right)\left(\frac{1}{8}\right) = \frac{1}{64}$

    b) $\left(-\frac{3}{4}\right)^2 = \left(-\frac{3}{4}\right)\left(-\frac{3}{4}\right) = \frac{9}{16}$

23. a) $1000^1 = 1000$

    b) $1^{1000} = 1$

25. a) $3^2 \cdot 3^3 = 3^{2+3} = 3^5 = 243$

    b) $(-3)^2 \cdot (-3)^3 = (-3)^{2+3} = (-3)^5 = -243$

27. a) $\frac{5^7}{5^5} = 5^{7-5} = 5^2 = 25$

    b) $\frac{(-5)^7}{(-5)^5} = (-5)^{7-5} = (-5)^2 = 25$

29. a) $6^0 = 1$

    b) $-6^0 = -1$

31. a) $(6x)^0 = 1$

    b) $6x^0 = 6$

33. a) $3^{-3} = \dfrac{1}{3^3} = \dfrac{1}{27}$

    b) $7^{-2} = \dfrac{1}{7^2} = \dfrac{1}{49}$

35. a) $-9^{-2} = -\dfrac{1}{9^2} = -\dfrac{1}{81}$

    b) $(-9)^{-2} = \dfrac{1}{(-9)^2} = \dfrac{1}{81}$

37. a) $\left(2^2\right)^3 = 2^{2\times3} = 2^6 = 64$

    b) $\left(2^3\right)^2 = 2^{3\times2} = 2^6 = 64$

39. a) $4^3 \cdot 4^{-2} = 4^{3-2} = 4^1 = 4$

    b) $2^{-2} \cdot 2^{-2} = 2^{-2-2} = 2^{-4} = \dfrac{1}{2^4} = \dfrac{1}{16}$

41. $175000 = 1.75 \times 10^5$

43. $0.00023 = 2.3 \times 10^{-4}$

45. $0.56 = 5.6 \times 10^{-1}$

47. $19000 = 1.9 \times 10^4$

49. $0.000186 = 1.86 \times 10^{-4}$

51. $0.00000423 = 4.23 \times 10^{-6}$

53. $711 = 7.11 \times 10^2$

55. $0.153 = 1.53 \times 10^{-1}$

57. $1.7 \times 10^2 = 170$

59. $1.097 \times 10^{-4} = 0.0001097$

61. $8.62 \times 10^{-5} = 0.0000862$

63. $3.12 \times 10^{-1} = 0.312$

65. $9.0 \times 10^6 = 9000000$

67. $2.31 \times 10^2 = 231$

69. $3.5 \times 10^4 = 35000$

71. $1.0 \times 10^4 = 10000$

73. $\left(3\times10^2\right)\left(2.5\times10^3\right) = 750,000$

75. $\left(5.1 \times 10^1\right)\left(3.0 \times 10^{-4}\right) = 15.3 \times 10^{-3} = 0.0153$

77. $\dfrac{7.5\times10^6}{3\times10^4} = 2.5\times10^2 = 250$

79. $\dfrac{8.4 \times 10^{-6}}{4.0 \times 10^{-3}} = 2.1 \times 10^{-3} = 0.0021$

81. $\dfrac{4.0 \times 10^5}{2.0 \times 10^4} = 2.0 \times 10^1 = 20$

83. $\left(5\times10^6\right)\left(2\times10^4\right) = 10\times10^{10} = 1\times10^{11}$

85. $\left(3.0 \times 10^{-3}\right)\left(1.5 \times 10^{-4}\right) = 4.5 \times 10^{-7}$

87. $\dfrac{5.6\times10^6}{8\times10^4} = 0.7\times10^2 = 7\times10^1$

89. $\dfrac{4.0\times10^{-5}}{2.0\times10^2} = 2.0 \times 10^{-7}$

91. $\dfrac{1.5\times10^5}{5.0\times10^{-4}} = 0.3 \times 10^9 = 3.0 \times 10^8$

93. $3.6\times10^{-3}, 1.7, 9.8\times10^2, 1.03\times10^4$

95. $8.3 \times 10^{-5}; 0.00079; 4.1 \times 10^3; 40,000;$ Note: $0.00079 = 7.9 \times 10^{-4}, 40,000 = 4 \times 10^4$

97. $\dfrac{2.990\times10^8}{6.522\times10^9} \approx 0.458\times10^{-1} \approx 0.046$

99. $\dfrac{1.095\times10^9}{6.522\times10^9} \approx 0.1679\times10^0 \approx 0.168$

101. $\dfrac{7.6\times10^{65}}{2.806\times10^{14}} \approx 2.708\times10^{51}$; about 2.7×10^{51} sec

103. $\dfrac{8.86\times10^{12}}{1.307\times10^9} \approx 6.779\times10^3 = 6779$ or \$6779

105. $t = \dfrac{d}{r} = \dfrac{239000 \text{ mi}}{20000 \text{ mph}} = 11.95$ 11.95 hrs

107. $(50)(5,800,000) = (5 \times 10^1)(5.8 \times 10^6) = 29 \times 10^7 = 2.9 \times 10^8$

   2.9×10^8 cells

109. (100,000 cu ft/sec) (60 sec/min) (60 min/hr) (24 hr) = 8,640,000,000 ft^3 or 8.64×10^9 cu ft

111. a) (0.60) (1,200,000,000) = \$720,000,000   b) (0.25) (1,200,000,000) = \$300,000,000

   c) (0.10) (1,200,000,000) = \$120,000,000   d) (0.05) (1,200,000,000) = \$60,000,000

113. 1,000 times, since 1 meter = 10^3 millimeters = 1,000 millimeters

115. $\dfrac{2 \times 10^{30}}{6 \times 10^{24}} = 0.\overline{3} \times 10^6 \approx 300,000$ times

117. a) $(1.86 \times 10^5$ mi/sec) (60 sec/min) (60 min/hr) (24 hr/day) (365 days/yr) (1 yr)

   $= (1.86 \times 10^5)(6 \times 10^1)(6 \times 10^1)(2.4 \times 10^1)(3.65 \times 10^2) = 586.5696 \times 10^{10} \approx 5.87 \times 10^{12}$ miles

   b) $t = \dfrac{d}{r} = \dfrac{9.3 \times 10^7}{1.86 \times 10^5} = 5.0 \times 10^2 = 500$ seconds or 8 min. 20 sec.

Exercise Set 5.7

1. A sequence is a list of numbers that are related to each other by a given rule. One example is 2, 4, 6, 8....

3. a) An arithmetic sequence is a sequence in which each term differs from the preceding term by a constant
       amount. One example is 1, 4, 7, 10,....

   b) A geometric sequence is one in which the ratio of any term to the term that directly precedes it is a constant.
       One example is 1, 3, 9, 27,....

5. a) $a_n = n^{th}$ term of the sequence   b) a_1 = first term of a sequence   c) d = common difference in a sequence

   d) s_n = the sum of the first n terms of the arithmetic sequence

7. $a_1 = 5, d = 1$ 5, 6, 7, 8, 9

9. $a_1 = 12, d = -2$ 12, 10, 8, 6, 4

11. $a_1 = 5, d = -2$ 5, 3, 1, -1, -3

13. $a_1 = \frac{3}{4}, d = \frac{1}{4}$ $\frac{3}{4}, 1, \frac{5}{4}, \frac{3}{2}, \frac{7}{4}$

15. $a_4 = 4 + (5-1)1 = 4 + 4 = 8$

17. $a_{10} = -5 + (10-1)(2) = -5 + 18 = 13$

19. $a_{20} = \dfrac{4}{5} + (19)(-1) = \dfrac{4}{5} - 19 = \dfrac{4}{5} - \dfrac{95}{5} = -\dfrac{91}{5}$

21. $a_{22} = -23 + (22-1)\left(\dfrac{5}{7}\right) = -23 + 15 = -8$

23. $a_n = n$ $(a_1 = 1, d = 1)$

25. $a_n = 2n$ $(a_1 = 2, d = 2)$

27. $a_1 = \dfrac{1}{4}, d = \dfrac{3}{4}$

   $a_n = \dfrac{1}{4} + (n-1)\left(\dfrac{3}{4}\right) = \dfrac{1}{4} + \dfrac{3}{4}n - \dfrac{3}{4} = \dfrac{3}{4}n - \dfrac{1}{2}$

29. $a_1 = -3, d = \dfrac{3}{2}$

   $a_n = -3 + (n-1)\left(\dfrac{3}{2}\right) = -3 + \dfrac{3}{2}n - \dfrac{3}{2} = \dfrac{3}{2}n = \dfrac{9}{2}$

31. $s_{50} = \dfrac{n(a_1 + a_n)}{2} = \dfrac{50(1+50)}{2} = \dfrac{50(51)}{2}$

$= (25)(51) = 1275$

33. $s_{50} = \dfrac{50(1+99)}{2} = \dfrac{50(100)}{2} = (25)(100) = 2500$

35. $s_8 = \dfrac{8(11 + (-24))}{2} = \dfrac{8 \cdot (-13)}{2} = -52$

37. $s_{24} = \dfrac{24\left(\dfrac{1}{5} + \dfrac{24}{5}\right)}{2} = \dfrac{24 \cdot \left(\dfrac{25}{5}\right)}{2} = 12 \cdot 5 = 60$

39. $a_1 = 1, \ r = 5 \qquad 1, 5, 25, 125, 625$

41. $a_1 = 2, r = -4 \qquad 2, -4, 8, -16, 32$

43. $a_1 = -3, r = -1 \qquad -3, 3, -3, 3, -3$

45. $a_1 = 81, r = -\dfrac{1}{3} \qquad 81, -27, 9, -3, 1$

47. $a_6 = 5(2)^5 = (5)(32) = 160$

49. $a_3 = 3\left(\dfrac{1}{2}\right)^2 = 3\left(\dfrac{1}{4}\right) = \dfrac{3}{4}$

51. $a_7 = (-5) \cdot 3^6 = (-5)(729) = -3645$

53. $a_{10} = (-2)(3)^9 = -39{,}366$

55. $1, 2, 4, 8 \quad a_n = 1(2)^{n-1} = 2^{n-1}$

57. $-1, 1, -1, 1 \quad a_n = (-1)(-1)^{n-1} = (-1)^n$

59. $2, 1, \dfrac{1}{2}, \dfrac{1}{4} \qquad a_n = (2)\left(\dfrac{1}{2}\right)^{n-1}$

61. $9, 3, 1, \dfrac{1}{3} \qquad a_n = (9)\left(\dfrac{1}{3}\right)^{n-1}$

63. $s_5 = \dfrac{a_1(1 - r^5)}{1 - r} = \dfrac{6(1 - 2^5)}{1 - 2} = \dfrac{6(-31)}{-1} = 186$

65. $s_6 = \dfrac{a_1(1 - r^6)}{1 - r} = \dfrac{(-3)(1 - 4^6)}{1 - 4} = \dfrac{(-3)(-4095)}{-3}$

$= -4095$

67. $s_{11} = \dfrac{a_1(1 - r^{11})}{1 - r} = \dfrac{-7(1 - 3^{11})}{1 - 3} = \dfrac{-7(-177{,}146)}{-2}$

$= -620{,}011$

69. $s_{13} = \dfrac{(-1)\left(1 - (-2)^{15}\right)}{1 - (-2)} = \dfrac{(-1)(1 + 32{,}768)}{3}$

$= \dfrac{(-1)(32{,}769)}{3} = -10{,}923$

71. $s_{100} = \dfrac{(100)(1 + 100)}{2} = \dfrac{(100)(101)}{2}$

$= 50(101) = 5050$

73. $s_{100} = \dfrac{(100)(2 + 200)}{2} = \dfrac{(100)(202)}{2}$

$= 50(202) = 10100$

75. a) $a_{12} = 96 + (11)(-3) = 96 - 33 = 63$ in.

   b) $\dfrac{[12(96 + 63)]}{2} = \dfrac{(12)(159)}{2} = (6)(159) = 954$ in.

77. a) Using the formula $a_n = a_1 + (n - 1)d$, we get

   $a_8 = 35{,}000 + (8 - 1)(1400) = \$44{,}800$

   b) $\dfrac{8(35000 + 44800)}{2} = \dfrac{8(79800)}{2} = \$319{,}200$

79. $1, 2, 3, \ldots \quad n = 31$

   $s_{31} = \dfrac{31(1 + 31)}{2} = \dfrac{31(32)}{2} = 31(16) = 496$ PCs

81. $a_6 = 200(0.8)^6 = 200(0.262144)^1 = 52.4288$ g

83. $a_{15} = 31{,}000(1.06)^{14} = \$70{,}088$

85. This is a geometric sequence where $a_1 = 2000$ and $r = 3$. In ten years the stock will triple its value 5 times.
$a_6 = a_1 r^{6-1} = 2000(3)^5 = \$486,000$

87. $\dfrac{82[1-(1/2)^6]}{1-(1/2)} = \dfrac{82[1-(1/64)]}{1/2} = \dfrac{82}{1} \bullet \dfrac{63}{64} \bullet \dfrac{2}{1} = 161.4375$

89. 12, 18, 24, ... ,1608 is an arithmetic sequence with $a_1 = 12$ and $d = 6$. Using the expression for the nth term of an arithmetic sequence $a_n = a_1 + (n-1)d$ or $1608 = 12 + (n-1)6$ and dividing both sides by 6 gives $268 = 2 + n - 1$ or $n = 267$

91. The total distance is 30 plus twice the sum of the terms of the geometric sequence having $a_1 = (30)(0.8) = 24$
and $r = 0.8$.   Thus $s_5 = \dfrac{24[1-(0.8)^5]}{(1-0.8)} = \dfrac{24[1-0.32768]}{0.2} = \dfrac{24(0.67232)}{0.2} = 80.6784$.
So the total distance is $30 + 2(80.6784) = 191.3568$ ft.

Exercise Set 5.8

1. Begin with the numbers 1, 1, then add 1 and 1 to get 2 and continue to add the previous two numbers in the sequence to get the next number in the sequence.

3. a) Golden number $= \dfrac{\sqrt{5}+1}{2}$

   b) 1.618 = golden ratio   When a line segment AB is divided at a point C, such that the ratio of the whole, AB, to the larger part, AC, is equal to the ratio of the larger part, AC, to the smaller part, CB, then each
   of the ratios $\dfrac{AB}{AC}$ and $\dfrac{AC}{CB}$ is known as the golden ratio.

   c) The golden proportion is: $\dfrac{AB}{AC} = \dfrac{AC}{CB}$

   d) The golden rectangle: $\dfrac{L}{W} = \dfrac{a+b}{a} = \dfrac{a}{b} = \dfrac{\sqrt{5}+1}{2} =$ golden number

5. a) Flowering head of a sunflower
   b) Great Pyramid

7. a) $\dfrac{\sqrt{5}+1}{2} = 1.618033989$ b) $\dfrac{\sqrt{5}-1}{2} = .618033989$
   c) Differ by 1

9. 1/1 = 1, 2/1 = 2, 3/2 = 1.5, 5/3 = 1.6, 8/5 = 1.6, 13/8 = 1.625, 21/13 = 1.6154, 34/21 = 1.619, 55/34 = 1.6176
   89/55 = 1.61818. The consecutive ratios alternate increasing then decreasing about the golden ratio.

11. If the first ten are selected; $\dfrac{1+1+2+3+5+8+13+21+34+55}{11} = \dfrac{143}{11} = 13$

13. If 2, 3, 5, and 8 are selected the result is $5^2 - 3^2 = 2 \cdot 8 \rightarrow 25 - 9 = 16 \rightarrow 16 = 16$

15. Answers will vary.   17. Answers will vary.   19. Answers will vary.
21. Answers will vary.

23. Not Fibonacci type; it is not true that each term is the sum of the two preceding terms.
25. Fibonacci type: $1 + 2 = 3$, $2 + 3 = 5$ Each term is the sum of the two preceding terms.
27. Fibonacci type; $40 + 65 = 105$; $65 + 105 = 170$
29. Fibonacci type; $-1 + 0 = -1$; $0 + (-1) = -1$

31. a) If 6 and 10 are selected the sequence is 6, 10, 16, 26, 42, 68, 110, ...
    b) $10/6 = 1.666$, $16/10 = 1.600$, $26/16 = 1.625$, $42/26 = 1.615$, $68/42 = 1.619$, $110/68 = 1.618$, ...

33. a) If 5, 8, and 13 are selected the result is $8^2 - (5)(13) = 64 - 65 = -1$.

    b) If 21, 34, and 55 are selected the result is $34^2 - (21)(55) = 1156 - 1155 = 1$.

    c) The square of the middle term of three consecutive terms in a Fibonacci sequence differs from the
       product of the 1st and 2nd term by 1.

35. a) Lucas sequence: 1, 3, 4, 7, 11, 18, 29, 47, ...   b) $8 + 21 = 29$; $13 + 34 = 47$
    c) The first column is a Fibonacci-type sequence.

37. $\dfrac{5}{x} = \dfrac{x}{5-x}$ $5(5-x) = x^2$ $25 - 5x = x^2$ $x^2 + 5x - 25 = 0$

    Solve for x using the quadratic formula,

    $$x = \frac{-b \pm \sqrt{b^2 - 4ac}}{2a} = \frac{-5 \pm \sqrt{25 - 4(1)(-25)}}{2(1)} = \frac{-5 \pm \sqrt{125}}{2} = 5\left(\frac{\sqrt{5}-1}{2}\right) \text{ since we want an}$$

    answer between 0 and 5.

39. Answers will vary.   {5, 12, 13}   {16, 30, 34}   {105, 208, 233}   {272, 546, 610}

Review Exercises

1. Use the divisibility rules in section 5.1.   2. Use the divisibility rules in section 5.1.
   158,340 is divisible by 2, 3, 4, 5, 6, and 10.   400,644 is divisible by 2, 3, 4, 6, and 9

3.
```
2 | 540
2 | 270
3 | 135
3 | 45
3 | 15
    5
```
$540 = 2^2 \cdot 3^3 \cdot 5$

4.
```
3 | 693
3 | 231
7 | 77
    11
```
$693 = 3^2 \cdot 7 \cdot 11$

5.
```
2 | 840
2 | 420
2 | 210
5 | 105
3 | 21
    7
```
$840 = 2^3 \cdot 3 \cdot 5 \cdot 7$

6.
$$882 = 2 \cdot 3^2 \cdot 7^2$$

7.
$$1452 = 2^2 \cdot 3 \cdot 11^2$$

8. $30 = 2 \cdot 3 \cdot 5, \ 105 = 3 \cdot 5 \cdot 7$
   gcd $= 15$ lcm $= 210$

9. $63 = 3 \cdot 3 \cdot 5, \ 108 = 3 \cdot 4 \cdot 9$
   gcd $= 9;$ lcm $= 756$

10. $45 = 3^2 \cdot 5, \ 250 = 2 \cdot 5^3;$ gcd $= 5;$ lcm $= 2 \cdot 3^2 \cdot 5^3 = 2250$

11. $90 = 2 \cdot 3^2 \cdot 5, \ 300 = 2^2 \cdot 3 \cdot 5^2;$ gcd $= 2 \cdot 3 \cdot 5 = 30;$ lcm $= 2^2 \cdot 3^2 \cdot 5^2 = 900$

12. $60 = 2^2 \cdot 3 \cdot 5, \ 40 = 2^3 \cdot 5, \ 96 = 2^5 \cdot 3;$ gcd $= 2^2 = 4;$ lcm $= 2^5 \cdot 3 \cdot 5 = 480$

13. $36 = 2^2 \cdot 3^2, \ 108 = 2^2 \cdot 3^3, \ 144 = 2^4 \cdot 3^2;$ gcd $= 2^2 \cdot 3^2 = 36;$ lcm $= 2^4 \cdot 3^3 = 432$

14. $15 = 3 \cdot 5, \ 9 = 3^2;$ lcm $= 3^2 \cdot 5 = 45.$ In 45 days the train stopped in both cities.

15. $4 + (-7) = -3$

16. $-2 + 5 = 3$

17. $(-2) + (-4) = -6$

18. $4 - 8 = 4 + (-8) = -4$

19. $-5 - 4 = -5 + (-4) = -9$

20. $-3 - (-6) = -3 + 6 = 3$

21. $(-3 + 7) - 4 = 4 + (-4) = 0$

22. $-1 + (9 - 4) = -1 + 5 = 4$

23. $-5 \cdot 3 = -15$

24. $(-2)(-12) = 24$

25. $14(-4) = -56$

26. $-35 / -7 = 5$

27. $12 / -6 = -2$

28. $[8 \div (-4)](-3) = (-2)(-3) = 6$

29. $-20 \div (-4 - 1) = -20 \div (-5) = 4$

30. $[-30 \div (10)] \div (-1) = -3 \div (-1) = 3$

31. $3/10 = 0.3$

32. $11/25 = 0.44$

33. $15/40 = 3/8 = 0.375$

34. $13/4 = 3.25$

35. $6/7 = 0.\overline{857142}$

36. $7/12 = 0.58\overline{3}$

37. $3/8 = 0.375$

38. $11/16 = 0.6875$

39. $5/7 = 0.\overline{714285}$

40. $0.225 = \dfrac{225}{1000} = \dfrac{45}{200} = \dfrac{9}{40}$

41. $1.4 = \dfrac{14}{10} = \dfrac{7}{5}$

42. $0.6666\ldots$ $10n = 6.6666\ldots.$
$$10n = 6.\overline{6} \qquad \dfrac{9n}{9} = \dfrac{6}{9}$$
$$\underline{-n = 0.\overline{6}} \qquad n = \dfrac{2}{3}$$
$$9n = 6.0$$

43. $0.5151\ldots$ $100n = 51.5151\ldots.$
$$100n = 51.\overline{51} \qquad \dfrac{99n}{99} = \dfrac{51}{99}$$
$$\underline{-n = \ 0.\overline{51}} \qquad n = \dfrac{51}{99} = \dfrac{17}{33}$$
$$99n = \ 51$$

44. $0.083 = \dfrac{83}{1000}$

45. $0.0073 = \dfrac{73}{10,000}$

46. $2.344444\ldots$ $100n = 234.444444\ldots$
$100n = 234.\overline{4}$
$\dfrac{-10n = \ \ 23.\overline{4}}{90n = 211.00}$ $\dfrac{90n}{90} = \dfrac{211}{90} = n$

47. $1\dfrac{3}{4} = \dfrac{(1)(4)+3}{4} = \dfrac{7}{4}$

48. $4\dfrac{1}{6} = \dfrac{(4)(6)+1}{6} = \dfrac{25}{6}$

49. $-3\,\tfrac{1}{4} = \dfrac{((-3)(4))-1}{4} = \dfrac{-13}{4}$

50. $-35\tfrac{3}{8} = \dfrac{((-35)(8))-3}{8} = \dfrac{-283}{8}$

51. $\dfrac{11}{5} = \dfrac{2\cdot 5+1}{5} = 2\dfrac{1}{5}$

52. $\dfrac{75}{8} = \dfrac{9\cdot 8+3}{8} = 9\dfrac{3}{8}$

53. $\dfrac{-12}{7} = \dfrac{(-1)(7)-5}{7} = -1\tfrac{5}{7}$

54. $\dfrac{-136}{5} = \dfrac{(-27)(5)-1}{5} = -27\tfrac{1}{5}$

55. $\dfrac{1}{3}+\dfrac{3}{4} = \dfrac{1}{3}\cdot\dfrac{4}{4}+\dfrac{3}{4}\cdot\dfrac{3}{3} = \dfrac{4}{12}+\dfrac{9}{12} = \dfrac{13}{12}$

56. $\dfrac{11}{12}-\dfrac{2}{3} = \dfrac{11}{12}-\dfrac{2}{3}\cdot\dfrac{4}{4} = \dfrac{11}{12}-\dfrac{8}{12} = \dfrac{3}{12} = \dfrac{1}{4}$

57. $\dfrac{1}{6}+\dfrac{5}{4} = \dfrac{1}{6}\cdot\dfrac{2}{2}+\dfrac{5}{4}\cdot\dfrac{3}{3} = \dfrac{2}{12}+\dfrac{15}{12} = \dfrac{17}{12}$

58. $\dfrac{7}{16}\cdot\dfrac{12}{21} = \dfrac{84}{336} = \dfrac{84}{4\cdot 84} = \dfrac{1}{4}$

59. $\dfrac{5}{9}\div\dfrac{6}{7} = \dfrac{5}{9}\div\dfrac{7}{6} = \dfrac{35}{54}$

60. $\left(\dfrac{4}{5}+\dfrac{5}{7}\right)\div\dfrac{4}{5} = \dfrac{28+25}{35}\cdot\dfrac{5}{4} = \dfrac{53}{35}\cdot\dfrac{5}{4} = \dfrac{53}{28}$

61. $\left(\dfrac{2}{3}\cdot\dfrac{1}{7}\right)\div\dfrac{4}{7} = \dfrac{2}{21}\cdot\dfrac{7}{4} = \dfrac{1}{6}$

62. $\left(\dfrac{1}{5}+\dfrac{2}{3}\right)\cdot\dfrac{3}{8} = \dfrac{3+10}{15}\cdot\dfrac{3}{8} = \dfrac{13}{15}\cdot\dfrac{3}{8} = \dfrac{13}{40}$

63. $\left(\dfrac{1}{5}\right)\left(\dfrac{2}{3}\right)+\left(\dfrac{1}{5}\div\dfrac{1}{2}\right) = \dfrac{2}{15}+\left(\dfrac{1}{5}\right)\left(\dfrac{2}{1}\right) = \dfrac{2}{15}+\dfrac{2}{5}$
$= \dfrac{2}{15}+\dfrac{6}{15} = \dfrac{8}{15}$

64. $\left(\dfrac{1}{8}\right)(17\tfrac{3}{4}) = \left(\dfrac{1}{8}\right)\left(\dfrac{71}{4}\right) = \dfrac{71}{32} = 2\tfrac{7}{32}$ teaspoons

65. $45 = \sqrt{9\cdot 5} = \sqrt{9}\cdot\sqrt{5} = 3\sqrt{5}$

66. $\sqrt{200} = \sqrt{100\cdot 2} = \sqrt{100}\cdot\sqrt{2} = 10\sqrt{2}$

67. $\sqrt{5}+7\sqrt{5} = 8\sqrt{5}$

68. $\sqrt{2}-4\sqrt{2} = (1-4)\sqrt{2} = -3\sqrt{2}$

69. $\sqrt{8}+6\sqrt{2} = 2\sqrt{2}+6\sqrt{2} = 8\sqrt{2}$

70. $\sqrt{3}-7\sqrt{27} = \sqrt{3}-21\sqrt{3} = -20\sqrt{3}$

71. $\sqrt{28}+\sqrt{63} = 2\sqrt{7}+3\sqrt{7} = 5\sqrt{7}$

72. $\sqrt{3}\cdot\sqrt{6} = \sqrt{18} = \sqrt{9\cdot 2} = \sqrt{9}\cdot\sqrt{2} = 3\sqrt{2}$

73. $\sqrt{8}\cdot\sqrt{6} = \sqrt{48} = \sqrt{16\cdot 3} = \sqrt{16}\cdot\sqrt{3} = 4\sqrt{3}$

74. $\dfrac{\sqrt{300}}{\sqrt{3}} = \sqrt{\dfrac{300}{3}} = \sqrt{100} = 10$

75. $\dfrac{\sqrt{56}}{\sqrt{2}} = \sqrt{\dfrac{56}{2}} = \sqrt{28} = 2\sqrt{7}$

76. $\dfrac{4}{\sqrt{3}}\cdot\dfrac{\sqrt{3}}{\sqrt{3}} = \dfrac{4\sqrt{3}}{3}$

77. $\dfrac{\sqrt{7}}{\sqrt{5}}\cdot\dfrac{\sqrt{5}}{\sqrt{5}} = \dfrac{\sqrt{35}}{5}$

78. $3(2+\sqrt{7}) = 6+3\sqrt{7}$

79. $\sqrt{3}(4+\sqrt{6})=4\sqrt{3}+\sqrt{18}=4\sqrt{3}+3\sqrt{2}$

80. $\sqrt{3}(\sqrt{6}+\sqrt{15})=\sqrt{18}+\sqrt{45}=3\sqrt{2}+3\sqrt{5}$

81. $x+2=2+x$ Commutative property of addition

82. $5\cdot m=m\cdot 5$ Commutative property of multiplication

83. Associative property of addition

84. Distributive property

85. Associative property of addition

86. Commutative property of addition

87. Associative property of multiplication

88. Commutative property of multiplication

89. Distributive property

90. Commutative property of multiplication

91. Natural numbers – not closed for subtraction
    $2-3=-1$ and -1 is not a natural number.

92. Whole numbers – closed for multiplication

93. Not closed; $1\div 2$ is not an integer

94. Closed

95. Not closed; $\sqrt{2}\cdot\sqrt{2}=2$ is not irrational

96. Not closed; $1\div 0$ is undefined

97. $5^2=5\bullet 5=25$

98. $5^{-2}=\dfrac{1}{5^2}=\dfrac{1}{5\bullet 5}=\dfrac{1}{25}$

99. $\dfrac{9^5}{9^3}=9^{5-3}=9^2=81$

100. $5^2\bullet 5^1=5^3=125$

101. $7^0=1$

102. $4^{-3}=\dfrac{1}{4^3}=\dfrac{1}{64}$

103. $(2^3)^2=2^{3\bullet 2}=2^6=64$

104. $(3^2)^2=3^{2\bullet 2}=3^4=81$

105. $8,200,000,000=8.2\times 10^9$

106. $0.0000158=1.58\times 10^{-5}$

107. $0.02309=2.309\times 10^{-2}$

108. $4,950,000=4.95\times 10^6$

109. $2.8\times 10^5=280,000$

110. $1.39\times 10^{-4}=0.000139$

111. $1.75\times 10^{-4}=0.000175$

112. $1\times 10^7=10,000,000$

113. a) $(3\times 10^4)(2\times 10^{-9})=$
     $6\times 10^{4-9}=6\times 10^{-5}$

114. a) $(5\times 10^6)(7.5\times 10^5)=$
     $(5)(7.5)\times (10^{6+5})=$
     $37.5\times 10^{11}=3.75\times 10^{12}$

115. $\dfrac{8.4\times 10^3}{4\times 10^2}=\dfrac{8.4}{4}\times\dfrac{10^3}{10^2}=2.1\times 10^1$

116. $\dfrac{1.5\times 10^{-3}}{5\times 10^{-4}}=\dfrac{1.5}{5}\times\dfrac{10^{-3}}{10^{-4}}=0.3\times 10^1=3.0\times 10^0$

117. a) $(550,000)(2,000,000)=(5.5\times 10^5)(2\times 10^6)$
     $=(5.5)(2)\times 10^{5+6}=11\times 10^{11}$
     $=1,100,000,000,000$

118. a) $(35,000)(0.00002)=(3.5\times 10^4)(2.0\times 10-^5)$
     $=(3.5)(2)\times 10^4\bullet 10-^5=7.0\times 10-^1=0.7$

119. $\dfrac{8,400,000}{70,000}=\dfrac{8.4\times 10^6}{7\times 10^4}=1.2\times 10^2=120$

120. $\dfrac{0.000002}{0.0000004}=\dfrac{2\times 10^{-6}}{4\times 10^{-7}}=0.5\times 10^1=5.0$

121. $\dfrac{1.49\times 10^{11}}{3.84\times 10^8}=.3880208333\times 10^3\approx 388.02$
     388 times

122. $\dfrac{20,000,000}{3,600}=\dfrac{2.0\times 10^7}{3.6\times 10^3}$
     $\approx 0.555556\times 10^4=\$5,555.56$

123. Arithmetic   17, 21

124. Geometric   8, 16

125. Arithmetic $-15, -18$

126. Geometric   1/32, 1/64

127. Arithmetic   16, 19

128. Geometric $\dfrac{1}{2}, -\dfrac{1}{2}$

129. $a_1 = 2,\ d = 5 \qquad a_6 = 2 + (6-1)(5) = 27$

130. $a_1 = -6, d = 2 \qquad a_9 = -6 + (9-1)2 = 10$

131. $a_{10} = -20 + 9(5) = -20 + 45 = 25$

132. 3, 6, 12, 24, 48 $a_4 = 48$

133. $a_5 = 4(1/2)^{5-1} = 4(1/2)^4 = 4(1/16) = 1/4$

134. $a_4 = -6(2)^{4-1} = -6(2)^3 = -6(8) = -48$

135. $s_{50} = \dfrac{50(3+150)}{2} = (25)(153) = 3825$

136. $s_8 = \dfrac{8(-4+(-2\frac{1}{4}))}{2} = \dfrac{(8)(-6\frac{1}{4})}{2} = -25$

137. $s_8 = \dfrac{8(100+58)}{2} = \dfrac{(8)(158)}{2} = 632$

138. $s_{20} = \dfrac{20(0.5+5.25)}{2} = \dfrac{(20)(5.75)}{2} = 57.5$

139. $s_5 = \dfrac{2(1-4^5)}{1-4} = \dfrac{(2)(-1023)}{-3} = 682$

140. $s_4 = \dfrac{3(1-2^4)}{1-2} = \dfrac{(3)(-15)}{-1} = 45$

141. $s_5 = \dfrac{3(1-(-2)^5)}{1-(-2)} = \dfrac{(3)(1+32)}{3} = \dfrac{(3)(33)}{3} = 33$

142. $s_6 = \dfrac{1(1-(-2)^6)}{1-(-2)} = \dfrac{(1)(1-64)}{3} = \dfrac{(1)(-63)}{3} = -21$

143. Arithmetic: $a_n = 3n$

144. Arithmetic: $a_n = 1 + (n-1)3 = 1 + 3n - 3$
$= 3n - 2$

145. Arithmetic: $a_n = -(3/2)n + (11/2)$

146. Geometric: $a_n = 3(2)^{n-1}$

147. Geometric: $a_n = 2(-1)^{n-1}$

148. Geometric: $a_n = 5(1/3)^{n-1}$

149. No; 13, 21

150. Yes; $-8, -13$

151. No

152. No

Chapter Test

1. 48,395 is divisible by: 5

2.

$$
\begin{array}{r|r}
2 & 414 \\
\hline
3 & 207 \\
\hline
3 & 69 \\
\hline
& 23
\end{array}
$$

$$414 = 2 \cdot 3^2 \cdot 23$$

3. $[(-3)+7]-(-4)=[4]+4=8$

4. $-7-13=-20$

5. $[(-70)(-5)] \div (8-10) = 350 \div [8+(-10)]$
   $=350 \div (-2) = -175$

6. $4\frac{5}{8} = \frac{(8)(4)+5}{8} = \frac{32+5}{8} = \frac{37}{8}$

7. $\frac{176}{9} = \frac{(19)(9)+5}{9} = 19\frac{5}{9}$

8. $\frac{5}{8} = 0.625$

9. $6.45 = \frac{645}{100} = \frac{129}{20}$

10. $\left(\frac{5}{16} \div 3\right) + \left(\frac{4}{5} \cdot \frac{1}{2}\right) = \left(\frac{5}{16} \cdot \frac{1}{3}\right) + \frac{4}{10}$

$$= \frac{5}{48} + \frac{4}{10} = \frac{50}{480} + \frac{192}{480} = \frac{242}{480} = \frac{121}{240}$$

11. $\frac{17}{24} - \frac{7}{12} = \frac{17}{24} - \left(\frac{7}{12}\right)\left(\frac{2}{2}\right) = \frac{17}{24} - \frac{14}{24} = \frac{3}{24} = \frac{1}{8}$

12. $\sqrt{75}+\sqrt{48}=\sqrt{25}\sqrt{3}+\sqrt{16}\sqrt{3}= 5\sqrt{3}+4\sqrt{3}=9\sqrt{3}$

13. $\frac{\sqrt{2}}{\sqrt{7}} = \frac{\sqrt{2}}{\sqrt{7}} \cdot \frac{\sqrt{7}}{\sqrt{7}} = \frac{\sqrt{14}}{\sqrt{49}} = \frac{\sqrt{14}}{7}$

14. The integers are closed under multiplication since the product of two integers is always an integer.

15. Associative property of addition

16. Distributive property

17. $\frac{4^5}{4^2} = 4^{5-2} = 4^3 = 64$

18. $4^3 \cdot 4^2 = 4^5 = 4 \cdot 4 \cdot 4 \cdot 4 \cdot 4 = 1024$

19. $3^{-4} = \frac{1}{3^4} = \frac{1}{81}$

20. $\frac{7.2 \times 10^6}{9.0 \times 10^{-6}} = 0.8 \times 10^{12} = 8.0 \times 10^{11}$

21. $a_n = -4n + 2$

22. $\frac{11[-2+(-32)]}{2} = \frac{11(-34)}{2} = -187$

23. $a_6 = 2 \cdot 3^{6-1} = 2 \cdot 243 = 486$

24. $s_8 = \frac{(-2)\left(1-2^8\right)}{1-2} = \frac{(-2)(-255)}{-1} = -510$

25. $a_n = 3 \cdot (2)^{n-1}$

26. 1, 1, 2, 3, 5, 8, 13, 21, 34, 55

CHAPTER SIX

ALGEBRA, GRAPHS, AND FUNCTIONS

Exercise Set 6.1

1. **Variables** are letters of the alphabet used to represent numbers.

3. The **solution** to an equation is the number or numbers that replace the variable to make the equation a true statement.

5. a) Base: 4, exponent: 5   b) Multiply 4 by itself 5 times.

7. a) $8 + 16 \div 4 = 8 + 4 = 12$ b) $9 + 6 \times 3 = 9 + 18 = 27$

9. $x = -5, \ x^2 = (-5)^2 = 25$

11. $x = -2, -x^2 = -(-2)^2 = -4$

13. $x = -7, -2x^3 = -2(-7)^3 = -2(-343) = 686$

15. $x = 4, x - 7 = 4 - 7 = -3$

17. $x = -2, \ -4x + 4 = -4(-2) + 4 = 8 + 4 = 12$

19. $x = -2, \ -x^2 + 3x - 10 = -(-2)^2 + 3(-2) - 10$
$$= -4 - 6 - 10 = -20$$

21. $x = \dfrac{2}{3}, \dfrac{1}{2}x^2 - 5x + 2 = \dfrac{1}{2}\left(\dfrac{2}{3}\right)^2 - 5\left(\dfrac{2}{3}\right) + 2$
$$= \dfrac{1}{2}\left(\dfrac{4}{9}\right) - \dfrac{10}{3} + 2$$
$$= \dfrac{4}{18} - \dfrac{60}{18} + \dfrac{36}{18} = -\dfrac{20}{18} = -\dfrac{10}{9}$$

23. $x = \dfrac{1}{2}, 8x^3 - 4x^2 + 7 = 8\left(\dfrac{1}{2}\right)^3 - 4\left(\dfrac{1}{2}\right)^2 + 7$
$$= 8\left(\dfrac{1}{8}\right) - 4\left(\dfrac{1}{4}\right) + 7$$
$$= 1 - 1 + 7 = 7$$

25. $x = -2, \ y = 1, \ 2x^2 + xy + 3y^2$
$$= 2(-2)^2 + (-2)(1) + 3(1)^2 = 8 - 2 + 3 = 9$$

27. $x = 2, \ y = -1, \ 4x^2 - 10xy + 3y^2$
$$= 4(2)^2 - 10(2)(-1) + 3(-1)^2 = 16 + 20 + 3 = 39$$

29. $8x + 3 = 23, \ x = 3$

    $8(3) + 3 = 24 + 3 = 27$

    $27 \neq 23, x = 3$ is not a solution.

31. $x - 3y = 0, \ x = 6, \ y = 3$

    $6 - 3(3) = 6 - 9 = -3$

    $-3 \neq 0, x = 6, y = 3$ is not a solution.

33. $x^2 - 3x + 6 = 5, x = 2$

    $(2)^2 - 3(2) + 6 = 4 - 6 + 6 = 4$

    $4 \neq 5, x = 2$ is not a solution.

35. $2x^2 + x = 28, x = -4$

    $2(-4)^2 + (-4) = 2(16) - 4 = 32 - 4 = 28$

    $28 = 28, x = -4$ is a solution.

37. $y = -x^2 + 3x - 1, \ x = 3, \ y = -1$

    $-(3)^2 + 3(3) - 1 = -9 + 9 - 1 = -1$

    $-1 = -1, x = 3, y = -1$ is a solution.

39. $d = \$899, 0.07d = 0.07(\$899) = \$62.93$

41. 2010 is 10 years since 2000.
$6.2(10) + 34.8 = 96.8$; 96.8 million

43. $2(0.60)^2 + 80(0.60) + 40 = 0.72 + 48 + 40 = 88.72$;
88.72 min

45. $R = 2, T = 70, 0.2R^2 + 0.003RT + 0.0001T^2 = 0.2(2)^2 + 0.003(2)(70) + 0.0001(70)^2 = 0.8 + 0.42 + 0.49 = 1.71$ in.

47.

x	y	$(x+y)^2$	$x^2 + y^2$
2	3	$5^2 = 25$	$4 + 9 = 13$
−2	−3	$(-5)^2 = 25$	$4 + 9 = 13$
−2	3	$1^2 = 1$	$4 + 9 = 13$
2	−3	$(-1)^2 = 1$	$4 + 9 = 13$

The two expressions are not equal.

Exercise Set 6.2

1. The parts that are added or subtracted in an algebraic expression are called **terms**.
   In $3x - 2y$, the $3x$ and $-2y$ are terms.

3. The numerical part of a term is called its **numerical coefficient.**
   For the term $3x$, 3 is the numerical coefficient.

5. To **simplify** an expression means to combine like terms by using the commutative, associative, and distributive properties. Example: $12 + x + 7 - 3x = x - 3x + 12 + 7 = -2x + 19$

7. If $a = b$, then $a \cdot c = b \cdot c$ for all real numbers a, b, and c, where $c \neq 0$. Example: If $\frac{x}{3} = 2$, then $3\left(\frac{x}{3}\right) = 3(2)$.

9. If $a = b$, then $\frac{a}{c} = \frac{b}{c}$ for all real numbers a, b, and c, where $c \neq 0$. Example: If $4x = 8$ then $\frac{4x}{4} = \frac{8}{4}$.

11. A **ratio** is a quotient of two quantities. Example: $\frac{7}{9}$

13. Yes. They have the same variable and the same exponent on the variable.

15. $4x + 6x = 10x$

17. $5x - 3x + 12 = 2x + 12$

19. $7x + 3y - 4x + 8y = 3x + 11y$

21. $-3x + 2 - 5x = -8x + 2$

23. $2 - 3x - 2x + 1 = -5x + 3$

25. $6.2x - 8.3 + 7.1x = 13.3x - 8.3$

27. $\frac{1}{5}x - \frac{1}{3}x - 4 = \frac{3}{15}x - \frac{5}{15}x - 4 = -\frac{2}{15}x - 4$

29. $6x - 4y - 5y + 4x + 3 = 6x + 4x - 4y - 5y + 3$
   $= 10x - 9y + 3$

31.

$2(s+3) + 6(s-4) + 1 = 2s + 6 + 6s - 24 + 1 = 8s - 17$

33. $0.2(x+4) + 1.2(x-3) = 0.2x + 0.8 + 1.2x - 3.6$
   $= 1.4x - 2.8$

35. $\frac{1}{4}x + \frac{4}{5} - \frac{2}{3}x = \frac{3}{12}x - \frac{8}{12}x + \frac{4}{5} = -\frac{5}{12}x + \frac{4}{5}$

37.

$0.5(2.6x - 4) + 2.3(1.4x - 5) = 1.3x - 2 + 3.22 - 11.5$

$= 4.52x - 13.5$

39.
$$y - 7 = 10$$
$$y - 7 + 7 = 10 + 7 \qquad \text{Add 7 to both sides of the equation.}$$
$$y = 17$$

41.
$$15 = 9 - 6$$
$$6 = -6x \qquad \text{Subtract 9 from both sides of the equation.}$$
$$\frac{-3}{-6} = \frac{-6x}{-6} \qquad \text{Divide both sides of the equation by } -6.$$
$$-1 = x$$

43.
$$\frac{3}{x} = \frac{7}{8}$$
$$3(8) = 7x \qquad \text{Cross multiplication}$$
$$24 = 7x$$
$$\frac{24}{7} = \frac{7x}{7} \qquad \text{Divide both sides of the equation by 7.}$$
$$\frac{24}{7} = x$$

45.
$$\frac{1}{2}x + \frac{1}{3} = \frac{2}{3}$$
$$6\left(\frac{1}{2}x + \frac{1}{3}\right) = 6\left(\frac{2}{3}\right) \qquad \text{Multiply both sides of the equation by the LCD.}$$
$$3x + 2 = 4 \qquad \text{Distributive Property}$$
$$3x + 2 - 2 = 4 - 2 \qquad \text{Subtract 2 from both sides of the equation.}$$
$$3x = 2$$
$$\frac{3x}{3} = \frac{2}{3} \qquad \text{Divide both sides of the equation by 3.}$$
$$x = \frac{2}{3}$$

47.
$$0.9x - 1.2 = 2.4$$
$$0.9x + 1.2 - 1.2 = 2.4 + 1.2 \qquad \text{Add 1.2 to both sides of the equation.}$$
$$0.9x = 3.6$$
$$\frac{0.9x}{0.9} = \frac{3.6}{0.9} \qquad \text{Divide both sides of the equation by 0.9.}$$
$$x = 4$$

49.
$$6t - 8 = 4t - 2$$
$$6t - 4t - 8 = 4t - 4t - 2 \qquad \text{Subtract } 4t \text{ from both sides of the equation.}$$
$$2t - 8 = -2$$
$$2t - 8 + 8 = -2 + 8 \qquad \text{Add 8 to both sides of the equation.}$$
$$2t = 6$$
$$\frac{2t}{2} = \frac{6}{2} \qquad \text{Divide both sides of the equation by 2.}$$
$$t = 3$$

51.

$$\frac{x-3}{2} = \frac{x+4}{3}$$

$3(x-3) = 2(x+4)$	Cross multiplication
$3x-9 = 2x+8$	Distributive Property
$3x-2x-9 = 2x-2x+8$	Subtract $2x$ from both sides of the equation.
$x-9 = 8$	
$x-9+9 = 8+9$	Add 9 to both sides of the equation.
$x = 17$	

53.

$6t-7 = 8t+9$	
$6t-6t-7 = 8t-6t+9$	Subtract $6t$ from both sides of the equation.
$-7 = 2t+9$	
$-7-9 = 2t+9-9$	Subtract 9 from both sides of the equation.
$-16 = 2t$	
$\dfrac{-16}{2} = \dfrac{2t}{2}$	Divide both sides of the equation by 2.
$-8 = t$	

55.

$2(x+3)-4 = 2(x-4)$	
$2x+6-4 = 2x-8$	Distributive Property
$2x+2 = 2x-8$	
$2x-2x+2 = 2x-2x-8$	Subtract $2x$ from both sides of the equation.
$2 = -8$	False

No solution

57.

$4(x-4)+12 = 4(x-1)$	
$4x-16+12 = 4x-4$	Distributive Property
$4x-4 = 4x-4$	

This equation is an identity. Therefore, the solution is all real numbers.

59.

$\dfrac{1}{3}(x+3) = \dfrac{2}{5}(x+2)$	
$15\left(\dfrac{1}{3}\right)(x+3) = 15\left(\dfrac{2}{5}\right)(x+2)$	Multiply both sides of the equation by the LCD.
$5(x+3) = 6(x+2)$	
$5x+15 = 6x+12$	Distributive Property
$5x-5x+15 = 6x-5x+12$	Subtract $5x$ from both sides of the equation.
$15 = x+12$	
$15-12 = x+12-12$	Subtract 12 from both sides of the equation.
$3 = x$	

61.
$$3x+2-6x=-x-15+8-5x$$
$$-3x+2=-6x-7$$
$$-3x+6x+2=-6x+6x-7 \qquad \text{Add } 6x \text{ to both sides of the equation.}$$
$$3x+2=-7$$
$$3x+2-2=-7-2 \qquad \text{Subtract 2 from both sides of the equation.}$$
$$3x=-9$$
$$\frac{3x}{3}=\frac{-9}{3} \qquad \text{Divide both sides of the equation by 3.}$$
$$x=-3$$

63.
$$4(t-3)+8=4(2t-6)$$
$$4t-12+8=8t-24 \qquad \text{Distributive Property}$$
$$4t-4=8t-24$$
$$4t-4t-4=8t-4t-24 \qquad \text{Subtract } 4t \text{ from both sides of the equation.}$$
$$-4=4t-24$$
$$-4+24=4t-24+24 \qquad \text{Add 24 to both sides of the equation.}$$
$$20=4t$$
$$\frac{20}{4}=\frac{4t}{4} \qquad \text{Divide both sides of the equation by 4.}$$
$$5=t$$

65.
$$\frac{7.75}{1000}=\frac{x}{27,000}$$
$$7.75(27,000)=1000x$$
$$\frac{7.75(27,000)}{1000}=\frac{1000x}{1000}$$
$$7.75(27)=x$$
$$x=\$209.25$$

67.
$$\frac{x}{1440}=\frac{1}{360}$$
$$360x=1440$$
$$\frac{360x}{360}=\frac{1440}{360}$$
$$x=4 \text{ gallons}$$

69.
$$\frac{1}{1,102,000}=\frac{14.4}{x}$$
$$x=1,102,000(14.4)$$
$$x=15,868,800 \text{ households}$$

71. a)
$$\frac{50}{80}=\frac{1}{x}$$
$$50x=80$$
$$\frac{50x}{50}=\frac{80}{50}$$
$$x=1.6 \text{ kph}$$

b)
$$\frac{50}{80}=\frac{x}{90}$$
$$80x=50(90)$$
$$80x=4500$$
$$\frac{80x}{80}=\frac{4500}{80}$$
$$x=56.25 \text{ mph}$$

73.
$$\frac{40}{1} = \frac{15}{x}$$
$$40x = 15$$
$$\frac{40x}{40} = \frac{15}{40}$$
$$x = 0.375 \text{ cc}$$

75. a) Answers will vary.

b)
$$2(x+3) = 4x+3-5x$$

$2x+6 = -x+3$	Distributive Property
$2x+x+6 = -x+x+3$	Add x to both sides of the equation.
$3x+6 = 3$	
$3x+6-6 = 3-6$	Subtract 6 from both sides of the equation.
$3x = -3$	
$\frac{3x}{3} = \frac{-3}{3}$	Divide both sides of the equation by 3.
$x = -1$	

77. a) An **inconsistent equation** is an equation that has no solution.

b) When solving an equation, if you obtain a false statement, then the equation is inconsistent.

79. a) 2:5; There are $2x$ males and a total of $2x + 3x = 5x$ students.

b) $m : m+n$

Exercise Set 6.3

1. A **formula** is an equation that typically has a real-life application.

3. **Subscripts** are numbers (or letters) placed below and to the right of variables. They are used to help clarify a formula.

5. An **exponential equation** is of the form $y = a^x, a > 0, a \neq 1$.

7. $A = lw = 4(14) = 56$

9. $P = 2l + 2w$
$$P = 2(12) + 2(16) = 24 + 32 = 56$$

11.
$$K = \frac{1}{2}mv^2$$
$$4500 = \frac{1}{2}m(30)^2$$
$$4500 = 450m$$
$$\frac{4500}{450} = \frac{450m}{450}$$
$$10 = m$$

13.
$$S = \pi r(r+h)$$
$$S = 3.14(8)(8+2)$$
$$S = 3.14(8)(10)$$
$$S = 3.14(80)$$
$$S = 251.2$$

15.
$$z = \frac{x - \mu}{\sigma}$$
$$\frac{2.5}{1} = \frac{42.1 - \mu}{2}$$
$$2.5(2) = 42.1 - \mu$$
$$5 = 42.1 - \mu$$
$$5 - 42.1 = 42.1 - 42.1 - \mu$$
$$-37.1 = -\mu$$
$$\frac{-37.1}{-1} = \frac{-\mu}{-1}$$
$$37.1 = \mu$$

17.
$$T = \frac{PV}{k}$$
$$\frac{80}{1} = \frac{P(20)}{0.5}$$
$$80(0.5) = 20P$$
$$40 = 20P$$
$$\frac{40}{20} = \frac{20P}{20}$$
$$2 = P$$

19.
$$A = P(1 + rt)$$
$$3600 = P(1 + 0.04(5))$$
$$3600 = P(1 + 0.2)$$
$$3600 = 1.2P$$
$$\frac{3600}{1.2} = \frac{1.2P}{1.2}$$
$$3000 = P$$

21.
$$V = \frac{1}{2}at^2$$
$$576 = \frac{1}{2}a(12)^2$$
$$\frac{576}{1} = \frac{144a}{2}$$
$$576(2) = 144a$$
$$1152 = 144a$$
$$\frac{1152}{144} = \frac{144a}{144}$$
$$8 = a$$

23.
$$C = \frac{5}{9}(F - 32)$$
$$C = \frac{5}{9}(77 - 32)$$
$$C = \frac{5}{9}(45) = 25$$

25.
$$m = \frac{y_2 - y_1}{x_2 - x_1}$$
$$m = \frac{8 - (-4)}{-3 - (-5)}$$
$$m = \frac{8 + 4}{-3 + 5} = \frac{12}{2} = 6$$

27.
$$S = R - rR$$
$$186 = 1R - 0.07R$$
$$186 = 0.93R$$
$$\frac{186}{0.93} = \frac{0.93R}{0.93}$$
$$200 = R$$

29.
$$E = a_1 p_1 + a_2 p_2 + a_3 p_3$$
$$E = 5(0.2) + 7(0.6) + 10(0.2)$$
$$E = 1 + 4.2 + 2 = 7.2$$

31.
$$s = -16t^2 + v_0 t + s_0$$
$$s = -16(4)^2 + 30(4) + 150$$
$$s = -16(16) + 120 + 150$$
$$s = -256 + 120 + 150 = 14$$

33.
$$P = \frac{f}{1 + i}$$
$$3000 = \frac{f}{1 + 0.08}$$
$$\frac{3000}{1} = \frac{f}{1.08}$$
$$3000(1.08) = f$$
$$3240 = f$$

35.
$$R_T = \frac{R_1 R_2}{R_1 + R_2}$$
$$R_T = \frac{(100)(200)}{100 + 200}$$
$$R_T = \frac{20,000}{300}$$
$$R_T = 66.67$$

37.
$$a_n = a_1 + (n-1)d$$
$$a_n = 15 + (4-1)(8)$$
$$a_n = 15 + 24$$
$$a_n = 39$$

39.
$$P = P_0 e^{kt}$$
$$P = 5000 e^{(0.06)(7)}$$
$$P = 5000 e^{0.42}$$
$$P \approx 7609.81$$

41.
$$4x - 9y = 14$$
$$4x - 4x - 9y = -4x + 14 \qquad \text{Subtract } 4x \text{ from both sides of the equation.}$$
$$-9y = -4x + 4$$
$$\frac{-9y}{-9} = \frac{-4x + 14}{-9} \qquad \text{Divide both sides of the equation by } -9.$$
$$y = \frac{-4x + 14}{-9} = \frac{-(-4x + 14)}{9}$$
$$= \frac{4x - 14}{9} = \frac{4}{9}x - \frac{14}{9}$$

43.
$$8x + 7y = 21$$
$$-8x + 8x + 7y = -8x + 21 \qquad \text{Subtract } 8x \text{ from both sides of the equation.}$$
$$7y = -8x + 21$$
$$\frac{7y}{7} = \frac{-8x + 21}{7} \qquad \text{Divide both sides of the equation by 7.}$$
$$y = \frac{-8x + 21}{7} = \frac{-8x}{7} + \frac{21}{7} = -\frac{8}{7}x + 3$$

45.
$$2x - 3y + 6 = 0$$
$$2x - 3y + 6 - 6 = 0 - 6 \qquad \text{Subtract 6 from both sides of the equation.}$$
$$2x - 3y = -6$$
$$-2x + 2x - 3y = -2x - 6 \qquad \text{Subtract } 2x \text{ from both sides of the equation.}$$
$$-3y = -2x - 6$$
$$\frac{-3y}{-3} = \frac{-2x - 6}{-3} \qquad \text{Divide both sides of the equation by -3.}$$
$$y = \frac{-2x - 6}{-3} = \frac{-(-2x - 6)}{3} = \frac{2x + 6}{3} = \frac{2x}{3} + \frac{6}{3} = \frac{2}{3}x + 2$$

47.
$$-2x+3y+z=15$$
$$-2x+2x+3y+z=2x+15 \qquad \text{Add } 2x \text{ to both sides of the equation.}$$
$$3y+z=2x+15$$
$$3y+z-z=2x-z+15 \qquad \text{Subtract } z \text{ from both sides of the equation.}$$
$$3y=2x-z+15$$
$$\frac{3y}{3}=\frac{2x-z+15}{3} \qquad \text{Divide both sides of the equation by 3.}$$
$$y=\frac{2x-z+15}{3}=\frac{2}{3}x-\frac{1}{3}z+5$$

49.
$$9x+4z=7+8y$$
$$9x+4z-7=7-7+8y \qquad \text{Subtract 7 from both sides of the equation.}$$
$$9x+4z-7=8y$$
$$\frac{9x+4z-7}{8}=\frac{8y}{8} \qquad \text{Divide both sides of the equation by 8.}$$
$$y=\frac{9x+4z-7}{8}=\frac{9}{8}x+\frac{1}{2}z-\frac{7}{8}$$

51.
$$d=rt$$
$$\frac{d}{t}=\frac{rt}{t} \qquad \text{Divide both sides of the equation by } t.$$
$$r=\frac{d}{t}$$

53.
$$p=a+b+c$$
$$p-b=a+b-b+c \qquad \text{Subtract } b \text{ from both sides of the equation.}$$
$$p-b=a+c$$
$$p-b-c=a+c-c \qquad \text{Subtract } c \text{ from both sides of the equation.}$$
$$a=p-b-c$$

55.
$$V=\frac{1}{3}Bh$$
$$3V=3\left(\frac{1}{3}Bh\right) \qquad \text{Multiply both sides of the equation by 3.}$$
$$3V=Bh$$
$$\frac{3V}{h}=\frac{Bh}{h} \qquad \text{Divide both sides of the equation by } h.$$
$$B=\frac{3V}{h}$$

57.
$$C=2\pi r$$
$$\frac{C}{2}=\frac{2\pi r}{2} \qquad \text{Divide both sides of the equation by 2.}$$
$$\frac{C}{2}=\pi r$$
$$\frac{C}{2\pi}=\frac{\pi r}{\pi} \qquad \text{Divide both sides of the equation by } \pi.$$
$$r=\frac{C}{2\pi}$$

59.
$$y = mx + b$$
$$y - mx = mx - mx + b$$ Subtract mx from both sides of the equation.
$$b = y - mx$$

61.
$$P = 2l + 2w$$
$$P - 2l = 2l - 2l + 2w$$ Subtract $2l$ from both sides of the equation.
$$P - 2l = 2w$$
$$\frac{P - 2l}{2} = \frac{2w}{2}$$ Divide both sides of the equation by 2.
$$w = \frac{P - 2l}{2}$$

63.
$$A = \frac{a + b + c}{3}$$
$$3A = 3\left(\frac{a + b + c}{3}\right)$$ Multiply both sides of the equation by 3.
$$3A = a + b + c$$
$$3A - a = a - a + b + c$$ Subtract a from both sides of the equation.
$$3A - a = b + c$$
$$3A - a - b = b - b + c$$ Subtract b from both sides of the equation.
$$c = 3A - a - b$$

65.
$$P = \frac{KT}{V}$$
$$PV = \left(\frac{KT}{V}\right)V$$ Multiply both sides of the equation by V.
$$PV = KT$$
$$\frac{PV}{K} = \frac{KT}{K}$$ Divide both sides of the equation by K.
$$T = \frac{PV}{K}$$

67.
$$F = \frac{9}{5}C + 32$$
$$F - 32 = \frac{9}{5}C + 32 - 32$$ Subtract 32 from both sides of the equation.
$$F - 32 = \frac{9}{5}C$$
$$\frac{5}{9}(F - 32) = \frac{5}{9}\left(\frac{9}{5}C\right)$$ Multiply both sides of the equation by $\frac{5}{9}$.
$$C = \frac{5}{9}(F - 32)$$

69.
$$S = 2\pi rh + 2\pi r^2$$
$$S - 2\pi r^2 = 2\pi rh + 2\pi r^2 - 2\pi r^2 \qquad \text{Subtract } 2\pi r^2 \text{ from both sides of the equation.}$$
$$S - 2\pi r^2 = 2\pi rh$$
$$\frac{S - 2\pi r^2}{2\pi r} = \frac{2\pi rh}{2\pi r} \qquad \text{Divide both sides of the equation by } 2\pi r.$$
$$\frac{S - 2\pi r^2}{2\pi r} = h$$

71. a) $i = prt$
$$i = 4500(0.025)(1) = \$112.50$$
   b) $\$4500 + \$112.50 = \$4612.50$

73. Radius $= 1$ in.
$$V = \frac{1}{3}\pi r^2 h$$
$$V = \frac{1}{3}\pi (1)^2 (4)$$
$$V = \frac{1}{3}\pi (4)$$
$$V \approx 4.19 \text{ in.}^3$$

75. $y = 2000(3)^x$
$$y = 2000(3)^5$$
$$y = 2000(243)$$
$$y = 486,000 \text{ bacteria}$$

77. $V = 24e^{0.08t}$
$$V = 24e^{0.08(382)}$$
$$V = 24e^{30.56}$$
$$V \approx \$4.49 \times 10^{14}$$

79. $V = lwh - \pi r^2 h$
$$V = 12(8)(12) - \pi (2)^2 (8)$$
$$V = 1152 - 100.5309649$$
$$V = 1051.469035 \text{ in.}^3 \approx 1051.47 \text{ in.}^3$$

Exercise Set 6.4

1. A **mathematical expression** is a collection of variables, numbers, parentheses, and operation symbols. An **equation** is two algebraic expressions joined by an equal sign.

3. $x+4$: 4 more than x

5. $2x-3$: 2 times x, decreased by 3

7. $8+x$

9. $3+2z$

11. $6w+9$

13. $4x+6$

15. $\dfrac{18-s}{4}$

17. $3(x+7)$

19. Let x = the number
   $x+5=5$ more than the number
   $x+5=11$
   $x+5-5=11-5$
   $x=6$

21. Let x = the number
   $x-4=$ the difference between the number and 4
   $x-4=20$
   $x-4+4=20+4$
   $x=24$

23. Let x = the number
   $4x-10=4$ times the number decreased by 10
   $4x-10=42$
   $4x-10+10=42+10$
   $4x=52$
   $\dfrac{4x}{4}=\dfrac{52}{4}$
   $x=13$

25. Let x = the number
   $4x+12=12$ more than 4 times the number
   $4x+12=32$
   $4x+12-12=32-12$
   $4x=20$
   $\dfrac{4x}{4}=\dfrac{20}{4}$
   $x=5$

27. Let x = the number
   $x+6=$ the number increased by 6
   $2x-3=3$ less than twice the number
   $x+6=2x-3$
   $x-x+6=2x-x-3$
   $6=x-3$
   $6+3=x-3+3$
   $9=x$

29. Let x = the number
   $x+10=$ the number increased by 10
   $2(x+3)=2$ times the sum of the number and 3
   $x+10=2(x+3)$
   $x+10=2x+6$
   $x-x+10=2x-x+6$
   $10=x+6$
   $10-6=x+6-6$
   $4=x$

31. Let x = the number of miles driven
   $0.42x=$ reimbursement for mileage
   $150+0.42x=207.54$
   $0.42x=57.54$
   $\dfrac{0.42x}{0.42}=\dfrac{57.54}{0.42}$
   $x=137$
   137 miles

33. Let x = the original cost before tax
   $0.05x=$ the amount of tax
   $x+0.05x=42$
   $1.05x=42$
   $\dfrac{1.05x}{1.05}=\dfrac{42}{1.05}$
   $x=40;\ \$40$ per half hour

35. Let x = the number of copies Ronnie
     must make

$0.08x$ = the amount spent on x copies

$0.08x = 250$

$\dfrac{0.08x}{0.08} = \dfrac{250}{0.08}$

$x = 3125$ copies

37. Let x = the amount donated for Business

$3x$ = the amount donated for
     Liberal Arts

$x + 3x = 1000$

$4x = 1000$

$\dfrac{4x}{4} = \dfrac{1000}{4}$

$x = \$250$ for Business

$3x = 3(250) = \$750$ for Liberal Arts

39. Let w = the width

$w + 3$ = the length

$2w + 2(w + 3) = P$

$2w + 2(w + 3) = 54$

$2w + 2w + 6 = 54$

$4w + 6 = 54$

$4w + 6 - 6 = 54 - 6$

$4w = 48$

$\dfrac{4w}{4} = \dfrac{48}{4}$

width = 12 ft, length = $w + 3 = 12 + 3 = 15$ ft

41. Let x = the number of wild mustangs
     in Utah

$6x + 2515$ = the the number of wild mustangs
     in Nevada

$x + 6x + 2515 = 21,730$

$7x = 19215$

$x = 2745$; Utah: 2745 mustangs

Nevada: $21,730 - 2745$
     $= 18,985$ mustangs

43. Let x = the typical number of tornados
     in December

$14x + 11$ = the typical number of tornados
     in May

$x + 14x + 11 = 341$

$15x + 11 = 330$

$x = 22$; December, 22 tornados

May, $(14)(22) + 11$
     $= 319$ tornados

45. Let w = width

$2w$ = length of entire enclosed region

$3w + 2(2w)$ = total amount of fencing

$3w + 2(2w) = 140$

$3w + 4w = 140$

$7w = 140$

$\dfrac{7w}{7} = \dfrac{140}{7}$

width = 20 ft

length = $2w = 2(20) = 40$ ft

47. Let x = the number of months

$70x$ = cost of laundry for x months

$70x = 760$

$\dfrac{70x}{70} = \dfrac{760}{70}$

$x = 10.85714286$ months ≈ 11 months

49.
Let r = regular fare

$\dfrac{r}{2}$ = half off regular fare

$0.07r$ = tax on regular fare

$$\dfrac{r}{2} + 0.07r = 257$$

$$2\left(\dfrac{r}{2} + 0.07r\right) = 2(257)$$

$$r + 0.14r = 514$$

$$1.14r = 514$$

$$\dfrac{1.14r}{1.14} = \dfrac{514}{1.14}$$

$$r = \$450.877193$$

$$\approx \$450.88$$

51.
Let x = amount of tax reduction to be deducted from Mr. McAdam's income

$3640 - x$ = amount of tax reduction to be deducted from Mrs. McAdam's income

$$24,200 - x = 26,400 - (3640 - x)$$

$$24,200 - x = 26,400 - 3640 + x$$

$$24,200 - x = 22,760 + x$$

$$24,200 - x + x = 22,760 + x + x$$

$$24,200 = 22,760 + 2x$$

$$24,200 - 22,760 = 22,760 - 22,760 + 2x$$

$$1440 = 2x$$

$$\dfrac{1440}{2} = \dfrac{2x}{2}$$

$$x = \$720 \text{ deducted from}$$

Mr. McAdam's income

$$3640 - x = 3640 - 720 = \$2920 \text{ deducted}$$

from Mrs. McAdam's income

53.
Let x = the first integer

$x + 1$ = the second integer

$x + 2$ = the third integer (the largest)

$$x + (x+1) + (x+2) = 3(x+2) - 3$$

$$3x + 3 = 3x + 6 - 3$$

$$3x + 3 = 3x + 3$$

55.
$$F = \dfrac{9}{5}C + 32$$

The thermometers will read the same when $F = C$.
Substitute C for F in the above equation.

$$C = \dfrac{9}{5}C + 32$$

$$5C = 5\left(\dfrac{9}{5}C + 32\right)$$

$$5C = 9C + 160$$

$$5C - 9C = 9C - 9C + 160$$

$$-4C = 160$$

$$\dfrac{-4C}{-4} = \dfrac{160}{-4}$$

$$C = -40°$$

Exercise Set 6.5

1. **Direct variation** — y varies directly with x if $y = kx$.

3. **Joint variation** — One quantity varies directly as the product of two or more other quantities.

5. Direct

7. Inverse

9. Direct

11. Inverse

13. Inverse

15. Inverse

17. Direct

19. Direct

21. Answers will vary.

23. a) $y = kx$

    b) $y = 8(15) = 120$

25. a) $m = \dfrac{k}{n^2}$

    b) $m = \dfrac{16}{(8)^2} = \dfrac{16}{64} = 0.25$

27. a) $A = \dfrac{kB}{C}$

29. a) $F = kDE$

    b) $F = 7(3)(10) = 210$

31. a) $t = \dfrac{kd^2}{f}$

    b) $192 = \dfrac{k(8)^2}{4}$

    $192 = \dfrac{64k}{4}$

    $768 = 64k$

    $\dfrac{768}{64} = \dfrac{64k}{64}$

    $k = 12$

    $t = \dfrac{12d^2}{f}$

    $t = \dfrac{12(10)^2}{6} = \dfrac{12(100)}{6} = \dfrac{1200}{6} = 200$

33. a) $Z = kWY$

    b) $12 = k(9)(4)$

    $12 = 36k$

    $\dfrac{12}{36} = \dfrac{36k}{36}$

    $k = \dfrac{1}{3}$

    $Z = \dfrac{1}{3}WY$

    $Z = \dfrac{1}{3}(50)(6) = \dfrac{300}{3} = 100$

35. a) $H = kL$

    b) $15 = k(50)$

    $\dfrac{15}{50} = \dfrac{50k}{50}$

    $k = 0.3$

    $H = 0.3L$

    $H = 0.3(10) = 3$

37. a) $A = kB^2$

    b) $245 = k(7)^2$

    $245 = 49k$

    $\dfrac{245}{49} = \dfrac{49k}{49}$

    $k = 5$

    $A = 5B^2$

    $A = 5(12)^2 = 5(144) = 720$

39. a) $F = \dfrac{kq_1 q_2}{d^2}$

    b) $80 = \dfrac{k(4)(16)}{(0.4)^2}$

    $80 = \dfrac{64k}{0.16}$

    $k = 0.2$

    $F = \dfrac{0.2 q_1 q_2}{d^2}$

    $F = \dfrac{0.2(12)(20)}{(0.2)^2} = \dfrac{48}{0.04} = 1200$

41. a) $t = kv$

    b) $2100 = k(140,000)$

    $\dfrac{2100}{140,000} = \dfrac{140,000k}{140,000}$

    $k = 0.015$

    $t = 0.015v$

    $t = 0.015(180,000) = \$2700$

43. a) $l = \dfrac{k}{d^2}$

    b) $20 = \dfrac{k}{(6)^2}$

    $k = 20(36) = 720$

    $l = \dfrac{720}{d^2}$

    $l = \dfrac{720}{(3)^2} = \dfrac{720}{9} = 80$ dB

45. a) $R = \dfrac{kA}{P}$

    b) $4800 = \dfrac{k(600)}{3}$

    $600k = 14,400$

    $k = \dfrac{14,400}{600} = 24$

    $R = \dfrac{24A}{P}$

    $R = \dfrac{24(700)}{3.50} = \dfrac{16,800}{3.50} = 4800$ tapes

47. a) $v = \dfrac{k\sqrt{t}}{l}$

    b) $5 = \dfrac{k\sqrt{225}}{0.60}$

    $k = \dfrac{5(0.60)}{\sqrt{225}} = \dfrac{3}{15} = 0.2$

    $v = \dfrac{0.2\sqrt{t}}{l}$

    $v = \dfrac{0.2\sqrt{196}}{0.70} = \dfrac{0.2(14)}{0.70}$

    $= 4$ vibrations per second

49. a) $N = \dfrac{kp_1 p_2}{d}$

    b) $100,000 = \dfrac{k(60,000)(200,000)}{300}$

    $12,000,000,000k = 30,000,000$

    $k = \dfrac{30,000,000}{12,000,000,000} = 0.0025$

    $N = \dfrac{0.0025 p_1 p_2}{d}$

    $N = \dfrac{0.0025(125,000)(175,000)}{450}$

    $N = \dfrac{54,687,500}{450}$

    $= 121,527.7778 \approx 121,528$ calls

51. a) $y = \dfrac{k}{x}$

    $y = \dfrac{0.3}{x}$

    $xy = 0.3$

    $\dfrac{xy}{y} = \dfrac{0.3}{y}$

    $x = \dfrac{0.3}{y}$

    Inversely

    b) k stays 0.3

53. $W = \dfrac{kTA\sqrt{F}}{R}$

    $72 = \dfrac{k(78)(1000)\sqrt{4}}{5.6}$

    $156,000k = 403.2$

    $k = \dfrac{403.2}{156,000} = 0.0025846154$

    $W = \dfrac{0.0025846154 TA\sqrt{F}}{R}$

    $W = \dfrac{0.0025846154(78)(1500)\sqrt{6}}{5.6}$

Exercise Set 6.6

1. $a < b$ means that a is less than b, $a \leq b$ means that a is less than or equal to b, $a > b$ means that a is greater than b, $a \geq b$ means that a is greater than or equal to b.

3. When both sides of an inequality are multiplied or divided by a negative number, the direction of the inequality symbol must be reversed.

5. Yes, the inequality symbol points to the -3 in both cases.

7. a) An inequality of the form $a < x < b$ is called a compound inequality.
    b) $-7 < x < 3$

9. $x \geq 4$

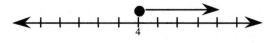

11. $x - 8 \geq 3$
     $x - 8 + 8 \geq 3 + 8$
     $x \geq 11$

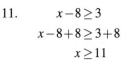

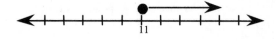

13. $-3x \leq 18$
     $\dfrac{-3x}{-3} \geq \dfrac{18}{-3}$
     $x \geq -6$

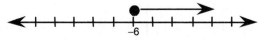

15. $\dfrac{x}{5} < 4$
     $5\left(\dfrac{x}{5}\right) < 5(4)$
     $x < 20$

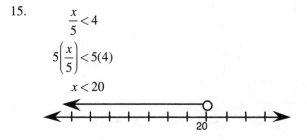

17. $\dfrac{-x}{3} \geq 3$
     $-3\left(\dfrac{-x}{3}\right) \leq -3(3)$
     $x \leq -9$

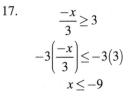

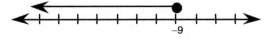

19. $2x + 6 \geq 14$
     $2x + 6 - 6 \geq 14 - 6$
     $2x \geq 8$
     $\dfrac{2x}{2} \geq \dfrac{8}{2}$
     $x \geq 4$

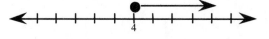

21. $4(2x - 1) < 2(4x - 3)$
     $8x - 4 < 8x - 6$
     $-4 < -6$
     False, no solution

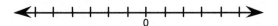

23. $-1 \leq x \leq 3$

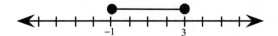

25.
$$3 < x - 7 \le 6$$
$$3 + 7 < x - 7 + 7 \le 6 + 7$$
$$10 < x \le 13$$

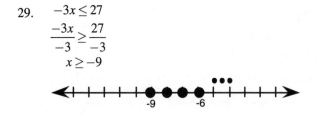

27. $x > 1$

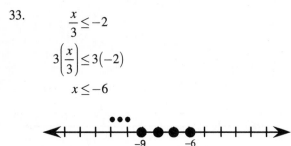

29.
$$-3x \le 27$$
$$\frac{-3x}{-3} \ge \frac{27}{-3}$$
$$x \ge -9$$

31.
$$x - 2 < 4$$
$$x - 2 + 2 < 4 + 2$$
$$x < 6$$

33.
$$\frac{x}{3} \le -2$$
$$3\left(\frac{x}{3}\right) \le 3(-2)$$
$$x \le -6$$

35.
$$-\frac{x}{4} \ge 2$$
$$(-4)\left(-\frac{x}{4}\right) \le (-4)(2)$$
$$x \le -8$$

37.
$$-15 < -4x - 3$$
$$-12 < -4x$$
$$\frac{-12}{-4} > \frac{-4x}{-4}$$
$$3 > x$$

39.
$$3(x + 4) \ge 4x + 13$$
$$3x + 12 \ge 4x + 13$$
$$3x - 4x + 12 \ge 4x - 4x + 13$$
$$-x + 12 \ge 13$$
$$-x + 12 - 12 \ge 13 - 12$$
$$-x \ge 1$$
$$\frac{-x}{-1} \le \frac{1}{-1}$$
$$x \le -1$$

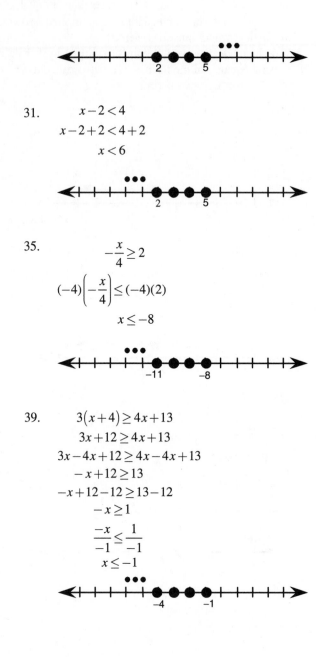

41. $5(x+4)-6 \le 2x+8$

$5x+20-6 \le 2x+8$

$5x+14 \le 2x+8$

$5x-2x+14 \le 2x-2x+8$

$3x+14 \le 8$

$3x+14-14 \le 8-14$

$3x \le -6$

$\dfrac{3x}{3} \le \dfrac{-6}{3}$

$x \le -2$

43. $1 > -x > -5$

$\dfrac{1}{-1} < \dfrac{-x}{-1} < \dfrac{-5}{-1}$

$-1 < x < 5$

45. $0.3 \le \dfrac{x+2}{10} \le 0.5$

$10(0.3) \le 10\left(\dfrac{x+2}{10}\right) \le 10(0.5)$

$3 \le x+2 \le 5$

$3-2 \le x+2-2 \le 5-2$

$1 \le x \le 3$

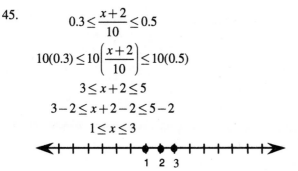

47.   a) 2009 through 2015
       b) 2005 through 2008
       c) 2006 through 2015
       d) 2005 and 2006

49.   Let x = the number of videos

No Fee Plan cost: $2.99x$

Annual Fee Plan: $30+1.49x$

$2.99x < 30+1.49x$

$2.99x-1.49x < 30+1.49x-1.49x$

$1.50x < 30$

$\dfrac{1.50x}{1.50} < \dfrac{30}{1.50}$

$x < 20$

The maximum number of videos that can be rented

for the No Fee Plan to cost less than the

Annual Fee Plan is 19.

51.   Let x = the number hours she works as cashier

$725+6.50x \le 3000$

$6.50x \le 2275$

$\dfrac{6.50x}{6.50} \le \dfrac{2275}{6.50}$

$x \le 350$

The maximum number of hours that

Samantha can work as a cashier is

350 hours.

53. Let $x =$ the length of time Tom can park.

$$1.75 + 0.50(x-1) \le 4.25$$

$$1.75 + 0.50x - 0.50 \le 4.25$$

$$0.50x \le 3$$

$$\frac{0.50x}{0.50} \le \frac{3}{0.50}$$

$$x \le 6$$

He can park for at most 6 hours.

55.
$$36 < 84 - 32t < 68$$

$$36 - 84 < 84 - 84 - 32t < 68 - 84$$

$$-48 < -32t < -16$$

$$\frac{-48}{-32} > \frac{-32t}{-32} > \frac{-16}{-32}$$

$$1.5 > t > 0.5$$

$$0.5 < t < 1.5$$

The velocity will be between $36\dfrac{\text{ft}}{\sec}$ and $68\dfrac{\text{ft}}{\sec}$ when t is between $0.5\sec$ and $1.5\sec$.

57. Let $x =$ Devon's grade on the fifth test

$$80 \le \frac{78 + 64 + 88 + 76 + x}{5} < 90$$

$$80 \le \frac{306 + x}{5} < 90$$

$$5(80) \le 5\left(\frac{306+x}{5}\right) < 5(90)$$

$$400 \le 306 + x < 450$$

$$400 - 306 \le 306 - 306 + x < 450 - 306$$

$$94 \le x < 144$$

Devon must have a score of $94 \le x \le 100$, assuming 100 is the highest grade possible.

59. Let $x =$ the number of gallons

$$250x = 2750 \text{ and } 400x = 2750$$

$$x = \frac{2750}{250} \text{ , } x = \frac{2750}{400}$$

$$x = 11, \ x = 6.875$$

$$6.875 \le x \le 11$$

61. Student's answer:
$$-\frac{1}{3}x \le 4$$

$$-3\left(-\frac{1}{3}x\right) \le -3(4)$$

$$x \le -12$$

Correct answer:
$$-\frac{1}{3}x \le 4$$

$$-3\left(-\frac{1}{3}x\right) \ge -3(4)$$

$$x \ge -12$$

Yes, -12 is in both solution sets.

Exercise Set 6.7

1. A **graph** is an illustration of all the points whose coordinates satisfy an equation.

3. To find the **y-intercept**, set $x = 0$ and solve the equation for y.

5. a) Divide the difference between the y-coordinates by the difference between the x-coordinates.

    b) $m = \dfrac{5-2}{-3-6} = \dfrac{3}{-9} = -\dfrac{1}{3}$

7. a) First

    b) Third

9. - 15.

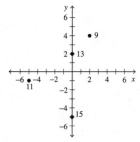

17. - 23.

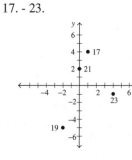

25. $(0, 2)$ 27. $(-2, 0)$ 29. $(-5, -3)$ 31. $(2, -3)$ 33. $(2, 2)$

35.   Substituting $(5, 2)$ into $x + 2y = 9$, we have

$$5 + 2(2) = 9$$
$$5 + 4 = 9$$
$$9 = 9$$

Therefore, $(5, 2)$ satisfies $x + 2y = 9$.

Substituting $(1, 4)$ into $x + 2y = 9$, we have

$$1 + 2(4) = 9$$
$$1 + 8 = 9$$
$$9 = 9$$

Therefore, $(1, 4)$ satisfies $x + 2y = 9$.

Substituting $(0, 5)$ into $x + 2y = 9$, we have

$$0 + 2(5) = 9$$
$$10 \neq 9$$

Therefore, $(0, 5)$ does not satisfy $x + 2y = 9$.

37.   Substituting $(8, 2)$ into $2x - 3y = 10$, we have

$$2(8) - 3(2) = 10$$
$$16 - 6 = 10$$
$$10 = 10$$

Therefore, $(8, 2)$ satisfies $2x - 3y = 10$.

Substituting $(-1, 4)$ into $2x - 3y = 10$, we have

$$2(-1) - 3(4) = 10$$
$$-2 - 12 = 10$$
$$-14 \neq 10$$

Therefore, $(-1, 4)$ does not satisfy $2x - 3y = 10$.

Substituting $\left(0, -\dfrac{10}{3}\right)$ into $2x - 3y = 10$, we have

$$2(0) - 3\left(-\dfrac{10}{3}\right) = 10$$
$$0 + 10 = 10$$
$$10 = 10$$

Therefore, $\left(0, -\dfrac{10}{3}\right)$ satisfies $2x - 3y = 10$.

39.   Substituting $(1, -1)$ into $7y = 3x - 5$, we have

$$7(-1) = 3(1) - 5$$
$$-7 = 3 - 5$$
$$-7 \neq -2$$

Therefore, $(1, -1)$ does not satisfy $7y = 3x - 5$.

Substituting $(-3, -2)$ into $7y = 3x - 5$, we have

$$7(-2) = 3(-3) - 5$$
$$-14 = -9 - 5$$
$$-14 = -14$$

Therefore, $(-3, -2)$ satisfies $7y = 3x - 5$.

Substituting $(2, 5)$ into $7y = 3x - 5$, we have

$$7(5) = 3(2) - 5$$
$$35 = 6 - 5$$
$$35 \neq 1$$

Therefore, $(2, 5)$ does not satisfy $7y = 3x - 5$.

41.   Substituting $(8, 0)$ into $\dfrac{x}{4} + \dfrac{2y}{3} = 2$, we have

$$\frac{8}{4} + \frac{2(0)}{3} = 2$$
$$2 + 0 = 2$$
$$2 = 2$$

Therefore, $(8, 0)$ satisfies $\dfrac{x}{4} + \dfrac{2y}{3} = 2$.

Substituting $\left(1, \dfrac{1}{2}\right)$ into $\dfrac{x}{4} + \dfrac{2y}{3} = 2$, we have

$$\frac{1}{4} + \frac{2}{3}\left(\frac{1}{2}\right) = 2$$
$$\frac{1}{4} + \frac{1}{3} = 2$$
$$\frac{7}{12} \neq 2$$

Therefore, $\left(1, \dfrac{1}{2}\right)$ does not satisfy $\dfrac{x}{4} + \dfrac{2y}{3} = 2$.

Substituting $(0, 3)$ into $\dfrac{x}{4} + \dfrac{2y}{3} = 2$, we have

$$\frac{0}{4} + \frac{3(2)}{3} = 2$$
$$2 = 2$$

Therefore, $(0, 3)$ satisfies $\dfrac{x}{4} + \dfrac{2y}{3} = 2$.

43.   Since the line is vertical, its slope is undefined.

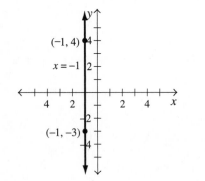

45.   Since the line is horizontal, its slope is 0.

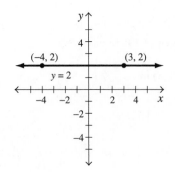

47.

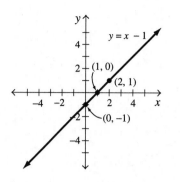

49.

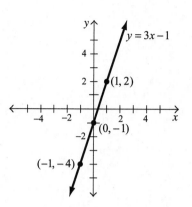

51.

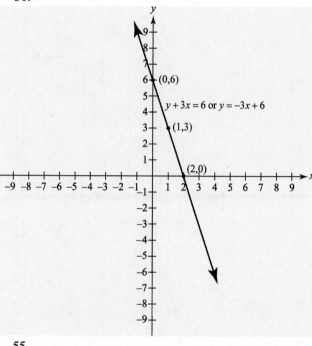

53.

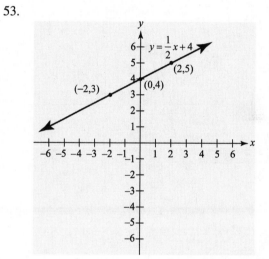

55.

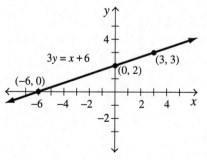

57.

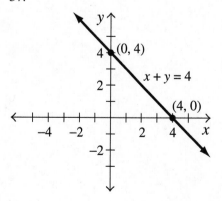

59.

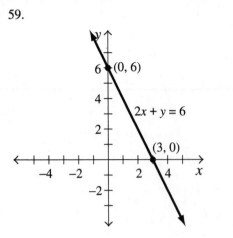

61.

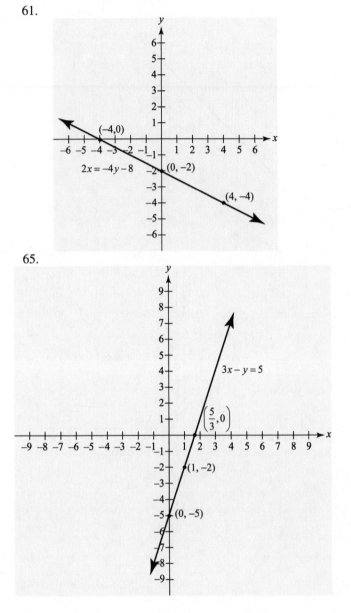

63.

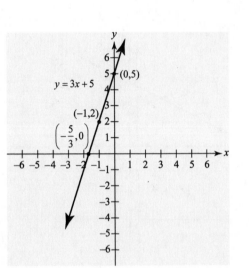

65.

67. $(6, 2), (8, 6)$ $m = \dfrac{6-2}{8-6} = \dfrac{4}{2} = 2$

69. $(-5, 6), (7, -9)$ $m = \dfrac{-9-6}{7-(-5)} = \dfrac{-15}{12} = -\dfrac{5}{4}$

71. $(5, 2), (-3, 2)$ $m = \dfrac{2-2}{-3-5} = \dfrac{0}{-8} = 0$

73. $(8, -3), (8, 3)$ $m = \dfrac{3-(-3)}{8-8} = \dfrac{6}{0}$ Undefined

77.

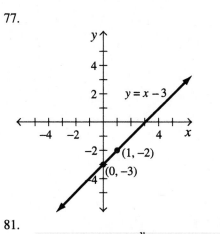

79.

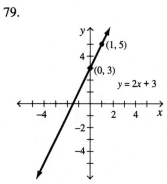

81.

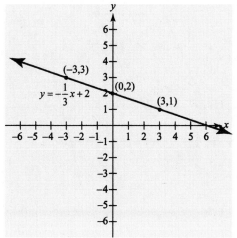

83.

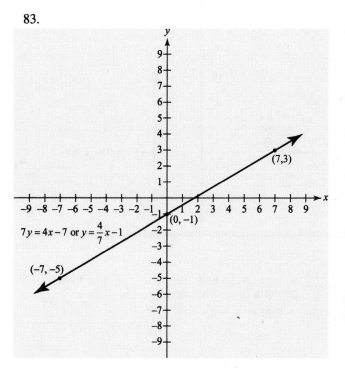

85.

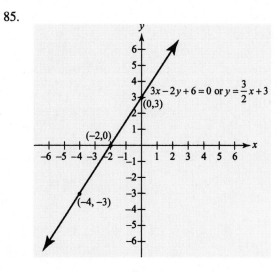

87. The y-intercept is 4; thus $b = 4$. The slope is negative since the graph falls from left to right. The change in y is 4, while the change in x is 2. Thus m, the slope, is $-\dfrac{4}{2} = -2$. The equation is $y = -2x + 4$.

89. The y-intercept is 2; thus $b = 2$. The slope is positive since the graph rises from left to right. The change in y is 3, while the change in x is 1. Thus m, the slope, is $\dfrac{3}{1} = 3$. The equation is $y = 3x + 2$.

91. a)

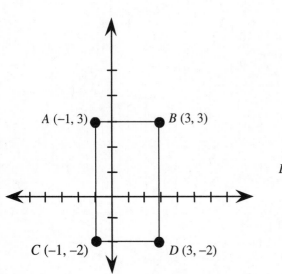

93.

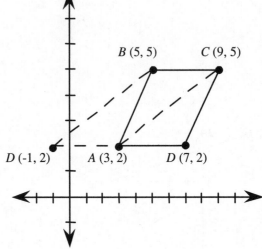

b) $A = lw = 5(4) = 20$ square units

95. For the line joining points P and Q to be parallel to the x-axis, both ordered pairs must have the same y-value. Thus, $b = -3$.

97. For the line joining points P and Q to be parallel to the x-axis, both ordered pairs must have the same y-value.
$$2b + 1 = 7$$
$$2b + 1 - 1 = 7 - 1$$
$$2b = 6$$
$$b = 3$$

b) \$130

c)
$$70 = 40 + 0.3s$$
$$70 - 40 = 40 - 40 + 0.3s$$
$$30 = 0.3s$$
$$s = 100 \text{ square feet}$$

99. a)

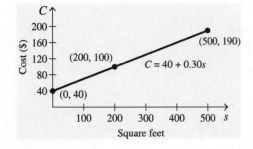

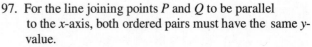

101. a)

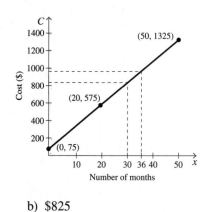

b) $825

c) 36 months

103. a) $m = \dfrac{19-9}{5-0} = \dfrac{10}{5} = 2$

b) $y = 2x + 9$

c) $y = 2(3) + 9 = 6 + 9 = 15$ defects

d) $17 = 2x + 9$

$17 - 9 = 2x + 9 - 9$

$8 = 2x$

$x = 4$ workers

105. a) slope $\approx \dfrac{2.41 - 2.60}{2004 - 2000} = \dfrac{-0.19}{4} \approx -0.05$

b) $y = -0.05x + 2.60$

c) $y = -0.05x + 2.60$

$y = -0.05(2) + 2.60 = 2.50$

$2.50 billion

d) $2.4 = -0.05x + 2.60$

$-0.20 = -0.05x$

$4 = x$

4 years after 2000, or in 2004

107. a) Solve the equations for y to put them in slope-intercept form. Then compare the slopes and y-intercepts. If the slopes are equal but the y-intercepts are different, then the lines are parallel.

b) $2x - 3y = 6$ $4x = 6y + 6$

$2x - 2x - 3y = -2x + 6$ $4x - 6 = 6y + 6 - 6$

$-3y = -2x + 6$ $4x - 6 = 6y$

$\dfrac{-3y}{-3} = \dfrac{-2x}{-3} + \dfrac{6}{-3}$ $\dfrac{4x}{6} - \dfrac{6}{6} = \dfrac{6y}{6}$

$y = \dfrac{2}{3}x - 2$ $\dfrac{2}{3}x - 1 = y$

Since the two equations have the same slope, $m = \dfrac{2}{3}$, the graphs of the equations are parallel lines.

Exercise Set 6.8

1. (1) Mentally substitute the equal sign for the inequality sign and plot points as if you were graphing the equation. (2) If the inequality is < or >, draw a dashed line through the points. If the inequality is ≤ or ≥, draw a solid line through the points. (3) Select a test point not on the line and substitute the x- and y- coordinates into the inequality. If the substitution results in a true statement, shade in the area on the same side of the line as the test point. If the substitution results in a false statement, shade in the area on the opposite side of the line as the test point.

3. A half-plane is the set of all the points in a plane on one side of a line.

5. a) $2x + 3y < 11$

   $2(1) + 3(3) < 11$

   $2 + 9 < 11$

   $11 < 11$ False

   No

   b) $2x + 3y \leq 11$

   $2(1) + 3(3) \leq 11$

   $2 + 9 \leq 11$

   $11 \leq 11$ True

   Yes

   c) $2x + 3y \geq 11$

   $2(1) + 3(3) \geq 11$

   $2 + 9 \geq 11$

   $11 \geq 11$ True

   Yes

   d) $2x + 3y > 11$

   $2(1) + 3(3) > 11$

   $2 + 9 > 11$

   $11 > 11$ False

   No

7. Graph $x = 3$. Since the original statement is less than or equal to, a solid line is drawn. Since the point (0, 0) satisfies the inequality $x \leq 3$, all points on the line and in the half-plane to the left of the line $x = 3$ are in the solution set.

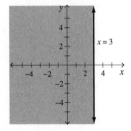

9. Graph $y = x - 4$. Since the original statement is greater than, a dashed line is drawn. Since the point (0, 0) satisfies the inequality $y > x - 4$, all points in the half-plane above the line $y = x - 4$ are in the solution set.

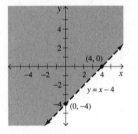

11. Graph $y = 2x - 6$. Since the original statement is greater than or equal to, a solid line is drawn. Since the point (0, 0) satisfies the inequality $y \geq 2x - 6$, all points on the line and in the half-plane above the line $y = 2x - 6$ are in the solution set.

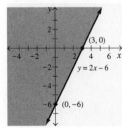

13. Graph $3x - 4y = 12$. Since the original statement is strictly greater than, a dashed line is drawn. Since the point (0, 0) does not satisfy the inequality $3x - 4y > 12$, all points in the half-plane below the line $3x - 4y = 12$ are in the solution set.

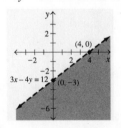

15. Graph $3x - 4y = 9$. Since the original statement is less than or equal to, a solid line is drawn. Since the point (0, 0) satisfies the inequality $3x - 4y \leq 9$, all points on the line and in the half-plane above the line $3x - 4y = 9$ are in the solution set.

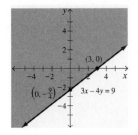

17. Graph $3x + 2y = 6$. Since the original statement is strictly less than, a dashed line is drawn. Since the point (0, 0) satisfies the inequality $3x + 2y < 6$, all points in the half-plane to the left of the line $3x + 2y = 6$ are in the solution set.

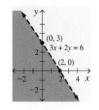

19. Graph $x + y = 0$. Since the original statement is strictly greater than, a dashed line is drawn. Since the point (1, 1) satisfies the inequality $x + y > 0$, all points in the half-plane above the line $x + y = 0$ are in the solution set.

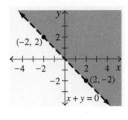

21. Graph $5x + 2y = 10$. Since the original statement is greater than or equal to, a solid line is drawn. Since the point (0, 0) does not satisfy the inequality $5x + 2y \geq 10$, all points on the line and in the half-plane above the line $5x + 2y = 10$ are in the solution set.

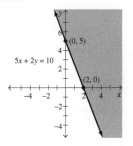

23. Graph $3x - 2y = 12$. Since the original statement is strictly less than, a dashed line is drawn. Since the point (0, 0) satisfies the inequality $3x - 2y < 12$, all points in the half-plane above the line $3x - 2y = 12$ are in the solution set.

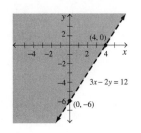

25. Graph $\frac{2}{5}x - \frac{1}{2}y = 1$. Since the original statement is less than or equal to, a solid line is drawn. Since the point (0, 0) satisfies the inequality $\frac{2}{5}x - \frac{1}{2}y \leq 1$, all points on the line and in the half-plane above the line $\frac{2}{5}x - \frac{1}{2}y = 1$ are in the solution set.

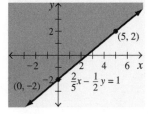

27.   Graph $0.1x + 0.3y = 0.4$. Since the original statement is less than or equal to, a solid line is drawn. Since the point (0, 0) satisfies the inequality $0.1x + 0.3y \le 0.4$, all points on the line and in the half-plane below the line $0.1x + 0.3y = 0.4$ are in the solution set.

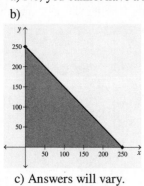

29.   a) $x + y \le 300$

b)

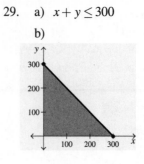

31.   a) No, you cannot have a negative number of shirts.

b)

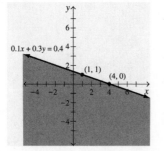

c) Answers will vary.

Exercise Set 6.9

1. A **binomial** is an expression that contains two terms in which each exponent that appears on the variable is a whole number. Examples: $2x+3$, $x-7$, x^2-9

3. The **FOIL method** is a method that obtains the products of the First, Outer, Inner, and Last terms of the binomials.

5. $ax^2+bx+c=0$, $a\neq 0$

7. $x^2+8x+15=(x+5)(x+3)$

9. $x^2-x-2=(x-2)(x+1)$

11. $x^2+2x-24=(x+6)(x-4)$

13. $x^2-2x-3=(x+1)(x-3)$

15. $x^2-10x+21=(x-7)(x-3)$

17. $x^2-49=(x-7)(x+7)$

19. $x^2+3x-28=(x+7)(x-4)$

21. $x^2+2x-63=(x+9)(x-7)$

23. $2x^2-x-6=(2x+3)(x-2)$

25. $5x^2+16x+3=(5x+1)(x+3)$

27. $5x^2+12x+4=(5x+2)(x+2)$

29. $4x^2+11x+6=(4x+3)(x+2)$

31. $4x^2-11x+6=(4x-3)(x-2)$

33. $8x^2-10x-3=(4x+1)(2x-3)$

35. $(x-3)(x+6)=0$
$x-3=0$ or $x+6=0$
$x=3$ $\qquad x=-6$

37. $(3x+4)(2x-1)=0$
$3x+4=0$ or $2x-1=0$
$3x=-4$ $\qquad 2x=1$
$x=-\dfrac{4}{3}$ $\qquad x=\dfrac{1}{2}$

39. $x^2+7x+10=0$
$(x+5)(x+2)=0$
$x+5=0$ or $x+2=0$
$x=-5$ $\qquad x=-2$

41. $x^2-7x+6=0$
$(x-6)(x-1)=0$
$x-6=0$ or $x-1=0$
$x=6$ $\qquad x=1$

43. $x^2-15=2x$
$x^2-2x-15=0$
$(x-5)(x+3)=0$
$x-5=0$ or $x+3=0$
$x=5$ $\qquad x=-3$

45. $x^2=4x-3$
$x^2-4x+3=0$
$(x-3)(x-1)=0$
$x-3=0$ or $x-1=0$
$x=3$ $\qquad x=1$

47. $x^2-81=0$
$(x-9)(x+9)=0$
$x-9=0$ or $x+9=0$
$x=9$ $\qquad x=-9$

49. $x^2+5x-36=0$
$(x+9)(x-4)=0$
$x+9=0$ or $x-4=0$
$x=-9$ $\qquad x=4$

51. $3x^2+10x=8$
$3x^2+10x-8=0$
$(3x-2)(x+4)=0$
$3x-2=0$ or $x+4=0$
$3x=2$ $\qquad x=-4$
$x=\dfrac{2}{3}$

53. $5x^2+11x=-2$
$5x^2+11x+2=0$
$(5x+1)(x+2)=0$
$5x+1=0$ or $x+2=0$
$5x=-1$ $\qquad x=-2$
$x=-\dfrac{1}{5}$

55. $3x^2 - 4x = -1$

    $3x^2 - 4x + 1 = 0$

    $(3x-1)(x-1) = 0$

    $3x - 1 = 0$ or $x - 1 = 0$

    $3x = 1$ $\qquad$ $x = 1$

    $x = \dfrac{1}{3}$

57. $6x^2 - 11x + 3 = 0$

    $(3x-1)(2x-3) = 0$

    $3x - 1 = 0$ or $2x - 3 = 0$

    $3x = 1$ $\qquad$ $2x = 3$

    $x = \dfrac{1}{3}$ $\qquad$ $x = \dfrac{3}{2}$

59. $x^2 + 2x - 15 = 0$

    $a = 1,\ \ b = 2,\ \ c = -15$

    $x = \dfrac{-2 \pm \sqrt{(2)^2 - 4(1)(-15)}}{2(1)}$

    $x = \dfrac{-2 \pm \sqrt{4 + 60}}{2} = \dfrac{-2 \pm \sqrt{64}}{2} = \dfrac{-2 \pm 8}{2}$

    $x = \dfrac{6}{2} = 3$ or $x = \dfrac{-10}{2} = -5$

61. $x^2 - 3x - 18 = 0$

    $a = 1,\ \ b = -3,\ \ c = -18$

    $x = \dfrac{-(-3) \pm \sqrt{(-3)^2 - 4(1)(-18)}}{2(1)}$

    $x = \dfrac{3 \pm \sqrt{9 + 72}}{2} = \dfrac{3 \pm \sqrt{81}}{2} = \dfrac{3 \pm 9}{2}$

    $x = \dfrac{12}{2} = 6$ or $x = \dfrac{-6}{2} = -3$

63. $x^2 - 8x = 9$

    $x^2 - 8x - 9 = 0$

    $a = 1, b = -8, c = -9$

    $x = \dfrac{-(-8) \pm \sqrt{(-8)^2 - 4(1)(-9)}}{2(1)}$

    $x = \dfrac{8 \pm \sqrt{64 + 36}}{2} = \dfrac{8 \pm \sqrt{100}}{2} = \dfrac{8 \pm 10}{2}$

    $x = \dfrac{18}{2} = 9$ or $x = \dfrac{-2}{2} = -1$

65. $x^2 - 2x + 3 = 0$

    $a = 1, b = -2, c = 3$

    $x = \dfrac{-(-2) \pm \sqrt{(-2)^2 - 4(1)(3)}}{2(1)}$

    $x = \dfrac{2 \pm \sqrt{4 - 12}}{2} = \dfrac{2 \pm \sqrt{-8}}{2}$

    No real solution

67. $x^2 - 4x + 2 = 0$

    $a = 1, b = -4, c = 2$

    $x = \dfrac{-(-4) \pm \sqrt{(-4)^2 - 4(1)(2)}}{2(1)}$

    $x = \dfrac{4 \pm \sqrt{16 - 8}}{2} = \dfrac{4 \pm \sqrt{8}}{2} = \dfrac{4 \pm 2\sqrt{2}}{2}$

    $x = 2 \pm \sqrt{2}$

69. $3x^2 - 8x + 1 = 0$

    $a = 3,\ \ b = -8,\ \ c = 1$

    $x = \dfrac{-(-8) \pm \sqrt{(-8)^2 - 4(3)(1)}}{2(3)}$

    $x = \dfrac{8 \pm \sqrt{64 - 12}}{6} = \dfrac{8 \pm \sqrt{52}}{6} = \dfrac{8 \pm 2\sqrt{13}}{6}$

    $x = \dfrac{4 \pm \sqrt{13}}{3}$

71. $3x^2 - 6x - 5 = 0$

     $a = 3,\ b = -6,\ c = -5$

$$x = \frac{-(-6) \pm \sqrt{(-6)^2 - 4(3)(-5)}}{2(3)}$$

$$x = \frac{6 \pm \sqrt{96}}{6} = \frac{6 \pm 4\sqrt{6}}{6} = \frac{3 \pm 2\sqrt{6}}{3}$$

73. $2x^2 + 7x + 5 = 0$

     $a = 2,\ b = 7,\ c = 5$

$$x = \frac{-7 \pm \sqrt{(7)^2 - 4(2)(5)}}{2(2)}$$

$$x = \frac{-7 \pm \sqrt{49 - 40}}{4} = \frac{-7 \pm \sqrt{9}}{4} = \frac{-7 \pm 3}{4}$$

$$x = \frac{-4}{4} = -1 \ \text{ or } \ x = \frac{-10}{4} = -\frac{5}{2}$$

75. $3x^2 - 10x + 7 = 0$

     $a = 3,\ b = -10,\ c = 7$

$$x = \frac{-(-10) \pm \sqrt{(-10)^2 - 4(3)(7)}}{2(3)}$$

$$x = \frac{10 \pm \sqrt{100 - 84}}{6} = \frac{10 \pm \sqrt{16}}{6} = \frac{10 \pm 4}{6}$$

$$x = \frac{14}{6} = \frac{7}{3} \ \text{ or } \ x = \frac{6}{6} = 1$$

77. $4x^2 + 6x + 5 = 0$

     $a = 4,\ b = 6,\ c = 5$

$$x = \frac{-6 \pm \sqrt{6^2 - 4(4)(5)}}{2(4)}$$

$$x = \frac{-6 \pm \sqrt{36 - 80}}{8} = \frac{6 \pm \sqrt{-44}}{8}$$

No real solution

79.   Area of backyard $= lw = 30(20) = 600 \ \text{m}^2$

Let x = width of grass around all sides of the flower garden

Width of flower garden $= 20 - 2x$

Length of flower garden $= 30 - 2x$

Area of flower garden $= lw = (30 - 2x)(20 - 2x)$

Area of grass $= 600 - (30 - 2x)(20 - 2x)$

$600 - (30 - 2x)(20 - 2x) = 336$

$600 - (600 - 100x + 4x^2) = 336$

$600 - 600 + 100x - 4x^2 = 336$

$-4x^2 + 100x = 336$

$4x^2 - 100x + 336 = 0$

$x^2 - 25x + 84 = 0$

$(x - 21)(x - 4) = 0$

$x - 21 = 0 \quad \text{or} \quad x - 4 = 0$

$x = 21 \qquad\qquad x = 4$

$x \neq 21$ since the width of the backyard is 20 m.

     Width of grass $= 4$ m

     Width of flower garden

     $= 20 - 2x = 20 - 2(4) = 20 - 8 = 12$ m

     Length of flower garden

     $= 30 - 2x = 30 - 2(4) = 30 - 8 = 22$ m

81. a) Since the equation is equal to 6 and not 0, the zero-factor property cannot be used.

b) $(x-4)(x-7)=6$

$x^2-11x+28=6$

$x^2-11x+22=0$

$a=1,\ b=-11,\ c=22$

$x=\dfrac{-(-11)\pm\sqrt{(-11)^2-4(1)(22)}}{2(1)}$

$x=\dfrac{11\pm\sqrt{121-88}}{2}=\dfrac{11\pm\sqrt{33}}{2}$

$x\approx 8.37$ or $x\approx 2.63$

83. $(x+1)(x-3)=0$

$x^2-2x-3=0$

Exercise Set 6.10

1. A **function** is a special type of relation where each value of the independent variable corresponds to a unique value of the dependent variable.

3. The **domain** of a function is the set of values that can be used for the independent variable.

5. The vertical line test can be used to determine if a graph represents a function. If a vertical line can be drawn so that it intersects the graph at more than one point, then each value of x does not have a unique value of y and the graph does not represent a function. If a vertical line cannot be made to intersect the graph in at least two different places, then the graph represents a function.

7. $x=-\dfrac{b}{2a}$

9. Not a function since $x=2$ is not paired with a unique value of y.

11. Function since each vertical line intersects the graph at only one point.
D: all real numbers   R: all real numbers

13. Not a function since $x=-1$ is not paired with a unique value of y.

15. Function since each vertical line intersects the graph at only one point.
D: all real numbers   R: $y\ge -4$

17. Not a function since it is possible to draw a vertical line that intersects the graph at more than one point.

19. Function since each vertical line intersects the graph at only one point.
D: $0\le x<12$ R: $y=1,\ 2,\ 3$

21. Not a function since it is possible to draw a vertical line that intersects the graph at more than one point.

23. Function since each vertical line intersects the graph at only one point.
D: all real numbers   R: $y>0$

25. Not a function since it is possible to draw a vertical line that intersects the graph at more than one point.

27. Function since each value of x is paired with a unique value of y.

29. Not a function since $x=6$ is paired with two different values of y.

31. Function since each value of x is paired with a unique value of y.

33. $f(x) = x + 5, \quad x = 3$

    $f(3) = 3 + 5 = 8$

35. $f(x) = -2x - 7, \ x = -4$

    $f(-4) = -2(-4) - 7 = 8 - 7 = 1$

37. $f(x) = 10x - 6, \ x = 0$

    $f(0) = 10(0) - 6 = 0 - 6 = -6$

39. $f(x) = x^2 - 3x + 1, \quad x = 4$

    $f(4) = (4)^2 - 3(4) + 1 = 16 - 12 + 1 = 5$

41. $f(x) = -x^2 - 2x + 1, \ x = 4$

    $f(4) = -(4)^2 - 2(4) + 1 = -16 - 8 + 1 = -23$

43. $f(x) = 4x^2 - 6x - 9, \ x = -3$

    $f(-3) = 4(-3)^2 - 6(-3) - 9 = 36 + 18 - 9 = 45$

45. $f(x) = -5x^2 + 3x - 9, \quad x = -1$

    $f(-1) = -5(-1)^2 + 3(-1) - 9 = -5 - 3 - 9 = -17$

47.

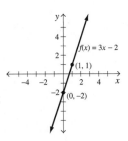

49.

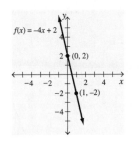

51.

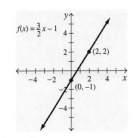

53. $y = x^2 - 9$

    a) $a = 1 > 0$, opens upward

    b) $x = 0$ c) $(0, -9)$ d) $(0, -9)$

    e) $(3, 0), (-3, 0)$

    f)

    ![graph of y = x^2 - 9 with points (-3,0), (3,0), (0,-9)]

    g) D: all real numbers   R: $y \geq -9$

55. $y = -x^2 + 4$

    a) $a = -1 < 0$, opens downward

    b) $x = 0$ c) $(0, 4)$ d) $(0, 4)$

    e) $(-2, 0), (2, 0)$

    f)

    ![graph of y = -x^2 + 4 with points (0,4), (-2,0), (2,0)]

    g) D: all real numbers   R: $y \leq 4$

57. $y = -2x^2 - 8$

a) $a = -2 < 0$, opens downward

b) $x = 0$ c) $(0, -8)$ d) $(0, -8)$

e) no x-intercepts

f)

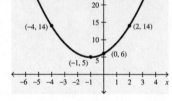

g) D: all real numbers   R: $y \leq -8$

59. $y = 2x^2 - 3$

a) $a = 2 > 0$, opens upward

b) $x = 0$ c) $(0, -3)$ d) $(0, -3)$

e) $(-1.22, 0), (1.22, 0)$

f)

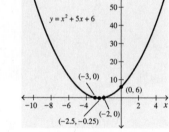

g) D: all real numbers   R: $y \geq -3$

61. $f(x) = x^2 + 2x + 6$

a) $a = 1 > 0$, opens upward

b) $x = -1$ c) $(-1, 5)$ d) $(0, 6)$

e) no x-intercepts

f)

g) D: all real numbers   R: $y \geq 5$

63. $y = x^2 + 5x + 6$

a) $a = 1 > 0$, opens upward

b) $x = -\dfrac{5}{2}$ c) $(-2.5, -0.25)$ d) $(0, 6)$

e) $(-3, 0), (-2, 0)$

f)

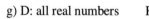

g) D: all real numbers   R: $y \geq -0.25$

65. $y = -x^2 + 4x - 6$

a) $a = -1 < 0$, opens downward

b) $x = 2$ c) $(2, -2)$ d) $(0, -6)$

e) no x-intercepts

f)

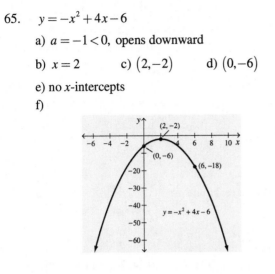

g) D: all real numbers   R: $y \leq -2$

67. $y = -2x^2 + 3x - 2$

a) $a = -2 < 0$, opens downward

b) $x = \dfrac{3}{4}$ c) $\left(\dfrac{3}{4}, -\dfrac{7}{8}\right)$ d) $(0, -2)$

e) None

f)

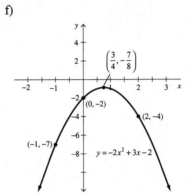

69.

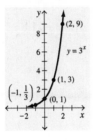

D: all real numbers   R: $y > 0$

71.

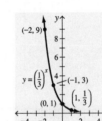

D: all real numbers   R: $y > 0$

73.

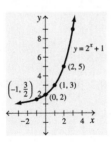

D: all real numbers   R: $y > 1$

75.

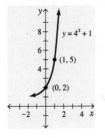

D: all real numbers   R: $y > 1$

77.

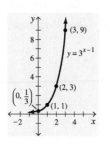

D: all real numbers   R: $y > 0$

79.

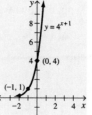

D: all real numbers   R: $y > 0$

81. $p(x) = 0.3x - 4000$

  $p(150,000) = 0.3(150,000) - 4000$
  $\qquad = 45,000 - 4000 = \$41,000$

83. a) $2004 \rightarrow x = 74$

  $f(74) = 0.005(74)^2 - 0.36(74) + 11.8$
  $\qquad = 12.54 \approx 13\%$

  b) 1970

  c) $x = -\dfrac{b}{2a} = -\dfrac{-0.36}{2(0.005)} = 36$

  $f(36) = 0.005(36)^2 - 0.36(36) + 11.8$
  $\qquad \approx 5.32 \approx 5\%$

85. $P(x) = 4000(1.3)^{0.1x}$

  a) $x = 10,\ P(10) = 4000(1.3)^{0.1(10)}$

  $\qquad = 4000(1.3) = 5200$ people

  b) $x = 50,\ P(50) = 4000(1.3)^{0.1(50)}$

  $\qquad = 4000(3.71293)$

  $\qquad = 14,851.72 \approx 14,852$ people

87. a) Yes

  b) $\approx \$59$

89. $d = (21.9)(2)^{(20-x)/12}$

  a) $x = 19,\ d = (21.9)(2)^{(20-19)/12}$

  $\qquad = (21.9)(1.059463094) \approx 23.2$ cm

  b) $x = 4,\ d = (21.9)(2)^{(20-4)/12}$

  $\qquad = (21.9)(2.519842099) \approx 55.2$ cm

  c) $x = 0,\ d = (21.9)(2)^{(20-0)/12}$

  $\qquad = (21.9)(3.174802105) \approx 69.5$ cm

91. $f(x) = -0.85x + 187$

  a) $f(20) = -0.85(20) + 187 = 170$ beats per minute

  b) $f(30) = -0.85(30) + 187 = 161.5 \approx 162$ beats per minute

  c) $f(50) = -0.85(50) + 187 = 144.5 \approx 145$ beats per minute

  d) $f(60) = -0.85(60) + 187 = 136$ beats per minute

  e) $-0.85x + 187 = 85$

  $\qquad -0.85x = -102$

  $\qquad x = 120$ years of age

Review Exercises

1. $x = 2$, $x^2 + 15 = (2)^2 + 15 = 4 + 15 = 19$

2. $x = -1$, $-x^2 - 7 = -(-1)^2 - 7 = -1 - 7 = -8$

3. $x = 2$, $4x^2 - 2x + 5 = 4(2)^2 - 2(2) + 5$
   $= 16 - 4 + 5 = 17$

4. $x = \dfrac{1}{2}$, $-x^2 + 7x - 3 = -\left(\dfrac{1}{2}\right)^2 + 7\left(\dfrac{1}{2}\right) - 3$
   $= -\dfrac{1}{4} + \dfrac{14}{4} - \dfrac{12}{4} = \dfrac{1}{4}$

5. $x = -1$, $4x^3 - 7x^2 + 3x + 1$
   $= 4(-1)^3 - 7(-1)^2 + 3(-1) + 1$
   $= -4 - 7 - 3 + 1 = -13$

6. $x = 1$, $y = -2$, $3x^2 - xy + 2y^2$
   $= 3(1)^2 - 1(-2) + 2(-2)^2$
   $= 3 + 2 + 8 = 13$

7. $3x - 2 + x + 7 = 4x + 5$

8. $2x + 5(x - 2) + 8x = 3x + 5x - 10 + 8x = 15x - 10$

9. $4(x - 1) + \dfrac{1}{3}(9x + 3) = 4x - 4 + 3x + 1 = 7x - 3$

10. $4s + 10 = -30$
    $4s + 10 - 10 = -30 - 10$
    $4s = -40$
    $\dfrac{4s}{4} = \dfrac{-40}{4}$
    $s = -10$

11. $3t + 14 = -6t - 13$
    $3t - 3t + 14 = -6t - 3t - 13$
    $14 = -9t - 13$
    $14 + 13 = -9t - 13 + 13$
    $27 = -9t$
    $\dfrac{27}{-9} = \dfrac{-9t}{-9}$
    $-3 = t$

12. $\dfrac{x + 5}{2} = \dfrac{x - 3}{4}$
    $4(x + 5) = 2(x - 3)$
    $4x + 20 = 2x - 6$
    $4x - 2x + 20 = 2x - 2x - 6$
    $2x + 20 = -6$
    $2x + 20 - 20 = -6 - 20$
    $2x = -26$
    $\dfrac{2x}{2} = \dfrac{-26}{2}$
    $x = -13$

13. $4(x - 2) = 3 + 5(x + 4)$
    $4x - 8 = 3 + 5x + 20$
    $4x - 8 = 5x + 23$
    $4x - 4x - 8 = 5x - 4x + 23$
    $-8 = x + 23$
    $-8 - 23 = x + 23 - 23$
    $-31 = x$

14. $\dfrac{x}{4} + \dfrac{3}{5} = 7$
    $20\left(\dfrac{x}{4} + \dfrac{3}{5}\right) = 20(7)$
    $5x + 12 = 140$
    $5x + 12 - 12 = 140 - 12$
    $5x = 128$
    $\dfrac{5x}{5} = \dfrac{128}{5}$
    $x = \dfrac{128}{5}$

15. $\dfrac{2}{\frac{1}{3}} = \dfrac{3}{x}$

$2x = 3\left(\dfrac{1}{3}\right)$

$2x = 1$

$\dfrac{2x}{2} = \dfrac{1}{2}$

$x = \dfrac{1}{2}$ cup

16. 1 hr 40 min $= 60$ min $+ 40$ min $= 100$ min

$\dfrac{120}{100} = \dfrac{300}{x}$

$120x = 100(300)$

$120x = 30{,}000$

$\dfrac{120x}{120} = \dfrac{30{,}000}{120}$

$x = 250$ min, or 4 hr 10 min

17. $A = bh$

$A = 12(4) = 48$

18. $V = 2\pi R^2 r^2$

$V = 2(3.14)(3)^2(1.75)^2$

$V = 2(3.14)(9)(3.0625)$

$V = 173.0925 \approx 173.1$

19. $z = \dfrac{\overline{x} - \mu}{\frac{\sigma}{\sqrt{n}}}$

$2 = \dfrac{\overline{x} - 100}{\frac{3}{\sqrt{16}}}$

$\dfrac{2}{1} = \dfrac{\overline{x} - 100}{\frac{3}{4}}$

$2\left(\dfrac{3}{4}\right) = 1(\overline{x} - 100)$

$\dfrac{3}{2} = \overline{x} - 100$

$\dfrac{3}{2} + 100 = \overline{x} - 100 + 100$

$\dfrac{3}{2} + \dfrac{200}{2} = \overline{x}$

$\dfrac{203}{2} = \overline{x}$

$101.5 = \overline{x}$

20. $E = mc^2$

$400 = m(4)^2$

$400 = 16m$

$\dfrac{400}{16} = \dfrac{16m}{16}$

$25 = m$

21.
$$8x - 4y = 24$$
$$8x - 8x - 4y = -8x + 24$$
$$-4y = -8x + 24$$
$$\frac{-4y}{-4} = \frac{-8x + 24}{-4}$$
$$y = 2x - 6$$

22.
$$2x + 7y = 15$$
$$2x - 2x + 7y = -2x + 15$$
$$7y = -2x + 15$$
$$\frac{7y}{7} = \frac{-2x + 15}{7} = -\frac{2}{7}x + \frac{15}{7}$$

23.
$$2x - 3y + 52 = 30$$
$$2x - 2x - 3y + 52 = -2x + 30$$
$$-3y + 52 = -2x + 30$$
$$-3y + 52 - 52 = -2x + 30 - 52$$
$$-3y = -2x - 22$$
$$\frac{-3y}{-3} = \frac{-2x - 22}{-3}$$
$$y = \frac{-2x - 22}{-3} = \frac{2x + 22}{3} = \frac{2}{3}x + \frac{22}{3}$$

24.
$$-3x - 4y + 5z = 4$$
$$-3x + 3x - 4y + 5z = 3x + 4$$
$$-4y + 5z = 3x + 4$$
$$-4y + 5z - 5z = 3x - 5z + 4$$
$$-4y = 3x - 5z + 4$$
$$\frac{-4y}{-4} = \frac{3x - 5z + 4}{-4}$$
$$y = \frac{3x - 5z + 4}{-4}$$
$$= \frac{-(3x - 5z + 4)}{4}$$
$$= \frac{-3x + 5z - 4}{4}$$
$$= -\frac{3}{4}x + \frac{5}{4}z - 1$$

25.
$$A = lw$$
$$\frac{A}{l} = \frac{lw}{l}$$
$$\frac{A}{l} = w$$

26.
$$P = 2l + 2w$$
$$P - 2l = 2l - 2l + 2w$$
$$P - 2l = 2w$$
$$\frac{P - 2l}{2} = \frac{2w}{2}$$
$$\frac{P - 2l}{2} = w$$

27.
$$L = 2(wh + lh)$$
$$L = 2wh + 2lh$$
$$L - 2wh = 2wh - 2wh + 2lh$$
$$L - 2wh = 2lh$$
$$\frac{L - 2wh}{2h} = \frac{2lh}{2h}$$
$$\frac{L - 2wh}{2h} = l \text{ or } l = \frac{L}{2h} - \frac{2wh}{2h} = \frac{L}{2h} - w$$

28.
$$a_n = a_1 + (n - 1)d$$
$$a_n - a_1 = a_1 - a_1 + (n - 1)d$$
$$a_n - a_1 = (n - 1)d$$
$$\frac{a_n - a_1}{n - 1} = \frac{(n - 1)d}{n - 1}$$
$$\frac{a_n - a_1}{n - 1} = d$$

29. $7 - 4x$

30. $2y + 7$

31. $10 + 3r$

32. $\dfrac{9}{q} - 15$

33.
$$\text{Let } x = \text{the number}$$
$$7x = 7 \text{ times the number}$$
$$3 + 7x = 3 \text{ increased by } 7 \text{ times the number}$$
$$3 + 7x = 17$$
$$3 - 3 + 7x = 17 - 3$$
$$7x = 14$$
$$\frac{7x}{7} = \frac{14}{7}$$
$$x = 2$$

34. Let $x =$ the number
$$3x = \text{the product of } 3 \text{ and a number}$$
$$3x + 8 = \text{the product of } 3 \text{ and a number}$$
$$\text{increased by } 8$$
$$x - 6 = 6 \text{ less than the number}$$
$$3x + 8 = x - 6$$
$$3x - x + 8 = x - x - 6$$
$$2x + 8 = -6$$
$$2x + 8 - 8 = -6 - 8$$
$$2x = -14$$
$$\frac{2x}{2} = \frac{-14}{2}$$
$$x = -7$$

35. Let $x =$ the number
$$x - 4 = \text{the difference of a number and } 4$$
$$5(x - 4) = 5 \text{ times the difference of a number}$$
$$\text{and } 4$$
$$5(x - 4) = 45$$
$$5x - 20 = 45$$
$$5x - 20 + 20 = 45 + 20$$
$$5x = 65$$
$$\frac{5x}{5} = \frac{65}{5}$$
$$x = 13$$

36. Let $x =$ the number
$$10x = 10 \text{ times a number}$$
$$10x + 14 = 14 \text{ more than } 10 \text{ times a number}$$
$$x + 12 = \text{the sum of a number and } 12$$
$$8(x + 12) = 8 \text{ times the sum of a number and } 12$$
$$10x + 14 = 8(x + 12)$$
$$10x + 14 = 8x + 96$$
$$10x - 8x + 14 = 8x - 8x + 96$$
$$2x + 14 = 96$$
$$2x + 14 - 14 = 96 - 14$$
$$2x = 82$$
$$\frac{2x}{2} = \frac{82}{2}$$
$$x = 41$$

37. Let $x =$ the amount invested in bonds
$$2x = \text{the amount invested in mutual funds}$$
$$x + 2x = 15,000$$
$$3x = 15,000$$
$$\frac{3x}{3} = \frac{15,000}{3}$$
$$x = \$5000 \text{ in bonds}$$
$$2x = 2(5000) = \$10,000 \text{ in mutual funds}$$

38. Let $x =$ number of lawn chairs
$$9.50x = \text{variable cost per lawn chair}$$
$$9.50x + 15,000 = 95,000$$
$$9.50x + 15,000 - 15,000 = 95,000 - 15,000$$
$$9.50x = 80,000$$
$$\frac{9.50x}{9.50} = \frac{80,000}{9.50}$$
$$x \approx 8421.05 \approx 8421 \text{ lawn chairs}$$

39.   Let $x =$ the number of species
                    of threatened mammals
        $6x + 2 =$ the number of species
                    of endangered mammals
$$6x + 2 + x = 79$$
$$7x + 2 = 79$$
$$7x + 2 - 2 = 79 - 2$$
$$7x = 77$$
$$\frac{7x}{7} = \frac{77}{7}$$
        $x = 11$ species of
            threatened mammals
$$6x + 2 = 6(11) + 2$$
$$= 68 \text{ species of}$$
            endangered mammals

40. Let $x =$ profit at restaurant B
     $x + 12{,}000 =$ profit at restaurant A
$$x + (x + 12{,}000) = 68{,}000$$
$$2x + 12{,}000 = 68{,}000$$
$$2x + 12{,}000 - 12{,}000 = 68{,}000 - 12{,}000$$
$$2x = 56{,}000$$
$$\frac{2x}{2} = \frac{56{,}000}{2}$$
$$x = \$28{,}000 \text{ for restaurant B}$$
$x + 12{,}000 = 28{,}000 + 12{,}000 = \$40{,}000$ for
restaurant A

41. $s = kt$
        $40 = k(5)$
        $k = \dfrac{40}{5} = 8$
        $s = 8t$
        $s = 8(8)$
        $s = 64$

42. $J = \dfrac{k}{A^2}$
        $8 = \dfrac{k}{3^2}$
        $k = 3^2(8)$
        $k = 72$
        $J = \dfrac{72}{A^2}$
        $J = \dfrac{72}{6^2}$
        $J = \dfrac{72}{36} = 2$

43. $W = \dfrac{kL}{A}$
        $80 = \dfrac{k(100)}{20}$
        $100k = 1600$
        $\dfrac{100k}{100} = \dfrac{1600}{100}$
        $k = 16$
        $W = \dfrac{16L}{A}$
        $W = \dfrac{16(50)}{40} = \dfrac{800}{40} = 20$

44. $z = \dfrac{kxy}{r^2}$
        $12 = \dfrac{k(20)(8)}{(8)^2}$
        $160k = 768$
        $\dfrac{160k}{160} = \dfrac{768}{160}$
        $k = 4.8$
        $z = \dfrac{4.8xy}{r^2}$
        $z = \dfrac{4.8(10)(80)}{(3)^2} = \dfrac{3840}{9} = 426.\overline{6} \approx 426.7$

45. a) $\dfrac{30 \text{ lb}}{2500 \text{ ft}^2} = \dfrac{x \text{ lb}}{12{,}500 \text{ ft}^2}$

$30(12{,}500) = 2500x$

$375{,}000 = 2500x$

$\dfrac{375{,}000}{2500} = \dfrac{2500x}{2500}$

$x = 150 \text{ lb}$

b) $\dfrac{150}{30} = 5 \text{ bags}$

46. $\dfrac{1 \text{ in.}}{30 \text{ mi}} = \dfrac{x \text{ in.}}{120 \text{ mi}}$

$30x = 120$

$\dfrac{30x}{30} = \dfrac{120}{30}$

$x = 4 \text{ in.}$

47. $\dfrac{1 \; kWh}{\$0.162} = \dfrac{740 \; kWh}{x}$

$x = \$119.88$

48. $S = kw$

$4.2 = k(60)$

$\dfrac{4.2}{60} = \dfrac{60k}{60}$

$k = 0.07$

$S = 0.07w$

$S = 0.07(25)$

$S = 1.75 \text{ in.}$

49. $8 + 7x \le 5x - 4$

$8 - 8 + 7x \le 5x - 4 - 8$

$7x \le 5x - 12$

$7x - 5x \le 5x - 5x - 12$

$2x \le -12$

$\dfrac{2x}{2} \le \dfrac{-12}{2}$

$x \le -6$

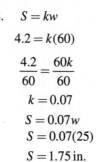

50. $2x + 8 \ge 5x + 11$

$2x - 5x + 8 \ge 5x - 5x + 11$

$-3x + 8 \ge 11$

$-3x + 8 - 8 \ge 11 - 8$

$-3x \ge 3$

$\dfrac{-3x}{-3} \le \dfrac{3}{-3}$

$x \le -1$

51. $3(x+9) \leq 4x+11$

$3x+27 \leq 4x+11$

$3x-4x+27 \leq 4x-4x+11$

$-x+27 \leq 11$

$-x+27-27 \leq 11-27$

$-x \leq -16$

$\dfrac{-x}{-1} \geq \dfrac{-16}{-1}$

$x \geq 16$

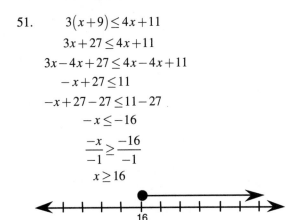

52. $-3 \leq x+1 < 7$

$-3-1 \leq x+1-1 < 7-1$

$-4 \leq x < 6$

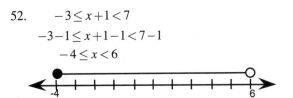

53. $2+4x > -10$

$2-2+4x > -10-2$

$4x > -12$

$\dfrac{4x}{4} > \dfrac{-12}{4}$

$x > -3$

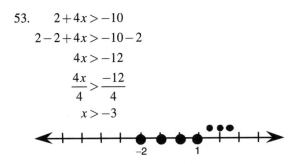

54. $5x+13 \geq -22$

$5x+13-13 \geq -22-13$

$5x \geq -35$

$\dfrac{5x}{5} \geq \dfrac{-35}{5}$

$x \geq -7$

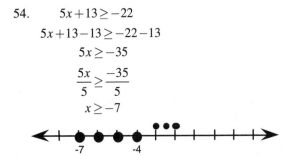

55. $-1 < x \leq 9$

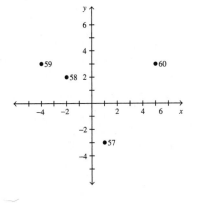

56. $-8 \leq x+2 \leq 7$

$-8-2 \leq x+2-2 \leq 7-2$

$-10 \leq x \leq 5$

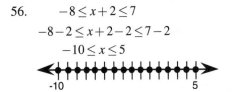

57. - 60.

61.

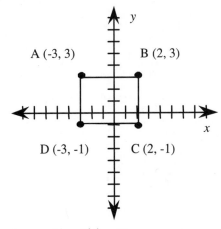

Area $= lw = 5(4) = 20$ square units

62.

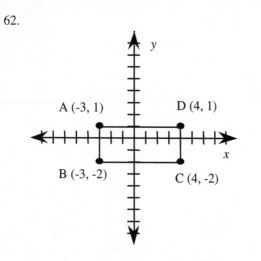

A (-3, 1)   D (4, 1)

B (-3, -2)   C (4, -2)

Area $= lw = 7(3) = 21$ square units

63.

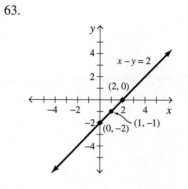

64.

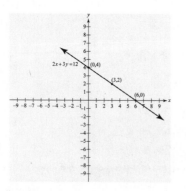

65.

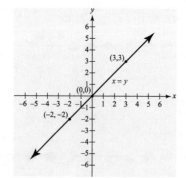

66.

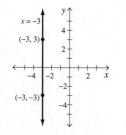

67.

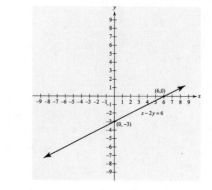

68.

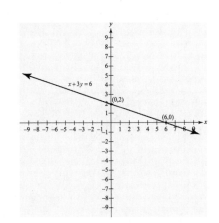

69.

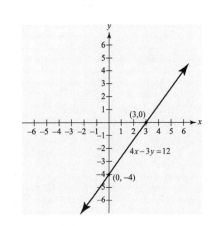

70.

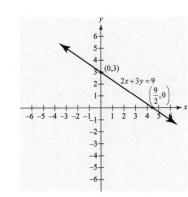

71. $m = \dfrac{5-3}{6-1} = \dfrac{2}{5}$

72. $m = \dfrac{-4-(-1)}{5-3} = \dfrac{-4+1}{5-3} = -\dfrac{3}{2}$

73. $m = \dfrac{3-(-4)}{2-(-1)} = \dfrac{3+4}{2+1} = \dfrac{7}{3}$

74. $m = \dfrac{-2-2}{6-6} = \dfrac{-4}{0}$ Undefined

75.

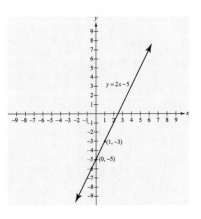

76.

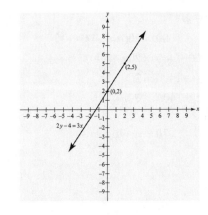

77.

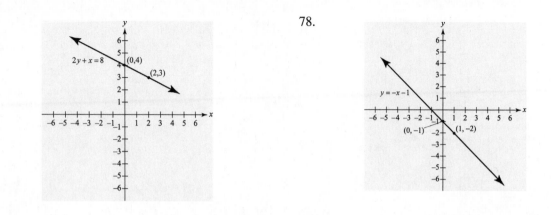

78.

79. The y-intercept is 4, thus $b = 4$. Since the graph rises from left to right, the slope is positive. The change in y is 4 units while the change in x is 2. Thus, m, the slope is $\frac{4}{2}$ or 2. The equation is $y = 2x + 4$.

80. The y-intercept is 1, thus $b = 1$. Since the graph falls from left to right, the slope is negative. The change in y is 3 units while the change in x is 3. Thus, m, the slope is $\frac{-3}{3}$ or -1. The equation is $y = -x + 1$.

81. a)

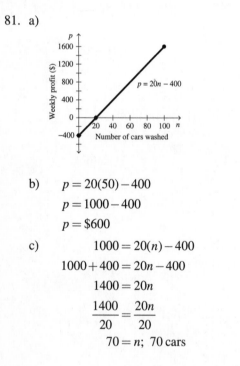

b) $p = 20(50) - 400$
     $p = 1000 - 400$
     $p = \$600$

c) $1000 = 20(n) - 400$
     $1000 + 400 = 20n - 400$
          $1400 = 20n$
     $\dfrac{1400}{20} = \dfrac{20n}{20}$
          $70 = n;\; 70$ cars

82. a)

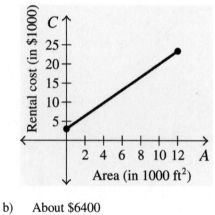

b)   About $6400

c)   About 4120 ft^2

83. Graph $2x + 3y = 12$. Since the original inequality is less than or equal to, a solid line is drawn. Since the point (0, 0) satisfies the inequality $2x + 3y \leq 12$, all points on the line and in the half-plane below the line $2x + 3y = 12$ are in the solution set.

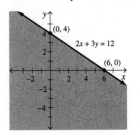

84. Graph $4x + 2y = 12$. Since the original inequality is greater than or equal to, a solid line is drawn. Since the point (0, 0) does not satisfy the inequality $4x + 2y \geq 12$, all points on the line and in the half plane above the line $4x + 2y = 12$ are in the solution set.

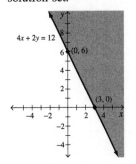

85. Graph $2x - 3y = 12$. Since the original inequality is strictly greater than, a dashed line is drawn. Since the point (0, 0) does not satisfy the inequality $2x - 3y > 12$, all points in the half-plane below the line $2x - 3y = 12$ are in the solution set.

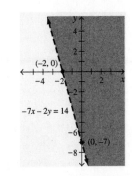

86. Graph $-7x - 2y = 14$. Since the original inequality is strictly less than, a dashed line is drawn. Since the point (0, 0) satisfies the inequality $-7x - 2y < 14$, all points in the half-plane to the right of the line $-7x - 2y = 14$ are in the solution set.

87. $x^2 + 11x + 18 = (x + 2)(x + 9)$

88. $x^2 + x - 30 = (x + 6)(x - 5)$

89. $x^2 - 10x + 24 = (x - 6)(x - 4)$

90. $x^2 - 9x + 20 = (x - 5)(x - 4)$

91. $6x^2 + 7x - 3 = (3x - 1)(2x + 3)$

92. $2x^2 + 13x - 7 = (2x - 1)(x + 7)$

93. $x^2 + 4x + 3 = 0$
    $(x + 1)(x + 3) = 0$
    $x + 1 = 0$ or $x + 3 = 0$
    $x = -1 \qquad x = -3$

94. $x^2 + 3x = 18$
    $x^2 + 3x - 18 = 0$
    $(x + 6)(x - 3) = 0$
    $x + 6 = 0$ or $x - 3 = 0$
    $x = -6 \qquad x = 3$

95. $3x^2 - 17x + 10 = 0$

$(3x-2)(x-5) = 0$

$3x - 2 = 0$ or $x - 5 = 0$

$3x = 2 \qquad x = 5$

$x = \dfrac{2}{3}$

96. $3x^2 = -7x - 2$

$3x^2 + 7x + 2 = 0$

$(x+2)(3x+1) = 0$

$x + 2 = 0$ or $3x + 1 = 0$

$x = -2 \qquad 3x = -1$

$\qquad\qquad x = -\dfrac{1}{3}$

97. $x^2 - 4x - 1 = 0$

$a = 1, \quad b = -4, \quad c = -1$

$x = \dfrac{-(-4) \pm \sqrt{(-4)^2 - 4(1)(-1)}}{2(1)}$

$x = \dfrac{4 \pm \sqrt{16+4}}{2} = \dfrac{4 \pm \sqrt{20}}{2} = \dfrac{4 \pm 2\sqrt{5}}{2} = 2 \pm \sqrt{5}$

98. $x^2 - 6x - 16 = 0$

$a = 1, b = -6, c = -6$

$x = \dfrac{-(-6) \pm \sqrt{(-6)^2 - 4(1)(-16)}}{2(1)}$

$x = \dfrac{6 \pm \sqrt{36+64}}{2} = \dfrac{6 \pm \sqrt{100}}{2} = \dfrac{6 \pm 10}{2}$

$x = \dfrac{16}{2} = 8$ or $x = \dfrac{-4}{2} = -2$

99. $2x^2 - 3x + 4 = 0$

$a = 2, b = -3, c = 4$

$x = \dfrac{-(-3) \pm \sqrt{(-3)^2 - 4(2)(4)}}{2(2)}$

$x = \dfrac{3 \pm \sqrt{9-32}}{4} = \dfrac{3 \pm \sqrt{-23}}{4}$

No real solution

100. $2x^2 - x - 3 = 0$

$a = 2, b = -1, c = -3$

$x = \dfrac{-(-1) \pm \sqrt{(-1)^2 - 4(2)(-3)}}{2(2)}$

$x = \dfrac{1 \pm \sqrt{1+24}}{4} = \dfrac{1 \pm \sqrt{25}}{4} = \dfrac{1 \pm 5}{4}$

$x = \dfrac{6}{4} = \dfrac{3}{2}$ or $x = \dfrac{-4}{4} = -1$

101. Function since each value of x is paired with a unique value of y.

D: $x = $ -2, -1, 2, 3 \qquad R: $y = $ -1, 0, 2

102. Not a function since it is possible to draw a vertical line that intersects the graph at more than one point.

103. Not a function since it is possible to draw a vertical line that intersects the graph at more than one point.

104. Function since each vertical line intersects the graph at only one point.

D: all real numbers \qquad R: all real numbers

105. $f(x) = 3x - 7, \quad x = 4$

$f(4) = 3(4) - 7 = 12 - 7 = 5$

106. $f(x) = -2x + 9, \quad x = -3$

$f(-3) = -2(-3) + 9 = 6 + 9 = 15$

107. $f(x) = 2x^2 - 3x + 4, \; x = 5$

$f(5) = 2(5)^2 - 3(5) + 4 = 50 - 15 + 4 = 39$

108. $f(x) = -4x^2 + 7x + 9, \; x = -2$

$f(-2) = -4(-2)^2 + 7(-2) + 9 = -16 - 14 + 9 = -21$

109. $y = -x^2 - 4x + 21$

    a) $a = -1 < 0$, opens downward

    b) $x = -2$ c) $(-2, 25)$ d) $(0, 21)$

    e) $(-7, 0), (3, 0)$

    f)

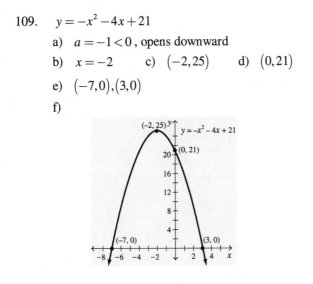

    g) D: all real numbers   R: $y \le 25$

110. $f(x) = 2x^2 + 8x + 6$

    a) $a = 2 > 0$, opens upward

    b) $x = -\dfrac{8}{2(2)} = -2$ c) $(-2, -2)$ d)

    $(0, 6)$

    e) $(-3, 0), (-1, 0)$

    f)

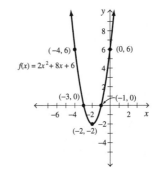

    g) D: all real numbers   R: $y \ge -2$

111.

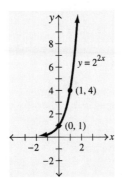

    D: all real numbers   R: $y > 0$

112.

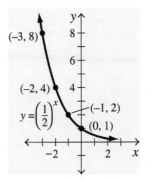

    D: all real numbers   R: $y > 0$

113. $m = 30 - 0.002n^2$, $n = 60$

    $m = 30 - 0.002(60)^2 = 30 - 0.002(3600)$

    $= 30 - 7.2 = 22.8$ mpg

114. $n = 2a^2 - 80a + 5000$

    a) $a = 18$

      $n = 2(18)^2 - 80(18) + 5000$

        $= 648 - 1440 + 5000 = 4208$

    b) $a = 25$

      $n = 2(25)^2 - 80(25) + 5000$

        $= 1250 - 2000 + 5000 = 4250$

115. $P = 100(0.92)^x$, $x = 4.5$

    $P = 100(0.92)^{4.5}$

    $= 100(0.6871399881) = 68.71399881 \approx 68.7\%$

Chapter Test

1. $4x^2 + 3x - 1, \quad x = -2$

    $4(-2)^2 + 3(-2) - 1 = 16 - 6 - 1 = 9$

2. $3x + 5 = 2(4x - 7)$

    $3x + 5 = 8x - 14$

    $3x - 8x + 5 = 8x - 8x - 14$

    $-5x + 5 = -14$

    $-5x + 5 - 5 = -14 - 5$

    $-5x = -19$

    $\dfrac{-5x}{-5} = \dfrac{-19}{-5}$

    $x = \dfrac{19}{5}$

3. $-2(x - 3) + 6x = 2x + 3(x - 4)$

    $-2x + 6 + 6x = 2x + 3x - 12$

    $4x + 6 = 5x - 12$

    $4x - 5x + 6 = 5x - 5x - 12$

    $-x + 6 = -12$

    $-x + 6 - 6 = -12 - 6$

    $-x = -18$

    $\dfrac{-x}{-1} = \dfrac{-18}{-1}$

    $x = 18$

4. Let $x =$ the number

    $3x =$ the product of the number and 3

    $3x + 5 =$ the product of the number and 3, increased by 5

    $3x + 5 = 17$

    $3x + 5 - 5 = 17 - 5$

    $3x = 12$

    $\dfrac{3x}{3} = \dfrac{12}{3}$

    $x = 4$

5. Let $x =$ Mary's weekly sales

    $0.06x =$ the amount of commission

    $350 + 0.06x = 710$

    $0.06x = 360$

    $\dfrac{0.06x}{0.06} = \dfrac{360}{0.06}$

    $x = \$6000$

6. $L = ah + bh + ch; \; a = 3, \, b = 4, \, c = 5, \, h = 7$

    $L = 2(7) + 5(7) + 4(7)$

    $= 14 + 35 + 28 = 77$

7. $3x + 5y = 11$

    $3x - 3x + 5y = -3x + 11$

    $5y = -3x + 11$

    $\dfrac{5y}{5} = \dfrac{-3x + 11}{5}$

    $y = \dfrac{-3x + 11}{5} = -\dfrac{3}{5}x + \dfrac{11}{5}$

8. $L = \dfrac{kMN}{P}$

    $12 = \dfrac{k(8)(3)}{2}$

    $24k = 24$

    $k = \dfrac{24}{24} = 1$

    $L = \dfrac{(1)MN}{P}$

    $L = \dfrac{(1)(10)(5)}{15} = \dfrac{50}{15} = 3.\overline{3} = 3\dfrac{1}{3}$

9. $l = \dfrac{k}{w}$

    $15 = \dfrac{k}{9}$

    $k = 15(9) = 135$

    $l = \dfrac{135}{w}$

    $l = \dfrac{135}{20} = 6.75 \text{ ft}$

10. $-3x + 11 \le 5x + 35$

      $-3x - 5x + 11 \le 5x - 5x + 35$

       $-8x + 11 \le 35$

      $-8x + 11 - 11 \le 35 - 11$

         $-8x \le 24$

         $\dfrac{-8x}{-8} \ge \dfrac{24}{-8}$

          $x \ge -3$

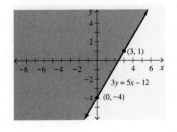

11. $m = \dfrac{14 - 8}{1 - (-2)} = \dfrac{6}{3} = 2$

12.

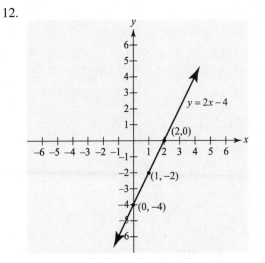

13.

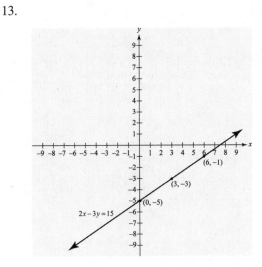

14. Graph $3y = 5x - 12$. Since the original statement is greater than or equal to, a solid line is drawn. Since the point (0, 0) satisfies the inequality $3y \ge 5x - 12$, all points on the line and in the half-plane above the line $3y = 5x - 12$ are in the solution set.

15. $x^2 + 5x = -4$

    $x^2 + 5x + 4 = 0$

    $(x+4)(x+1) = 0$

    $x + 4 = 0$ or $x + 1 = 0$

    $x = -4$ $x = -1$

16. $3x^2 + 2x = 8$

    $3x^2 + 2x - 8 = 0$

    $a = 3,\, b = 2,\, c = -8$

    $x = \dfrac{-2 \pm \sqrt{(2)^2 - 4(3)(-8)}}{2(3)}$

    $x = \dfrac{-2 \pm \sqrt{4+96}}{6} = \dfrac{-2 \pm \sqrt{100}}{6} = \dfrac{-2 \pm 10}{6}$

    $x = \dfrac{8}{6} = \dfrac{4}{3}$ or $x = \dfrac{-12}{6} = -2$

17. Function since each vertical line intersects the graph at only one point.

18. $f(x) = -3x^2 - 12x + 5,\ x = -2$

    $f(-2) = -3(-2)^2 - 12(-2) + 5$

    $\quad\quad = -12 + 24 + 5 = 17$

19. $y = x^2 - 2x + 4$

    a) $a = 1 > 0$, opens upward

    b) $x = 1$ c) $(1, 3)$ d) $(0, 4)$

    e) no x-intercepts

    f)

    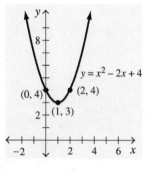

    g) D: all real numbers   R: $y \geq 3$

CHAPTER SEVEN

SYSTEMS OF LINEAR EQUATIONS AND INEQUALITIES

Exercise Set 7.1

1. Two or more linear equations form a system of linear equations.

3. An inconsistent system of equations is a system that has no solution.

5. A dependent system of equations is a system that has an infinite number of solutions.

7. The graphs of the system of equations are parallel and do not intersect.

9. The graphs of the system of equations are in fact the same line.

11. $(3, 5)$ $y = 3x - 4$ $y = -x + 8$

             $5 = 3(3) - 4$ $5 = -(3) + 8$

             $5 = 9 - 4$ $5 = 5$

               $5 = 5$

Therefore, $(3, 5)$ is a solution.

13.

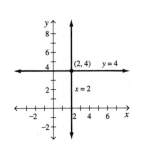

15.

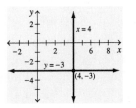

17.

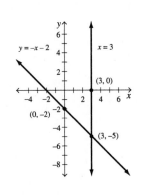

19.

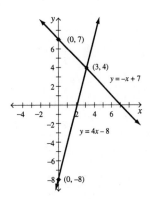

21.

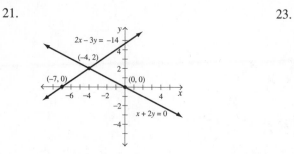

23.

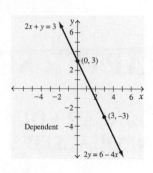

25.

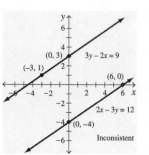

27.

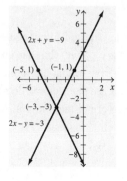

29.

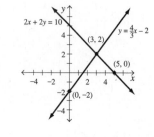

31.

33. a) Two lines with different slopes are not parallel, and therefore have exactly one point of intersection giving one solution.
   b) Two lines with the same slope and different y-intercepts are distinct parallel lines and have no solution.
   c) Two lines with the same slopes and y-intercepts have infinitely many solutions, each point on the line.

35. $3x + y = 9$ $y = -3x + 9$
   same slope, same y-intercept; infinite number of solutions

37. $3x + y = -6$ $4x - 2y = -8$
   different slopes, one solution

39. $3x + y = 7$ $y = -3x + 9$
   same slope, diff. y-intercepts; no solution

41. $2x - 3y = 6$ $x - (3/2)y = 3$

     same slopes, same y-intercepts;
     infinite number of solutions

43. $3x = 6y + 5$ $y = (1/2)x - 3$

     same slope, diff. y-intercepts; no solution

45. $4x + 7y = 2$ $4x = 6 + 7y$

     different slopes, one solution

47. $4y - 2x = 15$ $3y - 5x = 9$

     slopes are not negative reciprocals,
     not perpendicular

49. $2x + y = 3$ $2y - x = 5$

     slopes are negative reciprocals, perpendicular

51. a)

Let x equal the number of years, C equal cost

Cost (annual cost plus repair): $C_{\text{fix}} = 375x + 250$

Cost (annual cost plus

            replacement): $C_{\text{rep}} = 225x + 700$

     c) $375x + 250 = 225x + 700$

   $375x - 225x + 250 = 225x - 225x + 700$

         $150x + 250 = 700$

   $150x + 250 - 250 = 700 - 250$

            $150x = 450$

          $\dfrac{150x}{150} = \dfrac{450}{450}$

              $x = 3$

   3 years

b)

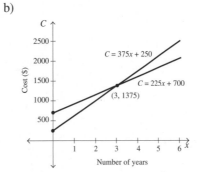

53. a)   Let C = cost , R = revenue

     $C(x) = 15x + 400$

     $R(x) = 25x$

   b)

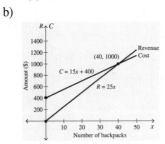

53. c)   The cost and revenue graphs intersect when
     $x = 40$ so 40 is the number of backpacks
     Benjamin's must sell to break even.

    d) $P = R(x) - C(x) = 25x - (15x + 400)$

        $P = 10x - 400$

    e) $P = 10(30) - 400 = 300 - 400 = -\$100$ (loss)

    f) $1000 = 10x - 400$ $\rightarrow$ $10x = 1400$

                          $x = 140$ BPs

55. a) Let $C(x) = $ cost , $R(x) = $ revenue
$C(x) = 230x + 8400$
$R(x) = 300x$

b)

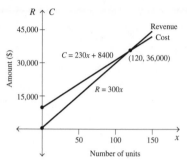

55. c) The cost and revenue graphs intersect when
$x = 120$ so 120 is the number of units the
manufacturer must sell to break even.

d) $P = R(x) - C(x) = 300x - (230x + 8400)$
$P = 70x - 8400$

e) $P = 70(100) - 8400 = 7000 - 8400$
$= -\$1400$ (loss)

f) $1260 = 70x - 8400$ → $70x = 9660$
$x = 138$ units

57. a) $s_1 = .15x + 500$
$s_2 = 650$

c) $.15x + 500 = 650$
$\underline{ -500 \quad -300}$
$.15x \qquad = 150$

$\dfrac{.15x}{.15} = \dfrac{150}{.15}$ so $x = \$1000$

b)

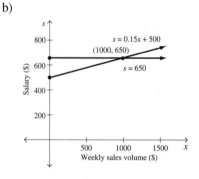

59. a) 1 point   b) 3 pts.   c) 6 pts.
d) 10 pts.
e) For n lines, the maximum number of intersections
is $\dfrac{n(n-1)}{2}$; for six lines,
there are 15 points.

Exercise Set 7.2

1. Solve one of the equations for one of the variables in terms of the other variable. Then substitute that
expression into the other equation and solve for the variable. Substitute the value found into one of the
original equations and solve for the other variable.

3. The system is dependent if the result is of the form a = a.

5. Solve one equation for the variable that is most readily manipulated, then substitute into the other equation.

$x + 3y = 3$
$\underline{ -3y \ -3y}$ → $3(3-3y) + 4y = 9$
$x = 3 - 3y$

7. $y = x + 8$

$y = -x + 4$

Substitute $(x + 8)$ in place of y in the second equation.

$x + 8 = -x + 4$ (solve for x)

$\underline{+x \qquad +x}$

$2x + 8 = 4$

$\underline{\quad -8 \ -8}$

$2x \quad = -4$

$\dfrac{2x}{2} = \dfrac{-4}{2} \qquad x = -2$

Now substitute -2 for x in an equation

$y = x + 8$

$y = (-2) + 8 = 6$

The solution is $(-2, 6)$. Consistent

9. $6x + 5y = 1$

$x - 3y = 4$ → $x = 3y + 4$

Substitute $(3y + 4)$ in place of x in the first equation.

$6(3y + 4) + 5y = 1$ (solve for y)

$18y + 24 + 5y = 1$

$23y = -23 \qquad y = -1$

Now substitute -1 for y in the 1st equation

$6x + 5(-1) = 1$

$6x = 6 \qquad x = 1$

The solution is $(1, -1)$. Consistent

11. $y - x = 4$

$x - y = 3$

Solve the first equation for y.

$y - x + x = x + 4$

$y = x + 4$

Substitute $(x + 4)$ for y in the second equation.

$x - (x + 4) = 3$ (combine like terms)

$-4 \qquad 3 \qquad$ False

Since -4 does not equal 3, there is no solution to this system. The equations are inconsistent.

13. $3y + 2x = 4$

$y = 6 - x$

Solve the second equation for x.

$3y = 6 - x$

$3y - 6 = 6 - 6 - x$

$3y - 6 = -x$

$-3y + 6 \qquad = x$

Now substitute $(-3y + 6)$ for x in the 1st eq'n.

$3y + 2(6 - 3y) = 4$ (solve for y)

$3y + 12 - 6y = 4$

$-3y = -8$ (div. by -3) $y = 8/3$

Substitute 8/3 for y in the 2nd eq'n.

$3(8/3)n = 6 - x$

$8 = 6 - x \qquad x = -2$

The solution is $(-2, 8/3)$.   Consistent

15. $y - 2x = 3$

$2y = 4x + 6$

Solve the first equation for y.

$y - 2x + 2x = 2x + 3$

$y = 2x + 3$

Now substitute $(2x + 3)$ for y in the 2nd eq'n.

$2(2x + 3) = 4x + 6$

$4x + 6 = 4x + 6$

$4x - 4x + 6 = 4x - 4x + 6$

$6 = 6$

This statement is true for all values of x.

The system is dependent.

17. $x = y + 3$

$x = -3$

Substitute -3 in place of x in the first equation.

$-3 = y + 3$

$-3 - 3 = y + 3 - 3$

$-6 = y$

The solution is $(-3, -6)$.   Consistent

19. $y + 3x - 4 = 0$
    $2x - y = 7$
    Solve the first equation for y.
    $y + 3x - 4 = 0$
    $y = 4 - 3x$
    Substitute $4 - 3x$ for y in the second eq.
    $2x - (4 - 3x) = 7$ (solve for x)
    $2x - 4 + 3x = 7$
    $5x = 11$ $x = 11/5$
    Substitute 11/5 for x in the second eq'n.
    $2(11/5) - y = 7$ (solve for y)
    $22/5 - y = 7$
    $-y = 13/5$ $y = -13/5$
    The solution is (11/5, –13/5).   Consistent

21. $x = 2y + 3$
    $y = 3x - 1$
    Substitute $(3x - 1)$ for y in the first equation.
    $x = 2(3x - 1) + 3$
    $x = 6x - 2 + 3$
    $x = 6x + 1$
    $x - 6x = 6x - 6x + 1$
    $-5x = 1$
    $\dfrac{-5x}{-5} = \dfrac{1}{-5}$ $x = -1/5$
    Substitute – 1/5 for x in the second equation.
    $y = 3(-1/5) - 1 = -3/5 - 5/5 = -8/5$
    The solution is (– 1/5, – 8/5).   Consistent

23. $y = -2x + 3$
    $4x + 2y = 12$
    Substitute $-2x + 3$ for y in the 2nd equation.
    $4x + 2(-2x + 3) = 12$
    $4x - 4x + 6 = 12$
    6   12   False
    Since 6 does not equal 12, there is no solution.
    The equations are inconsistent.

25. $3x + y = 9$
    $2x - y = 6$
    Add the equations to eliminate y.
    $5x = 15$ $x = 3$
    Substitute 3 for x in either eq'n.
    $3(3) + y = 9$ (solve for y)
    $9 + y = 9$ $y = 0$
    The solution is (3, 0)   Consistent

27. $x + y = 12$
    $x - 2y = -3$
    Multiply the 1st eq'n. by 2, then add the eq'ns.
    To eliminate y.
    $2x + 2y = 24$
    $x - 2y = -3$
    $3x = 21$ $x = 7$
    Substitute 7 for x in either eq'n.
    $(7) + y = 12$ (solve for y) $y = 5$

    The solution is (7, 5)   Consistent

29. $2x - y = -4$
    $-3x - y = 6$
    Multiply the second equation by – 1,
    $2x - y = -4$
    $3x + y = -6$ add the equations to eliminate y
    $5x = -10$ $x = -2$
    Substitute – 2 in place of x in the first equation.
    $2(-2) - y = -4$
    $-4 - y = -4$
    $-y = 0$ $y = 0$

31. $4x + 3y = -1$
    $2x - y = -13$
    Multiply the second equation by 3,
    $4x + 3y = -1$
    $6x - 3y = -39$ add the equations to eliminate y
    $10x = -40$ $x = -4$
    Substitute – 4 for x in the 2nd equation.
    $2(-4) - y = -13$
    $-8 - y = -13$ $y = 5$
    The solution is (– 4, 5).   Consistent

33. $2x + y = 11$
    $x + 3y = 18$
    Multiply the second equation by – 2,
    $2x + y = 11$
    $-2x - 6y = -36$ add the equations to elim. x
    $-5y = -25$ $y = 5$
    Substitute 5 for y in the 2nd equation.
    $x + 3(5) = 18$
    $x + 15 = 18$ $x = 3$
    The solution is (3, 5).

35. $3x - 4y = 11$
$3x + 5y = -7$
Multiply the first equation by (-1),
$-3x + 4y = -11$
$3x + 5y = -7$ add the equations to elim. x
$9y = -18$ $y = -2$
Substitute -2 for y in the first equation.
$3x - 4(-2) = 11$
$3x = 3$ $x = 1$
The solution is $(1, -2)$.   Consistent

37. $4x + y = 6$
$-8x - 2y = 13$
Multiply the first equation by 2,
$8x + 2y = 12$
$-8x - 2y = 13$ add the equations to elim. y
0 25 False
Since this statement is not true for any values
of x and y, the equations are inconsistent.

39. $3x - 4y = 2$
$4x + 3y = 11$
Multiply the first equation by 3, and the
second equation by 4,
$9x - 12y = 6$
$16x + 12y = 44$ add the equations to elim. y
$25x = 50$ $x = 2$
Substitute 2 for x in the second equation.
$4(2) + 3y = 11$
$8 + 3y = 11$
$3y = 3$ $y = 1$
The solution is $(2, 1)$.   Consistent

41. $S_1 = .15p + 12000$
$S_2 = .05p + 27000$

$.15p + 12000 = .05p + 27000$
$-.05p - 12000$ $-.05p - 12000$
$.10p = 15000$
$p = \$150,000.00$

43. Let x = # of medium pizzas
$50 - x$ = # of large pizzas

$10.95x + 14.95(50-x) = 663.50$
$10.95x + 747.50 - 14.95x = 663.50$
$-4.00x = -84.00$ $x = 21$
Substitute 21 for x in 2nd let statement
$50 - x = 50 - (21) = 29$

21 medium pizzas and 29 large pizzas

45. Let x = # of liters at 25%
$10 - x$ = # of liters at 50%

$.25x + .50(10 - x) = .40(10)$
$.25x + 5 - .50x = 4$
$-.25x = -1$ $x = 4$
Substitute 4 for x in 2nd let statement
$10 - x = 10 - (4) = 6$

4 liters of 25% solution and
6 liters of 50% solution

47. Let c = monthly cost
x = number of copies

Eco. Sales: $c = 18 + 0.02x$
Office Sup.: $c = 24 + 0.015x$ set eq'ns.
equal
$18 + 0.02x = 24 + 0.015x$
$0.005x = 6$ $x = 1200$

1200 copies per month

49. Let x = no. of pounds of nuts
y = no. of pounds of pretzels
$x + y = 20$ $y = -x + 20$
$3x + 1y = 30$
Substitute $(20 - x)$ for y in the 2nd equation.
$3x + (20 - x) = 30$
$3x + 20 - x = 30$
$2x = 10$ $x = 5$ Solve for y
$y = 20 - 5 = 15$

51. Let x = no. of inside tickets
         y = no. of lawn tickets
     x + y = 4600   x = 4600 - y
     27x + 14y = 104,700
     Substitute (4600 - y) for x in the 2nd
      equation.
     27(4600 - y) + 14y = 104,700
     124,200 – 27y + 14y = 104,700
     -13y = -19,500   y = 1500
     Substitute 1500 for y in the 1st eq'n.
     x + (1500) = 4600   x = 3100

     3100 inside tickets and 1500 lawn tickets

53. y_{GM} = -0.83x + 28.33
     y_T = 0.89x + 9.11

     -0.83x + 28.33 = 0.89x + 9.11
     <u>0.83x   -9.11   0.83x - 9.11</u>
         19.22 = 1.72x
     $\dfrac{19.22}{1.72} = \dfrac{1.72x}{1.72}$

         $x = 11.17... \approx 11.2$
     11.2 years after 2000, or in 2011

55. (1/u) + (2/v) = 8
     (3/u) – (1/v) =3
     Substitute x for $\dfrac{1}{u}$ and y for $\dfrac{1}{v}$.
     (1) x + 2y = 8
     (2) 3x – y = 3
     Multiply eq'n. (2) by 2,
     x + 2y = 8
     6x – 2y = 6   add to eliminate y
     7x = 14   x= 2, thus u = ½
     Substitute 2 for x in eq. (1).
     (2) + 2y = 8
     2y = 6   y = 3, thus v = 1/3

     Answer: (1/2, 1/3)

57. a) (2) + (1) + (4) = 7   (2) – (1) + 2(4) = 9
            7 = 7   9 = 9
        -(2) + 2(1) + (4) = 4
            4 = 4   (2,1,4) is a solution.
    b) Add eq'ns. 1 and 2 to yield eq'n. 4
        Multiply eq'n. 2 by 2, then add eq'ns. 2 and 3
        to yield eq'n. 5
        Combine eq'ns. 4 and 5 to find one variable.
        Substitute back into various equations to find
        the other 2 variables.

59. y = 3x + 3
     (1/3)y = x + 1
     If we multiply the 2nd eq'n. by 3, we get the eq'n.
     y = 3x + 1, the same as eq'n. # 1.
     2 lines that lie on top of one another have an
     infinite number of solutions.

61. a) (0, 0)   b) (1, 0)   c) (0, 1)   d) (1, 1)

Exercise Set 7.3

1. A matrix is a rectangular array of elements.
3. A square matrix contains the same number of rows as columns.
5. A 4 x 3 matrix has 4 rows.
7. a) Add numbers in the same positions to produce an entry in that position.

    b) $\begin{bmatrix} 5 & 4 & -1 \\ 3 & 9 & 5 \end{bmatrix} + \begin{bmatrix} 2 & 5 & -6 \\ -1 & 7 & 4 \end{bmatrix} = \begin{bmatrix} 5+2 & 4+5 & -1+(-6) \\ 3+(-1) & 9+7 & 5+4 \end{bmatrix} = \begin{bmatrix} 7 & 9 & -7 \\ 2 & 16 & 9 \end{bmatrix}$

9. a) The number of columns of the first matrix must be the same as the number of rows of the second matrix.

b) The dimensions of the resulting matrix will have the same number of rows as the first matrix and the same number of columns as the second matrix. The product of a 2 x 2 with a 2 x 3 matrix will yield a 2 x 3 matrix.

11. a) Identity matrix for 2x2 $\begin{bmatrix} 1 & 0 \\ 0 & 1 \end{bmatrix}$ b) Identity matrix for 3x3 $\begin{bmatrix} 1 & 0 & 0 \\ 0 & 1 & 0 \\ 0 & 0 & 1 \end{bmatrix}$

13. $A = \begin{bmatrix} 1 & 8 \\ 2 & 7 \end{bmatrix}$ $B = \begin{bmatrix} -4 & 1 \\ 7 & 2 \end{bmatrix}$ $A + B = \begin{bmatrix} 1+(-4) & 8+1 \\ 2+7 & 7+2 \end{bmatrix} = \begin{bmatrix} -3 & 9 \\ 9 & 9 \end{bmatrix}$

15. $A + B = \begin{bmatrix} 2 & 1 \\ -1 & 4 \\ 6 & 0 \end{bmatrix} + \begin{bmatrix} -3 & 3 \\ -4 & 0 \\ 1 & 6 \end{bmatrix} = \begin{bmatrix} 2+(-3) & 1+3 \\ -1+(-4) & 4+0 \\ 6+1 & 0+6 \end{bmatrix} = \begin{bmatrix} -1 & 4 \\ -5 & 4 \\ 7 & 6 \end{bmatrix}$

17. $A - B = \begin{bmatrix} 4 & -2 \\ -3 & 5 \end{bmatrix} - \begin{bmatrix} -2 & 5 \\ 9 & 1 \end{bmatrix} = \begin{bmatrix} 4-(-2) & -2-(-5) \\ -3-(9) & 5-1 \end{bmatrix} = \begin{bmatrix} 6 & -7 \\ -12 & 4 \end{bmatrix}$

19. $A - B = \begin{bmatrix} -5 & 1 \\ 8 & 6 \\ 1 & -5 \end{bmatrix} - \begin{bmatrix} -6 & -8 \\ -10 & -11 \\ 3 & -7 \end{bmatrix} = \begin{bmatrix} -5+6 & 1+8 \\ 8+10 & 6+11 \\ 1-3 & -5+7 \end{bmatrix} = \begin{bmatrix} 1 & 9 \\ 18 & 17 \\ -2 & 2 \end{bmatrix}$

21. $2B = 2\begin{bmatrix} 3 & 2 \\ 5 & 0 \end{bmatrix} = \begin{bmatrix} 2(3) & 2(2) \\ 2(5) & 2(0) \end{bmatrix} = \begin{bmatrix} 6 & 4 \\ 10 & 0 \end{bmatrix}$

23. $2B + 4C = 2\begin{bmatrix} 3 & 2 \\ 5 & 0 \end{bmatrix} + 4\begin{bmatrix} -2 & 3 \\ 4 & 0 \end{bmatrix} = \begin{bmatrix} 6 & 4 \\ 10 & 0 \end{bmatrix} + \begin{bmatrix} -8 & 12 \\ 16 & 0 \end{bmatrix} = \begin{bmatrix} 6-8 & 4+12 \\ 10+16 & 0+0 \end{bmatrix} = \begin{bmatrix} -2 & 16 \\ 26 & 0 \end{bmatrix}$

25. $4B - 2C = 4\begin{bmatrix} 3 & 2 \\ 5 & 0 \end{bmatrix} - 2\begin{bmatrix} -2 & 3 \\ 4 & 0 \end{bmatrix} = \begin{bmatrix} 12 & 8 \\ 20 & 0 \end{bmatrix} - \begin{bmatrix} -4 & 6 \\ 8 & 0 \end{bmatrix} = \begin{bmatrix} 12+4 & 8-6 \\ 20-8 & 0-0 \end{bmatrix} = \begin{bmatrix} 16 & 2 \\ 12 & 0 \end{bmatrix}$

27. $A \times B = \begin{bmatrix} 1 & 3 \\ 0 & 6 \end{bmatrix}\begin{bmatrix} 2 & 6 \\ 8 & 4 \end{bmatrix} = \begin{bmatrix} 1(2)+3(8) & 1(6)+3(4) \\ 0(2)+6(8) & 0(6)+6(4) \end{bmatrix} = \begin{bmatrix} 26 & 18 \\ 48 & 24 \end{bmatrix}$

29. $A \times B = \begin{bmatrix} 2 & 3 & -1 \\ 0 & 4 & 6 \end{bmatrix}\begin{bmatrix} 2 \\ 4 \\ 1 \end{bmatrix} = \begin{bmatrix} 2(2)+3(4)-1(1) \\ 0(2)+4(4)+6(1) \end{bmatrix} = \begin{bmatrix} 15 \\ 22 \end{bmatrix}$

31. $A \times B = \begin{bmatrix} 4 & 7 & 6 \\ -2 & 3 & 1 \\ 5 & 1 & 2 \end{bmatrix}\begin{bmatrix} 1 & 0 & 0 \\ 0 & 1 & 0 \\ 0 & 0 & 1 \end{bmatrix} = \begin{bmatrix} 4+0+0 & 0+7+0 & 0+0+6 \\ -2+0+0 & 0+3+0 & 0+0+1 \\ 5+0+0 & 0+1+0 & 0+0+2 \end{bmatrix} = \begin{bmatrix} 4 & 7 & 6 \\ -2 & 3 & 1 \\ 5 & 1 & 2 \end{bmatrix}$

33. $A + B = \begin{bmatrix} 2 & 3 & 5 \\ 4 & 0 & 3 \end{bmatrix} + \begin{bmatrix} 7 & -2 & 3 \\ 2 & -1 & 1 \end{bmatrix} = \begin{bmatrix} 2+7 & 3+(-2) & 5+3 \\ 4+2 & 0+(-1) & 3+1 \end{bmatrix} = \begin{bmatrix} 9 & 1 & 8 \\ 6 & -1 & 4 \end{bmatrix}$

$A \times B = \begin{bmatrix} 2 & 3 & 5 \\ 4 & 0 & 3 \end{bmatrix} \times \begin{bmatrix} 7 & -2 & 3 \\ 2 & -1 & 1 \end{bmatrix}$

Operation cannot be performed because number of columns of A is not equal to number of rows of B.

35. Matrices A and B cannot be added because they do not have the same dimensions.

$A \times B = \begin{bmatrix} 4 & 5 & 3 \\ 6 & 2 & 1 \end{bmatrix} \times \begin{bmatrix} 3 & 2 \\ 4 & 6 \\ -2 & 0 \end{bmatrix} = \begin{bmatrix} 4(3)+5(4)+3(-2) & 4(2)+5(6)+3(0) \\ 6(3)+2(4)+1(-2) & 6(2)+2(6)+1(0) \end{bmatrix} = \begin{bmatrix} 26 & 38 \\ 24 & 24 \end{bmatrix}$

37. A and B cannot be added because they do not have the same dimensions.

$A \times B = \begin{bmatrix} 1 & 2 \\ 3 & 4 \end{bmatrix} \begin{bmatrix} -3 \\ 2 \end{bmatrix} = \begin{bmatrix} 1(-3)+2(2) \\ 3(-3)+4(2) \end{bmatrix} = \begin{bmatrix} 1 \\ -1 \end{bmatrix}$

39. $A + B = \begin{bmatrix} 3 & 5 \\ -2 & -3 \end{bmatrix} + \begin{bmatrix} 4 & 5 \\ 6 & 7 \end{bmatrix} = \begin{bmatrix} 3+4 & 5+5 \\ -2+6 & -3+7 \end{bmatrix} = \begin{bmatrix} 7 & 10 \\ 4 & 4 \end{bmatrix}$

$B + A = \begin{bmatrix} 4 & 5 \\ 6 & 7 \end{bmatrix} + \begin{bmatrix} 3 & 5 \\ -2 & -3 \end{bmatrix} = \begin{bmatrix} 4+3 & 5+5 \\ 6+(-2) & 7+(-3) \end{bmatrix} = \begin{bmatrix} 7 & 10 \\ 4 & 4 \end{bmatrix}$ Thus $A + B = B + A$.

41. $A + B = \begin{bmatrix} 0 & -1 \\ 3 & -4 \end{bmatrix} + \begin{bmatrix} 8 & 1 \\ 3 & -4 \end{bmatrix} = \begin{bmatrix} 0+8 & -1+1 \\ 3+3 & -4+(-4) \end{bmatrix} = \begin{bmatrix} 8 & 0 \\ 6 & -8 \end{bmatrix}$

$B + A = \begin{bmatrix} 8 & 1 \\ 3 & -4 \end{bmatrix} + \begin{bmatrix} 0 & -1 \\ 3 & -4 \end{bmatrix} = \begin{bmatrix} 8+0 & 1+(-1) \\ 3+3 & -4+(-4) \end{bmatrix} = \begin{bmatrix} 8 & 0 \\ 6 & -8 \end{bmatrix}$ Thus $A + B = B + A$.

43. $(A + B) + C = \left(\begin{bmatrix} 5 & 2 \\ 3 & 6 \end{bmatrix} + \begin{bmatrix} 3 & 4 \\ -2 & 7 \end{bmatrix} \right) + \begin{bmatrix} -1 & 4 \\ 5 & 0 \end{bmatrix} = \begin{bmatrix} 8 & 6 \\ 1 & 13 \end{bmatrix} + \begin{bmatrix} -1 & 4 \\ 5 & 0 \end{bmatrix} = \begin{bmatrix} 7 & 10 \\ 6 & 13 \end{bmatrix}$

$A + (B + C) = \begin{bmatrix} 5 & 2 \\ 3 & 6 \end{bmatrix} + \left(\begin{bmatrix} 3 & 4 \\ -2 & 7 \end{bmatrix} + \begin{bmatrix} -1 & 4 \\ 5 & 0 \end{bmatrix} \right) = \begin{bmatrix} 5 & 2 \\ 3 & 6 \end{bmatrix} + \begin{bmatrix} 2 & 8 \\ 3 & 7 \end{bmatrix} = \begin{bmatrix} 7 & 10 \\ 6 & 13 \end{bmatrix}$

Thus, $(A + B) + C = A + (B + C)$.

45. $(A + B) + C = \left(\begin{bmatrix} 7 & 4 \\ 9 & -36 \end{bmatrix} + \begin{bmatrix} 5 & 6 \\ -1 & -4 \end{bmatrix} \right) + \begin{bmatrix} -7 & -5 \\ -1 & 3 \end{bmatrix} = \begin{bmatrix} 12 & 10 \\ 8 & -40 \end{bmatrix} + \begin{bmatrix} -7 & -5 \\ -1 & 3 \end{bmatrix} = \begin{bmatrix} 5 & 5 \\ 7 & -37 \end{bmatrix}$

$A + (B + C) = \begin{bmatrix} 7 & 4 \\ 9 & -36 \end{bmatrix} + \left(\begin{bmatrix} 5 & 6 \\ -1 & -4 \end{bmatrix} + \begin{bmatrix} -7 & -5 \\ -1 & 3 \end{bmatrix} \right) = \begin{bmatrix} 7 & 4 \\ 9 & -36 \end{bmatrix} + \begin{bmatrix} -2 & 1 \\ -2 & -1 \end{bmatrix} = \begin{bmatrix} 5 & 5 \\ 7 & -37 \end{bmatrix}$

Thus, $(A + B) + C = A + (B + C)$.

47. $A \times B = \begin{bmatrix} 1 & -2 \\ 4 & -3 \end{bmatrix}\begin{bmatrix} -1 & -3 \\ 2 & 4 \end{bmatrix} = \begin{bmatrix} 1(-1)-2(2) & 1(-3)-2(4) \\ 4(-1)+(-3)(2) & 4(-3)+(-3)(4) \end{bmatrix} = \begin{bmatrix} -5 & -11 \\ -10 & -24 \end{bmatrix}$

$B \times A = \begin{bmatrix} -1 & -3 \\ 2 & 4 \end{bmatrix}\begin{bmatrix} 1 & -2 \\ 4 & -3 \end{bmatrix} = \begin{bmatrix} -1(1)+(-3)4 & -1(-2)+(-3)(-3) \\ 2(1)+4(4) & 2(-2)+4(-3) \end{bmatrix} = \begin{bmatrix} -13 & 11 \\ 18 & -16 \end{bmatrix}$

Thus, $A \times B \neq B \times A$.

49. $A \times B = \begin{bmatrix} 4 & 2 \\ 1 & -3 \end{bmatrix}\begin{bmatrix} 2 & 4 \\ -3 & 1 \end{bmatrix} = \begin{bmatrix} 4(2)+2(-3) & 4(4)+2(1) \\ 1(2)+(-3)(-3) & 1(4)+(-3)(1) \end{bmatrix} = \begin{bmatrix} 2 & 18 \\ 11 & 1 \end{bmatrix}$

$B \times A = \begin{bmatrix} 2 & 4 \\ -3 & 1 \end{bmatrix}\begin{bmatrix} 4 & 2 \\ 1 & -3 \end{bmatrix} = \begin{bmatrix} 2(4)+4(1) & 2(2)+4(-3) \\ -3(4)+1(1) & -3(2)+1(-3) \end{bmatrix} = \begin{bmatrix} 12 & -8 \\ -11 & -9 \end{bmatrix}$ Thus, $A \times B \neq B \times A$.

51. Since $B = I$, (the identity matrix), and $A \times I = I \times A = A$, we can conclude that $A \times B = B \times A$.

53. $(A \times B) \times C = \left(\begin{bmatrix} 1 & 3 \\ 4 & 0 \end{bmatrix}\begin{bmatrix} 4 & 2 \\ 3 & 1 \end{bmatrix}\right)\begin{bmatrix} 2 & 1 \\ 3 & 0 \end{bmatrix} = \begin{bmatrix} 13 & 5 \\ 16 & 8 \end{bmatrix}\begin{bmatrix} 2 & 1 \\ 3 & 0 \end{bmatrix} = \begin{bmatrix} 41 & 13 \\ 56 & 16 \end{bmatrix}$

$A \times (B \times C) = \begin{bmatrix} 1 & 3 \\ 4 & 0 \end{bmatrix}\left(\begin{bmatrix} 4 & 2 \\ 3 & 1 \end{bmatrix}\begin{bmatrix} 2 & 1 \\ 3 & 0 \end{bmatrix}\right) = \begin{bmatrix} 1 & 3 \\ 4 & 0 \end{bmatrix}\begin{bmatrix} 14 & 4 \\ 9 & 3 \end{bmatrix} = \begin{bmatrix} 41 & 13 \\ 56 & 16 \end{bmatrix}$ Thus, $(A \times B) \times C = A \times (B \times C)$.

55. $(A \times B) \times C = \left(\begin{bmatrix} 4 & 3 \\ -6 & 2 \end{bmatrix}\begin{bmatrix} 1 & 2 \\ 0 & 1 \end{bmatrix}\right)\begin{bmatrix} 4 & 3 \\ 0 & -2 \end{bmatrix} = \begin{bmatrix} 4 & 11 \\ -6 & -10 \end{bmatrix}\begin{bmatrix} 4 & 3 \\ 0 & -2 \end{bmatrix} = \begin{bmatrix} 16 & -10 \\ -24 & 2 \end{bmatrix}$

$A \times (B \times C) = \begin{bmatrix} 4 & 3 \\ -6 & 2 \end{bmatrix}\left(\begin{bmatrix} 1 & 2 \\ 0 & 1 \end{bmatrix}\begin{bmatrix} 4 & 3 \\ 0 & -2 \end{bmatrix}\right) = \begin{bmatrix} 4 & 3 \\ -6 & 2 \end{bmatrix}\begin{bmatrix} 4 & -1 \\ 0 & -2 \end{bmatrix} = \begin{bmatrix} 16 & -10 \\ -24 & 2 \end{bmatrix}$

Thus, $(A \times B) \times C = A \times (B \times C)$.

57. $(A \times B) \times C = \left(\begin{bmatrix} 3 & 4 \\ -1 & -2 \end{bmatrix}\begin{bmatrix} 0 & 1 \\ 1 & 0 \end{bmatrix}\right)\begin{bmatrix} 2 & 0 \\ 3 & 0 \end{bmatrix} = \begin{bmatrix} 4 & 3 \\ -2 & -1 \end{bmatrix}\begin{bmatrix} 2 & 0 \\ 3 & 0 \end{bmatrix} = \begin{bmatrix} 17 & 0 \\ -7 & 0 \end{bmatrix}$

$A \times (B \times C) = \begin{bmatrix} 3 & 4 \\ -1 & -2 \end{bmatrix}\left(\begin{bmatrix} 0 & 1 \\ 1 & 0 \end{bmatrix}\begin{bmatrix} 2 & 0 \\ 3 & 0 \end{bmatrix}\right) = \begin{bmatrix} 3 & 4 \\ -1 & -2 \end{bmatrix}\begin{bmatrix} 3 & 0 \\ 2 & 0 \end{bmatrix} = \begin{bmatrix} 17 & 0 \\ -7 & 0 \end{bmatrix}$

Thus, $(A \times B) \times C = A \times (B \times C)$.

59. $A + B = \begin{bmatrix} 40 & 22 & 31 \\ 38 & 25 & 34 \end{bmatrix} + \begin{bmatrix} 48 & 36 & 39 \\ 40 & 29 & 37 \end{bmatrix}$

$= \begin{bmatrix} 40+48 & 22+36 & 31+39 \\ 38+40 & 25+29 & 34+37 \end{bmatrix} = \begin{bmatrix} 88 & 58 & 70 \\ 78 & 54 & 71 \end{bmatrix}$

$\qquad\qquad\quad$ T $\quad$ O $\quad$ C

Total weight: $\begin{bmatrix} 88 & 58 & 70 \\ 78 & 54 & 71 \end{bmatrix}$ Chase's $\quad$ Gro-More

61. $A \times B = \begin{bmatrix} 7 & 8.5 & 10 \\ 7.5 & 8 & 11 \end{bmatrix} \times \begin{bmatrix} 3 \\ 2 \\ 1.5 \end{bmatrix} = \begin{bmatrix} 7(3)+8.5(2)+10(1.5) \\ 7.5(3)+8(2)+11(1.5) \end{bmatrix} = \begin{bmatrix} 53 \\ 55 \end{bmatrix}$

Total cost: $\begin{bmatrix} 53 \\ 55 \end{bmatrix}$ Java's / Spot

63. $A \times B = \begin{bmatrix} 2 & 2 & .5 & 1 \\ 3 & 2 & 1 & 2 \\ 0 & 1 & 0 & 3 \\ .5 & 1 & 0 & 0 \end{bmatrix} \begin{bmatrix} 10 & 12 \\ 5 & 8 \\ 8 & 8 \\ 4 & 6 \end{bmatrix} = \begin{bmatrix} 2 \cdot 10+2 \cdot 5+.5 \cdot 8+1 \cdot 4 & 2 \cdot 12+2 \cdot 8+.5 \cdot 8+1 \cdot 6 \\ 3 \cdot 10+2 \cdot 5+1 \cdot 8+2 \cdot 4 & 3 \cdot 12+2 \cdot 8+1 \cdot 8+2 \cdot 6 \\ 0 \cdot 10+1 \cdot 5+0 \cdot 8+3 \cdot 4 & 0 \cdot 12+1 \cdot 8+0 \cdot 8+3 \cdot 6 \\ .5 \cdot 10+1 \cdot 5+0 \cdot 8+0 \cdot 4 & .5 \cdot 12+1 \cdot 8+0 \cdot 8+0 \cdot 6 \end{bmatrix} = \begin{bmatrix} 38 & 50 \\ 56 & 72 \\ 17 & 26 \\ 10 & 14 \end{bmatrix}$

65. $C(A \times B) = [40 \ 30 \ 12 \ 20] \begin{bmatrix} 38 & 50 \\ 56 & 72 \\ 17 & 26 \\ 10 & 14 \end{bmatrix} = [36.04 \ 47.52]$ cents   small $36.04, large $47.52

67. a) A must have 3 rows and B must have one column.

b) $A = \begin{bmatrix} 1 & -2 \\ 0 & 1 \\ 1 & 3 \end{bmatrix}$ $B = \begin{bmatrix} 6 \\ 7 \end{bmatrix}$ $A \times B = \begin{bmatrix} -8 \\ 7 \\ 27 \end{bmatrix}$

69. $A \times B = \begin{bmatrix} 5 & -2 \\ -2 & 1 \end{bmatrix} \begin{bmatrix} 1 & 2 \\ 2 & 5 \end{bmatrix} = \begin{bmatrix} 5(1)-2(2) & 5(2)-2(5) \\ -2(1)+1(2) & -2(2)+1(5) \end{bmatrix} = \begin{bmatrix} 1 & 0 \\ 0 & 1 \end{bmatrix}$

$B \times A = \begin{bmatrix} 1 & 2 \\ 2 & 5 \end{bmatrix} \begin{bmatrix} 5 & -2 \\ -2 & 1 \end{bmatrix} = \begin{bmatrix} 1(5)+2(-2) & 1(-2)+2(1) \\ 2(5)+5(-2) & 2(-2)+5(1) \end{bmatrix} = \begin{bmatrix} 1 & 0 \\ 0 & 1 \end{bmatrix}$

Thus, A and B are multiplicative inverses.

71. False. Let $A = [1 \ 3]$ and $B = [2 \ 1]$. Then $A - B = [-1, 2]$ and $B - A = [1, -2]$ $A - B \neq B - A$.

73. a) $1.4(14) + 0.7(10) + 0.3(7) = \$28.70$
b) $2.7(12) + 2.8(9) + 0.5(5) = \$60.10$

c) $L \times C = \begin{bmatrix} 28.7 & 24.6 \\ 41.3 & 35.7 \\ 69.3 & 60.1 \end{bmatrix}$ small / medium / large (Ames Bay)

This array shows the total cost of each sofa at each plant.

75. A + B cannot be calculated because the # of columns of A ≠ # of rows of B and the # of rows of A ≠ # of rows of B.

$A \times B = \begin{bmatrix} 1 & 2 & 3 \\ 3 & 2 & 1 \end{bmatrix} \begin{bmatrix} 0 & 4 \\ 1 & 5 \\ 2 & 1 \end{bmatrix} = \begin{bmatrix} 1(0)+2(1)+3(2) & 1(4)+2(5)+3(1) \\ 3(0)+2(1)+1(2) & 3(4)+2(5)+1(1) \end{bmatrix} = \begin{bmatrix} 8 & 17 \\ 4 & 23 \end{bmatrix}$

Exercise Set 7.4

1. a) An augmented matrix is a matrix formed with the coefficients of the variables and the constants. The coefficients are separated from the constants by a vertical bar.

   b) $\begin{bmatrix} 1 & 3 & | & 7 \\ 2 & -1 & | & 4 \end{bmatrix}$

3. If you obtain an augmented matrix in which a 0 appears across an entire row, the system of equations is dependent.

5. 1) Multiply the 2nd row by -1/2; 2) add -3 times the 2nd row to the first row; and 3) identify the values of x and y.

1) $\begin{bmatrix} 1 & 3 & | & 5 \\ 0 & 1 & | & (-1/2) \end{bmatrix}$ 2) $\begin{bmatrix} 1+0 & 3+(-3) & | & 5+(3/2) \\ 0 & 1 & | & (-1/2) \end{bmatrix} = \begin{bmatrix} 1 & 0 & | & 13/2 \\ 0 & 1 & | & -1/2 \end{bmatrix}$ 3) $(x, y) = \left(\dfrac{13}{2}, \dfrac{-1}{2} \right)$

7. $x + 3y = 7 \qquad -x + y = 1$

$\begin{bmatrix} 1 & 3 & | & 7 \\ -1 & 1 & | & 1 \end{bmatrix} \rightarrow \begin{bmatrix} 1 & 3 & | & 7 \\ -1+1 & 1+3 & | & 7+1 \end{bmatrix} = \begin{bmatrix} 1 & 3 & | & 7 \\ 0 & 4 & | & 8 \end{bmatrix} \rightarrow$

$\begin{bmatrix} 1 & 3 & | & 7 \\ 0 & 1 & | & 2 \end{bmatrix} \rightarrow \begin{bmatrix} 1+0 & 3+(-3) & | & 7+(-6) \\ 0 & 1 & | & 2 \end{bmatrix} = \begin{bmatrix} 1 & 0 & | & 1 \\ 0 & 1 & | & 2 \end{bmatrix} \rightarrow (1, 2)$

9. $x - 2y = -1 \qquad 2x + y = 8$

$\begin{bmatrix} 1 & -2 & | & -1 \\ 2 & 1 & | & 8 \end{bmatrix} \rightarrow \begin{bmatrix} 1 & -2 & | & -1 \\ 2-2 & 1+4 & | & 8+2 \end{bmatrix} = \begin{bmatrix} 1 & -2 & | & -1 \\ 0 & 5 & | & 10 \end{bmatrix} \rightarrow \begin{bmatrix} 1+0 & -2+2 & | & -1+4 \\ 0 & 5 & | & 10 \end{bmatrix} \rightarrow$

$\begin{bmatrix} 1 & 0 & | & 3 \\ 0 & 5 & | & 10 \end{bmatrix} = \begin{bmatrix} 1 & 0 & | & 3 \\ 0 & 1 & | & 2 \end{bmatrix} \rightarrow (3, 2)$

11. $\begin{bmatrix} 2 & -5 & | & -6 \\ -4 & 10 & | & 12 \end{bmatrix} (r_2 + 2r_1) \begin{matrix} = \\ \\ \end{matrix} \begin{bmatrix} 2 & -5 & | & -6 \\ 0 & 0 & | & 0 \end{bmatrix} \Rightarrow$ Dependent system

   The solution is all points on the line $2x - 5y = -6$.

13. $\begin{bmatrix} 2 & -3 & | & 10 \\ 2 & 2 & | & 5 \end{bmatrix} \begin{matrix} (r_1 \div 2) \\ (r_2 - 2r_1) \end{matrix} \begin{bmatrix} 1 & -\frac{3}{2} & | & 5 \\ 0 & 5 & | & -5 \end{bmatrix} \begin{matrix} = \\ (r_2 \div (5)) \end{matrix} \begin{bmatrix} 1 & -\frac{3}{2} & | & 5 \\ 0 & 1 & | & -1 \end{bmatrix} \begin{matrix} (r_1 + \frac{3}{2} r_2) \\ = \end{matrix} \begin{bmatrix} 1 & 0 & | & \frac{7}{2} \\ 0 & 1 & | & -1 \end{bmatrix}$ The solution is $(7/2, -1)$.

15. $\begin{bmatrix} 4 & 2 & | & 6 \\ 5 & 4 & | & 9 \end{bmatrix} \begin{matrix} (r_1 \div 4) \\ = \end{matrix} \begin{bmatrix} 1 & \frac{1}{2} & | & \frac{3}{2} \\ 5 & 4 & | & 9 \end{bmatrix} \begin{matrix} = \\ (r_2 - 5r_1) \end{matrix} \begin{bmatrix} 1 & \frac{1}{2} & | & \frac{3}{2} \\ 0 & \frac{3}{2} & | & \frac{3}{2} \end{bmatrix} \begin{matrix} = \\ (\frac{2}{3} r_2) \end{matrix} \begin{bmatrix} 1 & \frac{1}{2} & | & \frac{3}{2} \\ 0 & 1 & | & 1 \end{bmatrix} \begin{matrix} (r_1 - \frac{1}{2} r_2) \\ = \end{matrix} \begin{bmatrix} 1 & 0 & | & 1 \\ 0 & 1 & | & 1 \end{bmatrix}$

   The solution is $(1, 1)$.

17. $\begin{bmatrix} -3 & 6 & | & 5 \\ 2 & -4 & | & 8 \end{bmatrix} \begin{matrix} (r_1 \div (-3)) \\ = \end{matrix} \begin{bmatrix} 1 & -2 & | & \frac{-5}{3} \\ 2 & -4 & | & 8 \end{bmatrix} \begin{matrix} = \\ (r_2 - 2r_1) \end{matrix} \begin{bmatrix} 1 & -2 & | & \frac{-5}{3} \\ 0 & 0 & | & \frac{34}{3} \end{bmatrix} \Rightarrow$ Inconsistent system   No solution.

19. $\begin{bmatrix} 3 & 1 & | & 13 \\ 1 & 3 & | & 15 \end{bmatrix} \begin{matrix} (r_1 \div 3) \\ = \end{matrix} \begin{bmatrix} 1 & \frac{1}{3} & | & \frac{13}{3} \\ 1 & 3 & | & 15 \end{bmatrix} \begin{matrix} \\ (r_2 - r_1) \end{matrix} = \begin{bmatrix} 1 & \frac{1}{3} & | & \frac{13}{3} \\ 0 & \frac{8}{3} & | & \frac{32}{3} \end{bmatrix} \begin{matrix} \\ (\frac{3}{8}r_2) \end{matrix} \begin{bmatrix} 1 & \frac{1}{3} & | & \frac{13}{3} \\ 0 & 1 & | & 4 \end{bmatrix} \begin{matrix} (r_1 - \frac{1}{3}r_2) \\ = \end{matrix} \begin{bmatrix} 1 & 0 & | & 3 \\ 0 & 1 & | & 4 \end{bmatrix}$

The solution is (3, 4).

21. F + S = 32   35F + 25S = 980

$\begin{bmatrix} 1 & 1 & | & 32 \\ 35 & 25 & | & 980 \end{bmatrix} \begin{matrix} \\ (r_2 - 35r_1) \end{matrix} = \begin{bmatrix} 1 & 1 & | & 32 \\ 0 & -10 & | & -140 \end{bmatrix} \begin{matrix} \\ (r_2 \div -10) \end{matrix} = \begin{bmatrix} 1 & 1 & | & 32 \\ 0 & 1 & | & 14 \end{bmatrix} \begin{matrix} (r_1 - r_2) \\ = \end{matrix} \begin{bmatrix} 1 & 0 & | & 18 \\ 0 & 1 & | & 14 \end{bmatrix}$

The solution is (18, 14); 18 fitted caps and 14 stretch-fit caps.

23. Let T = # of hours for truck driver   L = # of hours for laborer

10T + 8L = 144   L = T + 2   →   T = L − 2

$\begin{bmatrix} 10 & 8 & | & 144 \\ 1 & -1 & | & -2 \end{bmatrix} \begin{matrix} r_1 + 8r_2 \\ = \end{matrix} \begin{bmatrix} 18 & 0 & | & 128 \\ 1 & -1 & | & -2 \end{bmatrix} \begin{matrix} (r_1 \div 18) \\ = \end{matrix} \begin{bmatrix} 1 & 0 & | & 64/9 \\ 1 & -1 & | & -2 \end{bmatrix} \begin{matrix} \\ (r_2 - r_1) \end{matrix} \begin{bmatrix} 1 & 0 & | & 64/9 \\ 0 & -1 & | & -82/9 \end{bmatrix}$

$\begin{bmatrix} 1 & 0 & | & 64/9 \\ 0 & -1 & | & -82/9 \end{bmatrix} \begin{matrix} \\ (r_2 \bullet -1) \end{matrix} = \begin{bmatrix} 1 & 0 & | & 64/9 \\ 0 & 1 & | & 82/9 \end{bmatrix}$ (64/9, 82/9)

7 1/9 hours for the truck driver and 9 1/9 hours for the laborer.

25. Let x = amount of $6/ream paper ordered and y = amount of $7.50/ream paper ordered.

6x + 7.5y = 1275   x + y = 200

$\begin{bmatrix} 6 & 7.5 & | & 1275 \\ 1 & 1 & | & 200 \end{bmatrix} = \begin{bmatrix} 1 & 1.25 & | & 212.5 \\ 1 & 1 & | & 200 \end{bmatrix} = \begin{bmatrix} 1 & 1.25 & | & 212.5 \\ 0 & -.25 & | & -12.5 \end{bmatrix} = \begin{bmatrix} 1 & 1.25 & | & 212.5 \\ 0 & 1 & | & 50 \end{bmatrix} = \begin{bmatrix} 1 & 0 & | & 150 \\ 0 & 1 & | & 50 \end{bmatrix}$

The solution is 150 reams at $6/ream and 50 reams at $7.50/ream.

Exercise Set 7.5

1. The solution set of a system of linear inequalities is the set of points that satisfy all inequalities in the system.

3. Yes; a point of intersection satisfies both inequalities and is therefore a solution.

5.   7.

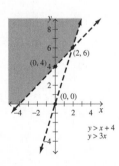

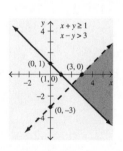

9.

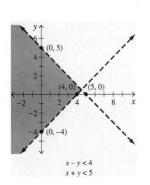

$x - y < 4$
$x + y < 5$

11.

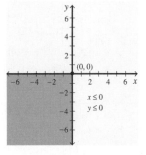

$x + 2y \geq 4$
$3x - y \geq -6$

13.

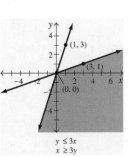

$y \leq 3x$
$x \geq 3y$

15.

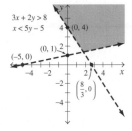

$x \leq 0$
$y \leq 0$

17.

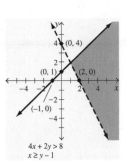

$4x + 2y > 8$
$x \geq y - 1$

19.

$3x + 2y > 8$
$x < 5y - 5$

21. a) $20x + 30y \leq 600$, $x \geq 2y$, $x \geq 10$, $y \geq 5$

    c) One example is 15 small bowls, 7 large
       bowls.

b)

23. $x \leq 0$, $y \geq 0$

25. Yes. One example is
    $x \leq 0$, $y \leq 0$, $x \geq 0$, $y \geq 0$.

27. $y \leq x$, $y \geq x$, $y \leq 0$, $y \geq 0$

Exercise Set 7.6

1. Constraints are restrictions that are represented as linear inequalities.

3. Vertices

5. If a linear equation of the form K = Ax + By is evaluated at each point in a closed polygonal region, the maximum and minimum values of the equation occur at a corner.

7. At (0, 0), K= 6(0) + 4(0) = 0
   At (0, 4), K= 6(0) + 4(4) = 16
   At (2, 3), K= 6(2) + 4(3) = 24
   At (5, 0), K= 6(5) + 4(0) = 30
   The maximum value is 30 at (5, 0); the minimum
   value is 0 at (0, 0).

9. At (10, 20), K = 2(10) + 3(20) = 80
   At (10, 40), K = 2(10) + 3(40) = 140
   At (50, 30), K = 2(50) + 3(30) = 190
   At (50, 10), K = 2(50) + 3(10) = 130
   At (20, 10), K = 2(20) + 3(10) = 70

   The maximum value is 190 at (50, 30); the minimum value is 70 at (20, 10).

11. a)

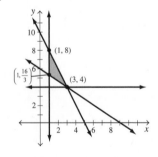

   b) $x + y \leq 6$ $2x + y \leq 8$ $x \geq 0$ $y \geq 0$
   $P = 4x + 5y$
   At (0,0), P = 4(0) + 5(0) = 0   min. at (0, 0)
   At (0,4), P = 4(0) + 5(4) = 20
   At (2, 4), P = 4(2) + 5(4) = 28
   At (0,6), P = 4(0) + 5(6) = 30   max. at (0, 6)

13. a)

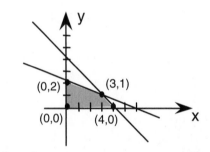

   b) P = 7x + 6y
   At (0,0), P = 7(0) + 6(0) = 0   min. at (0, 0)
   At (0,2), P = 7(0) + 6(2) = 12
   At (3,1), P = 7(3) + 6(1) = 27
   At (4,0), P = 7(4) + 6(0) = 28   max. at (4, 0)

15. a)

   b) P = 2.20x + 1.65y
   At (3, 4), P = 2.20(3)+1.65(4) = 13.2
   At (1, 8), P = 2.20(1)+1.65(8) = 15.4
   At (1, 16/3), P = 2.20(1)+1.65(16/3) = 11
   Maximum of 15.4 at (1, 8) and
   Minimum of 11 at (1, 16/3)

17. a) $x + y \le 24$, $x \ge 2y$, $y \ge 4$, $x \ge 0$, $y \ge 0$

   b) $P = 40x + 55y$

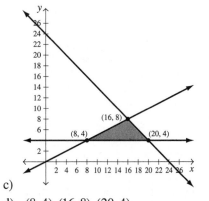

   c)

   d) $(8, 4), (16, 8), (20, 4)$

   e) At $(8, 4)$, $P = 40(8) + 55(4) = 540$
      At $(16, 8)$, $P = 40(16) + 55(8) = 1080$
      At $(20, 4)$, $P = 40(20) + 55(4) = 1020$
      16 Kodak cameras and 8 Canon cameras

   f) The maximum profit is \$1080

19. Let x = gallons of indoor paint
      y = gallons of outdoor paint
      $x \ge 60$ $y \ge 100$

   (a) $3x + 4y \ge 60$ $x \ge 0$
       $10x + 5y \ge 100$ $y \ge 0$

   (b) C = 28x + 33y

   (c)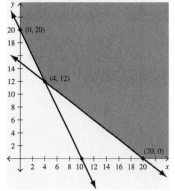

   d) At $(0, 20)$, C = 28(0) + 33(20) = 660
      At $(20, 0)$, C = 28(20) + 33(0) = 560
      At $(4, 12)$, C = 28(4) + 33(12) = 508

   e) 4 hours on Mach. 1 and 12 hours on Mach. 2

   f) Min. profit = \$ 508.00

21. Let x = # of car seats
      y = # of strollers
      $x + 3y \le 24$ $2x + y \le 16$ $x + y \le 10$

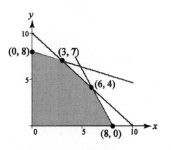

   P = 25x + 35y
   At $(0, 8)$, P = 25(0) + 35(8) = 280
   At $(3, 7)$, P = 25(3) + 35(7) = 320
   At $(6, 4)$, P = 25(6) + 35(4) = 290
   At $(8, 0)$, P = 25(8) + 35(0) = 200
   3 car seats and 7 strollers
   Max. profit = \$ 320.00

Review Exercises

1.

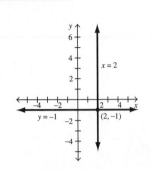

2.

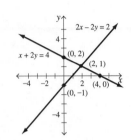

3.

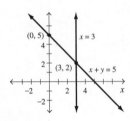

4.

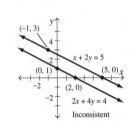

5. $y = (1/3)x + 5$

   $y = (1/3)x + 5$

   Same slope and y-intercept. Infinite # of solutions.

6. $y = -2x + 4$

   $y = -2x + 6$

   Same slope but different y-intercepts. No solution.

7. $6y - 2x = 20$ becomes $y = (1/3)x + 10/3$

   $4y + 2x = 10$ becomes $y = -(1/2)x + 5/2$

   Different slopes. One solution.

8. $y = (1/2)x - 2$

   $y = 2x + 6$

   Different slopes. One solution.

9. (1) $-x + y = -2$

   (2) $\underline{x + 2y = 5}$ (add)

   $\qquad 3y = 3 \qquad y = 1$

   Substitute 1 in place of y in the first equation.

   $-x + 1 = -2$

   $-x = -3 \qquad x = 3$

   The solution is (3, 1).

10. $x - 2y = 9$

    $y = 2x - 3$

    Substitute $(2x - 3)$ in place of y in the 1st equation.

    $x - 2(2x - 3) = -11$ (solve for x)

    $x - 4x - 6 = -11$

    $5x - 6 = -1$

    $5x = -5 \qquad x = -1$

    Substitute (-1) in place of x in the 2nd equation.

    $y = 2(-1) - 3 = -2 - 3 = -5$

    The solution is (-1, -5).

11. $2x - y = 4 \qquad y = 2x - 4$

    $3x - y = 2$

    Substitute $2x - 4$ for y in the second equation.

    $3x - (2x - 4) = 2$ (solve for x)

    $3x - 2x + 4 = 2$

    $x + 4 = 2 \qquad x = -2$

    Substitute -2 for x in an equation.

    $2(-2) - y = 4$

    $-4 - y = 4 \qquad y = -8$

    The solution is (-2, -8).

12. $3x + y = 1 \qquad y = -3x + 1$

    $3y = -9x - 4$

    Substitute $-3x + 1$ for y in the second equation.

    $3(-3x + 1) = -9x - 4$ (solve for x)

    $-9x + 3 = -9x - 4$

    $3 \neq 4$ False There is no solution to this system.

    The equations are inconsistent.

13. (1) $x - 2y = 8$
    (2) $2x + y = 6$
    Multiply the second equation by 2.
    $x - 2y = 8$
    $\underline{4x + 2y = 12}$
    $5x \quad\;\; = 20$
    $\qquad x = 4$
    Substitute 4 in place of x in the equation (2).
    $2(4) + y = 6$
    $\qquad y = 6 - 8$
    $\qquad y = -2$
    The solution is (4, −2).

14. (1) $2x + y = 2$
    (2) $\underline{-3x - y = 5}$ (add)
    $\;\;-x \;= 7 \qquad\qquad x = -7$
    Substitute (− 7) in place of x in the 1st equation.
    $2(-7) + y = 2$
    $-14\,y = 2 \qquad\qquad y = 16$
    The solution is (− 7, 16).

15. (1) $x + y = 2$
    (2) $x + 3y = -2$
    Multiply the first equation by −1.
    $-x - y = -2$
    $\underline{x + 3y = -2}$ (add)
    $2y = -4 \qquad\qquad y = -2$
    Substitute (−2) for y in equation (2).
    $x + 3(-2) = -2$
    $x - 6 = -2 \qquad x = 4$
    The solution is (4,−2).

16. (1) $4x - 8y = 16$
    (2) $x - 2y = 4 \qquad x = 2y + 4$
    Substitute 2y + 4 for x in the first equation.
    $4(2y + 4) - 8y = 16$
    $8y + 16 - 8y = 16$
    $16 = 16$ True
    There are an infinite number of solutions.
    The system is dependent.

17. (1) $3x - 4y = 10$
    (2) $5x + 3y = 7$
    Multiply the first equation by 3, and the 2nd equation by 4.
    $9x - 12y = 30$
    $\underline{20x + 12y = 28}$ (add)
    $29x = 58 \qquad\qquad x = 2$
    Substitute 2 for x in the second equation.
    $5(2) + 3y = 7$
    $3y = -3 \quad y = -1$
    The solution is (2, −1).

18. (1) $3x + 4y = 6$
    (2) $2x - 3y = 4$
    Multiply the first equation by 2, and the second equation by − 3.
    $6x + 8y = 12$
    $\underline{-6x + 9y = -12}$ (add)
    $17y = 0 \qquad y = 0$
    Substitute 0 for y in the first equation.
    $3x + 4(0) = 6$
    $3x = 6 \qquad\qquad x = 2$
    The solution is (2,0).

19. $A + B = \begin{bmatrix} 1 & -3 \\ 2 & 4 \end{bmatrix} + \begin{bmatrix} -2 & -5 \\ 6 & 3 \end{bmatrix} = \begin{bmatrix} 1+(-2) & -3+(-5) \\ 2+6 & 4+3 \end{bmatrix} = \begin{bmatrix} -1 & -8 \\ 8 & 7 \end{bmatrix}$

20. $A - B = \begin{bmatrix} 1 & -3 \\ 2 & 4 \end{bmatrix} - \begin{bmatrix} -2 & -5 \\ 6 & 3 \end{bmatrix} = \begin{bmatrix} 1-(-2) & -3-(-5) \\ 2-6 & 4-3 \end{bmatrix} = \begin{bmatrix} 3 & 2 \\ -4 & 1 \end{bmatrix}$

21. $2A = 2\begin{bmatrix} 1 & -3 \\ 2 & 4 \end{bmatrix} = \begin{bmatrix} 2(1) & 2(-3) \\ 2(2) & 2(4) \end{bmatrix} = \begin{bmatrix} 2 & -6 \\ 4 & 8 \end{bmatrix}$

22. $2A - 3B = 2\begin{bmatrix} 1 & -3 \\ 2 & 4 \end{bmatrix} - 3\begin{bmatrix} -2 & -5 \\ 6 & 3 \end{bmatrix} = \begin{bmatrix} 2 & -6 \\ 4 & 8 \end{bmatrix} + \begin{bmatrix} 6 & 15 \\ -18 & -9 \end{bmatrix} = \begin{bmatrix} 2+6 & -6+15 \\ 4-18 & 8-9 \end{bmatrix} = \begin{bmatrix} 8 & 9 \\ -14 & -1 \end{bmatrix}$

23. $A \times B = \begin{bmatrix} 1 & -3 \\ 2 & 4 \end{bmatrix} \times \begin{bmatrix} -2 & -5 \\ 6 & 3 \end{bmatrix} = \begin{bmatrix} 1(-2)+(-3)6 & 1(-5)+(-3)3 \\ 2(-2)+4(6) & 2(-5)+4(3) \end{bmatrix} = \begin{bmatrix} -20 & -14 \\ 20 & 2 \end{bmatrix}$

24. $B \times A = \begin{bmatrix} -2 & -5 \\ 6 & 3 \end{bmatrix} \times \begin{bmatrix} 1 & -3 \\ 2 & 4 \end{bmatrix} = \begin{bmatrix} (-2)1+(-5)2 & (-2)(-3)+(-5)4 \\ 6(1)+3(2) & 6(-3)+3(4) \end{bmatrix} = \begin{bmatrix} -12 & -14 \\ 12 & -6 \end{bmatrix}$

25. $\begin{bmatrix} 1 & 3 & | & 8 \\ 1 & 1 & | & 4 \end{bmatrix} \underset{(r_2-r_1)}{=} \begin{bmatrix} 1 & 3 & | & 8 \\ 0 & -2 & | & -4 \end{bmatrix} \underset{-\frac{1}{2}_2}{=} \begin{bmatrix} 1 & 3 & | & 8 \\ 0 & 1 & | & 2 \end{bmatrix} \overset{(r_1-3r_2)}{=} \begin{bmatrix} 1 & 0 & | & 2 \\ 0 & 1 & | & 2 \end{bmatrix}$ The solution is (2, 2).

26. $\begin{bmatrix} -1 & 1 & | & 4 \\ 1 & 3 & | & 4 \end{bmatrix} = \begin{bmatrix} 1 & -1 & | & 4 \\ 0 & 4 & | & 8 \end{bmatrix} = \begin{bmatrix} -1 & 1 & | & 4 \\ 0 & 1 & | & 2 \end{bmatrix} = \begin{bmatrix} 1 & -1 & | & -4 \\ 0 & 1 & | & 2 \end{bmatrix} = \begin{bmatrix} 1 & 0 & | & -2 \\ 0 & 1 & | & 2 \end{bmatrix}$ The solution is (– 2, 2).

27. $\begin{bmatrix} 2 & 1 & | & 3 \\ 3 & -1 & | & 12 \end{bmatrix} \overset{(r_1 \div 2)}{=} \begin{bmatrix} 1 & \frac{1}{2} & | & \frac{3}{2} \\ 3 & -1 & | & 12 \end{bmatrix} \underset{(r_2-3r_1)}{=} \begin{bmatrix} 1 & \frac{1}{2} & | & \frac{3}{2} \\ 0 & -\frac{5}{2} & | & \frac{15}{2} \end{bmatrix} \overset{(-\frac{2}{5}r_2)}{=} \begin{bmatrix} 1 & \frac{1}{2} & | & \frac{3}{2} \\ 0 & 1 & | & -3 \end{bmatrix} \overset{(-\frac{1}{2}r_2+r_1)}{=} \begin{bmatrix} 1 & 0 & | & 3 \\ 0 & 1 & | & -3 \end{bmatrix}$

The solution is (3,– 3).

28. $\begin{bmatrix} 2 & 3 & | & 2 \\ 4 & -9 & | & 4 \end{bmatrix} = \begin{bmatrix} 1 & \frac{3}{2} & | & 1 \\ 0 & -15 & | & 0 \end{bmatrix} = \begin{bmatrix} 1 & \frac{3}{2} & | & 1 \\ 0 & 1 & | & 0 \end{bmatrix} = \begin{bmatrix} 1 & 0 & | & 1 \\ 0 & 1 & | & 0 \end{bmatrix}$ The solution is (1,0)

29. $\begin{bmatrix} 1 & 3 & | & 3 \\ 3 & -2 & | & 2 \end{bmatrix} = \begin{bmatrix} 1 & 3 & | & 3 \\ 0 & -11 & | & -7 \end{bmatrix} = \begin{bmatrix} 1 & 3 & | & 3 \\ 0 & 1 & | & \frac{7}{11} \end{bmatrix} = \begin{bmatrix} 1 & 0 & | & \frac{12}{11} \\ 0 & 1 & | & \frac{7}{11} \end{bmatrix}$ The solution is $\left(\dfrac{12}{11}, \dfrac{7}{11} \right)$

30. $\begin{bmatrix} 3 & -6 & | & -9 \\ 4 & 5 & | & 14 \end{bmatrix} \overset{(r_1 \bullet -1)}{=} \begin{bmatrix} -3 & 6 & | & 9 \\ 4 & 5 & | & 14 \end{bmatrix} \overset{(r_2+r_1)}{=} \begin{bmatrix} 1 & 11 & | & 23 \\ 4 & 5 & | & 14 \end{bmatrix} \overset{(r_1 \bullet -4)}{\underset{(r_2+r_1)}{=}} \begin{bmatrix} 1 & 11 & | & 23 \\ 0 & -39 & | & -78 \end{bmatrix}$

$\begin{bmatrix} 1 & 11 & | & 23 \\ 0 & -39 & | & -78 \end{bmatrix} \overset{(r_1 \bullet 11/39)}{\underset{(r_2+r_1)}{=}} \begin{bmatrix} 1 & 0 & | & 1 \\ 0 & -39 & | & -78 \end{bmatrix} \overset{(r_2 \div -2)}{=} \begin{bmatrix} 1 & 0 & | & 1 \\ 0 & 1 & | & 2 \end{bmatrix}$ The solution is (1, 2).

31. Let x = amount borrowed at 4%   y = amount borrowed at 6%

    .04x + .06y = 29000   x + y = 600000

$\begin{bmatrix} .04 & .06 & | & 29000 \\ 1 & 1 & | & 600000 \end{bmatrix} \overset{(r_2 \bullet 25)}{=} \begin{bmatrix} 1 & 1.5 & | & 725000 \\ 1 & 1 & | & 600000 \end{bmatrix} \overset{(r_2-r_1)}{=} \begin{bmatrix} 1 & 1.5 & | & 725000 \\ 0 & -0.5 & | & -125000 \end{bmatrix}$

$\underset{(-2 \bullet r_2)}{=} \begin{bmatrix} 1 & 1.5 & | & 725000 \\ 0 & 1 & | & 250000 \end{bmatrix} \overset{(r_1-1.5r_1)}{=} \begin{bmatrix} 1 & 0 & | & 350000 \\ 0 & 1 & | & 250000 \end{bmatrix}$

$350,000 borrowed at 4% and $ 250,000 borrowed at 6%

32. Let s = liters of 80% acid solution
    w = liters of 50% acid solution

    s + w = 100

    0.80s + 0.50w = 100(0.75)

    0.80s + 0.50w = 75

    s = 100 – w

    0.80(100 – w) + 0.50w = 75

    80 – 0.80w + 0.50w = 75

    – 0.30w = – 5

    w = – 5/(– 0.30) = 16 2/3 liters

    s = 100 – 16 2/3 = 83 1/3 liters

    Mix 83 1/3 liters of 80% solution with
    16 2/3 liters of 50% solution.

33. Let t = price per ton of topsoil
    m = price per ton of mulch

    (1) 4t + 3m = 1529

    (2) 2t + 5m = 1405

    Subtract 2 times eq'n. (2) from eq'n. (1):

    -7m = -1281

    m = 183

    Substitute 183 for m in eq'n. 1.

    4t = 1529 – 3(183) = 980 so t = 245

    Topsoil costs $245 per ton and mulch
    costs $183 per ton.

34. Let c = total cost   x = no. of months to
operate

    a) model 1600A: c_A = 950 + 32x
        model 6070B: c_B = 1275 + 22x
        950 + 32x= 1275 + 22x
        10x = 325   x = 32.5 months
        After 32.5 months of operation the total cost
        of the units will be equal.

    b) After 32.5 months or 2.7 years, the most cost
        effective unit is the unit with the lower per
        month to operate cost. Thus, model 6070B is
        the better deal in the long run.

35. a) Let C = total cost for parking
        x = number of additional hours
        All-Day: C = 5 + 0.50x
        Sav-A-Lot: C = 4.25 + 0.75x
        5 + 0.50x= 4.25 + 0.75x
        0.75 = 0.25x   3 = x
        The total cost will be the same after 3
        additional hours or 4 hours total.

    b) After 5 hours or x = 4 additional hours:
        All-Day: C = 5 + 0.50(4) = $7.00
        Sav-A-Lot: C = 4.25 + 0.75(4) = $7.25
        All-Day would be less expensive.

36.

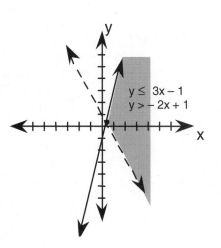

y ≤ 3x – 1
y > – 2x + 1

37.

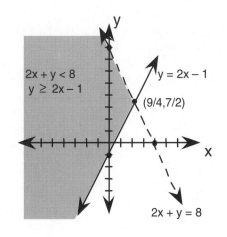

2x + y < 8
y ≥ 2x – 1

y = 2x – 1

(9/4,7/2)

2x + y = 8

38.

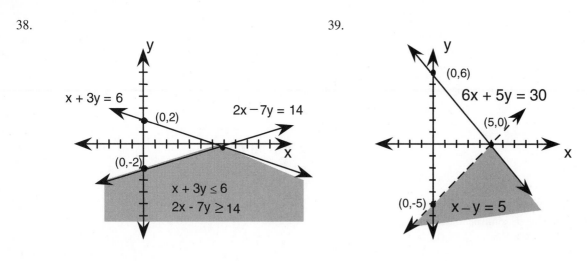

39.

40. a)

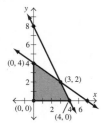

b) $P = 5x + 3y$

At $(0, 0)$ $P = 5(0) + 3(0) = 0$

At $(0, 4)$, $P = 5(0) + 3(4) = 12$

At $(3, 2)$, $P = 5(3) + 3(2) = 21$

At $(4, 0)$, $P = 5(4) + 3(0) = 20$

The maximum is 21 at $(3, 2)$ and the minimum is 0 at $(0, 0)$.

Chapter Test

1. If the lines do not intersect (parallel) the system of equations is inconsistent. The system of equations is consistent if the lines intersect only once. If both equations represent the same line then the system of equations is dependent.

2.

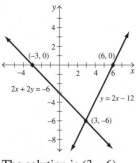

The solution is $(3, -6)$.

3. Write each equation in slope intercept form, then compare slopes and intercepts.

$4x + 5y = 6$ $-3x + 5y = 13$

$5y = -4x + 6$ $5y = 3x + 13$

$y = -(4/5)x + 6/5$

$y = (3/5)x + 13/5$

The slopes are different so there is only one solution.

4. $x + y = -1$ 　　$x = -y - 1$
  $2x + 3y = -5$
  Substitute $(-y - 1)$ for x in the second equation.
  $2(-y - 1) + 3y = -5$ (solve for y)
  $-2y - 2 + 3y = -5$
  　$y = -3$
  Substitute (-3) for y in the equation $x = -y - 1$.
  $x = -(-3) - 1 = 2$ 　　　　The solution is $(2, -3)$.

5. $y = 3x - 7$ 　　$y = 5x - 3$
  Substitute $(3x - 7)$ for y in the second equation.
  $3x - 7 = 5x - 3$ (solve for x)
  $2x = -4$ 　　　$x = -2$
  Substitute -2 for x in the first equation.
  　$y = 3(-2) - 7 = -13$
  The solution is $(-2, -13)$.

6. $x - y = 4$
  $\underline{2x + y = -10}$ (add)
  $3x = -6$ 　　　　$x = -2$
  Substitute -2 for x in the 2nd equation.
  $2(-2) + y = -10$
  $-4 + y = -10$ 　　　$y = -6$
  　The solution is $(-2, -6)$.

7. $4x + 3y = 5$
  $2x + 4y = 10$
  Multiply the second equation by (-2).
  $4x + 3y = 5$
  $\underline{-4x - 8y = -20}$ (add)
  $-5y = -15$ 　　　　$y = 3$
  Substitute 3 for y in the first equation.
  $4x + 3(3) = 5$
  $4x + 9 = 5$
  $4x = -4$ 　　　$x = -1$
  The solution is $(-1, 3)$.

8. $3x + 4y = 6$
  $2x - 3y = 4$
  Multiply the 1st eq'n. by 3 and the 2nd eq'n. by 4.
  $9x + 12y = 18$
  $\underline{8x - 12y = 16}$
  $17x = 34$ 　　　　$x = 2$

Substitute 2 for x in an equation.
  $2(2) - 3y = 4$ (solve for y)
  $-3y = 0$ 　　　　$y = 0$

The solution is $(2, 0)$.

9. $\begin{bmatrix} 1 & 3 & | & 4 \\ 5 & 7 & | & 4 \end{bmatrix} \begin{matrix} \\ (-5r_1 + r_2) \end{matrix} = \begin{bmatrix} 1 & 3 & | & 4 \\ 0 & -8 & | & -16 \end{bmatrix} \begin{matrix} \\ (r_2 \div (-8)) \end{matrix} = \begin{bmatrix} 1 & 3 & | & 4 \\ 0 & 1 & | & 2 \end{bmatrix} \begin{matrix} (r_1 - 3r_2) \\ = \end{matrix} \begin{bmatrix} 1 & 0 & | & -2 \\ 0 & 1 & | & 2 \end{bmatrix}$

The solution is $(-2, 2)$.

10. $A + B = \begin{bmatrix} 2 & -5 \\ 1 & 3 \end{bmatrix} + \begin{bmatrix} -1 & -3 \\ 5 & 2 \end{bmatrix} = \begin{bmatrix} 2+(-1) & -5-3 \\ 1+5 & 3+2 \end{bmatrix} = \begin{bmatrix} 1 & -8 \\ 6 & 5 \end{bmatrix}$

11. $A - 2B = \begin{bmatrix} 2 & -5 \\ 1 & 3 \end{bmatrix} - 2\begin{bmatrix} -1 & -3 \\ 5 & 2 \end{bmatrix} = \begin{bmatrix} 2-2(-1) & -5-2(-3) \\ 1-2(5) & 3-2(2) \end{bmatrix} = \begin{bmatrix} 4 & 1 \\ -9 & -1 \end{bmatrix}$

12. $A \times B = \begin{bmatrix} 2 & -5 \\ 1 & 3 \end{bmatrix}\begin{bmatrix} -1 & -3 \\ 5 & 2 \end{bmatrix} = \begin{bmatrix} 2(-1)+(-5)(5) & 2(-3)+(-5)(2) \\ 1(-1)+(3)(5) & 1(-3)+3(2) \end{bmatrix} = \begin{bmatrix} -27 & -16 \\ 14 & 3 \end{bmatrix}$

13. $y < -2x + 2$ $y > 3x + 2$

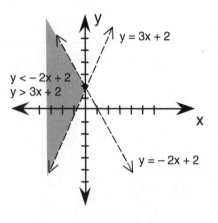

14. Let x = daily fee
      y = mileage charge
      $3x + 150y = 132$
      $2x + 400y = 142$ $x = 71 - 200y$
      Substitute $(71 - 200y)$ for x in the 1st equation.
      $3(71 - 200y) + 150y = 132$
      $213 - 600y + 150y = 132$
        $-450y = -81$
      $y = 0.18$
      Substitute 0.18 for y in the first equation.
      $3x + 150(0.18) = 132$
      $3x + 27 = 132$
      $3x = 105$ so $x = 35$
      The daily fee is \$35 and the mileage charge
      is 18 cents per mile.

15. (a) Let x = no. of checks written in one month.
        Cost at Union Bank: $6 + .10x$
        Cost at Citrus Bank: $2 + .20x$
        These are equal when:
            $2 + .2x = 6 + .1x$
                $.1x = 4$
                  $x = 40$

   (b) Since $14 < 40$, the bank with the lower
        monthly fee is better, which is Citrus Bank.

16. a)

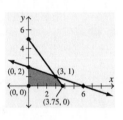

   b) $P = 6x + 4y$
        At $(0, 0)$ $P = 6(0) + 4(0) = 0$
        At $(0, 2)$ $P = 6(0) + 4(2) = 8$
        At $(3, 1)$ $P = 6(3) + 4(1) = 22$
        At $(3.75, 0)$ $P = 6(3.75) + 4(0) = 22.5$
        Max. is 22.5 at $(3.75, 0)$
        and min. is 0 at $(0, 0)$

CHAPTER EIGHT

THE METRIC SYSTEM

Exercise Set 8.1

1. The metric system.

3. It is the worldwide accepted standard of measurement. There is only 1 basic unit of measurement for each quantity. It is based on the number 10, which makes many calculations easier.

5. a) Move the decimal point one place for each change in unit of measure.

   b) $214.6 \text{ cm} = \dfrac{214.6}{10^5} \text{ km} = 214.6 \times 10^{-5} \text{ km} = 0.002146 \text{ km}$

   c) $60.8 \text{ hm} = (60.8)(1000) \text{ dm} = 60800 \text{ dm}$

7. kilo   1000 times the base unit   k
   hecto   100 times the base unit   h
   deka   10 times the base unit   da
   deci   1/10 times the base unit   d
   centi   1/100 times the base unit   c
   milli   1/1000 times the base unit   m

9. a) 10,000 times greater
   b) 1 h = 10,000 cm
   c) 1 cm = 0.0001 hm

11. 1 yard

13. 5 grams

15. 22° C

17. milli b

19. hecto c

21. deci f

23. a) 100 grams   b) 0.001 gram   c) 1000 grams   d) 0.01 gram   e) 10 grams   f) 0.1 gram

25. cg   1/100 gm

27. dg   1/10 gm

29. hg   100 gm

31. Max. mass 320 kg = (320 x 1000) g = 320 000 g

33. 5 m = (5 x 1000) mm = 5000 mm

35. 0.085 hℓ = (0.085)(0.1) kℓ = 0.0085 kℓ

37. 242.6 cm = (242.6)(0.0001) hm = 0.02426 hm

39. 22435 mg = (2435)(0.00001) hg = 0.02435 hg

41. 1.34 hm = (1.34)(10000) cm = 13,400 cm

43. 32.5 kg = 325 hg

45. 895 ℓ = 895,000 mℓ

47. 140 cg = 1.40 g

49. 40,302 mℓ = 4.0302 daℓ

51. 590 cm, 2.3 dam, 0.47 km

53. 1.4 kg, 1600 g, 16,300 dg

55. 203,000 mm, 2.6 km, 52.6 hm

57. Jim, since a meter is longer than a yard.

59. The pump that removes 1 daℓ of water per min.
    1 dekaliter > 1 deciliter

61. a) Perimeter= 2l + 2w= 2(74) + 2(99)= 346 cm
    b) 346 cm = (346 x 10) mm = 3,460 mm

63. a) 1,200 km / 187 ℓ = 6.417 km/ℓ
    b) 1,200,00 / 187 ℓ = 6,417 m/ℓ

65. a) 6(360) ml = 2,160 mℓ
    b) 2160(1000) = 2.16 ℓ
    c) 2.45 / 2.16 = $1.13 per liter

67. a) $(4)(27 \text{ m}) = 108 \text{ m}$ b) $108 \text{ m} = 0.108 \text{ km}$ 69. a) $(16950 - 5830) \text{ km} = 11,120 \text{ km}$

    c) $108 \text{ m} = 108\ 000 \text{ mm}$ b) $11,120,000 \text{ m}$

71. 1 gigameter = 1000 megameters   73. 1 teraliter = 1×10^{24} picoliters   75. $0.8/.027 = 29.6 \approx 30$ eggs

77. $195 \text{ mg} = 0.195 \text{ g}$ 79. $7000 \text{ cm} = 7 \text{ dam}$ 81. $0.00006 \text{ hg} = 6 \text{ mg}$

    $0.8/0.195 = 4.1$ cups

83. $0.02 \text{ k}\ell = 2 \text{ da}\ell$ 85. magr   gram   87. rteli   liter

89. terem   meter   91. reketolim   kilometer   93. greseed sulesic   degrees celsius

Exercise Set 8.2

1. volume   3. area

5. volume   7. volume

9. area   11. length

13. Answers will vary.   15. Answers will vary.
    (AWV)

17. Answers will vary.   19. 1 cubic decimeter

21. 1 cubic centimeter   23. area

25. centimeters   27. cm or mm

29. centimeters   31. kilometers

33. centimeters   35. kilometers

37. c   36 m   39. c   130 km

41. b   4 cm   43. c   375 m

45. mm   AWV   47. cm or m   AWV

49. mm or cm   AWV   51. cm, km

53. m   55. cm

57. sq. cm   59. sq. cm or sq. m

61. sq. m or hectares   63. sq. cm

65. sq. m   67. a 5 cm^2

69. b   1/8 ha   71. a 100 cm^2

73. c 4900 km^2 75. AWV

77. AWV   79. AWV

81. liters   83. kiloliters

85. cubic meters   87. liters
    or cubic centimeters

89. cubic meters   91. c 7780 cm^3

93. c $55 \text{ k}\ell$ 95. c 0.04 m^3

97. a 30 m^3 99. a) AWV
                                         b) $152,561 \text{ cm}^3$

101. a) AWV   b) $v \approx (3.14)(0.25)^2 (1) = 0.20 \text{ m}^3$ 103. a) AWV   b) $A = lw = (4)(2.2) = 8.8 \text{ cm}^2$

105. $A = \pi r^2 \approx 3.14\left(10.2^2\right) \approx 326.7 \text{ m}^3$ 107. a) $(73)(53) = 3869 \text{ m}^2$
                                                    b) $3869 - (70)(50) = 3869 - 3500 = 369 \text{ m}^2$

109. a) $(62.4)(50.5) = 3151.2 \text{ m}^2$

    b) $(3151.2)(0.0001) \text{ ha} = 0.31512 \text{ ha}$

111. Total Surface Area of 4 walls $= 2lh + 2wh = 2(20)(6) + 2(12)(6) = 384 \text{ m}^2$

    Liters for first coat $= (384 \text{ m}^2)\left(\dfrac{1 \, \ell}{10 \text{ m}^2}\right) = 38.4 \, 1$ Liters for second coat $= (384 \text{ m}^2)\left(\dfrac{1 \, \ell}{15 \text{ m}^2}\right) = 25.6 \, 1$

    Total liters $= 38.4 + 25.6 = 64 \, \ell$ Total cost $= (64)(\$4.75) = \$304$

113. a) $V = lwh = (70)(40)(20) = 56{,}000 \text{ cm}^3$ b) $56{,}000 \text{ cm}^3 = 56{,}000 \text{ m}\ell$ c) $56 \, 000 \text{ m}\ell = \left(\dfrac{56000}{1000}\right)\ell = 56 \, \ell$

115. $10^2 = 100$ times larger

117. $10^3 = 1000$ times larger   119. 100 mm^2 121. 100 hm^2

123. 0.000001 dm^3 125. $1{,}000{,}000 \text{ cm}^3$ 127. $218 \text{ cm}^3 = 218 \text{ m}\ell$

129. $76 \text{ k}\ell = 76 \text{ m}^3$ 131. $60 \text{ m}^3 = 60 \text{ k}\ell$ 133. Answers will vary

135. $6.7 \text{ kl} = 6.7 \text{ m}^3 = (6.7 \times 10^3) \text{ dm}^3 = 6{,}700 \text{ dm}^3$

137. a) $1 \text{ sq mi} = (1 \text{ mi}^2)(5280)^2\,\dfrac{\text{ft}^2}{\text{mi}^2} = 27{,}878{,}400 \text{ ft}^2$ b) It is easier to convert in the metric system

                                                           because it is a base 10 system.

        $27{,}878{,}400 \text{ ft}^2 \times (12)^2\,\dfrac{\text{in}^2}{\text{ft}^2} = 4{,}014{,}489{,}600 \text{ in}^2$

139. a) AWV; the average use is 5150.7 liters / day

    b) AWV; the average use is 493.2 liters / day

Exercise Set 8.3

1. kilogram   3. 2 lb

5. approx. 35° C AWV   7. Answers will vary

9. kilograms   11. grams

13. grams   15. kilograms

17. kilograms   19. b   2.26 kg

    or metric tonnes

21. b   1.4 kg   23. b   2800 kg

25. AWV   27. AWV

29. c 0° C   31. b   Dress warmly

                                         and walk.

33. c 40° C   35. c 177° C

37. c 40° C

39. $F = \dfrac{9}{5}(25) + 32 = 45 + 32 = 77^\circ \text{ F}$ 41. $C = \dfrac{5}{9}(92 - 32) = \dfrac{5}{9}(60) = 33.3^\circ \text{ C}$

43. $C = \dfrac{5}{9}(0-32) = \dfrac{5}{9}(-32) \approx -17.8^\circ\, C$

45. $F = \dfrac{9}{5}(37) + 32 = 66.6 + 32 = 98.6^\circ\, F$

47. $C = \dfrac{5}{9}(13-32) = \dfrac{5}{9}(-19) = -10.6^\circ\, C$

49. $F = \dfrac{9}{5}(0) + 32 = 0 + 32 = 32^\circ\, F$

51. $C = \dfrac{5}{9}(-20-32) = \dfrac{5}{9}(-52) = -28.9^\circ\, C$

53. $C = \dfrac{5}{9}(165-32) = \dfrac{5}{9}(133) = 73.9^\circ\, F$

55. $F = \dfrac{9}{5}(22) + 32 = 39.6 + 32 = 71.6^\circ\, F$

57. $F = \dfrac{9}{5}(35.1) + 32 = 63.2 + 32 = 95.2^\circ\, F$

59. low: $F = \dfrac{9}{5}(17.8) + 32 = 32 + 32 = 64.04^\circ\, F$

61. cost = $(8.1)(2.50) = \$20.25$

high: $F = \dfrac{9}{5}(23.5) + 32 = 42.3 + 32 = 74.3^\circ\, F$

Range = $74.30 - 64.04 = 10.26^\circ\, F$

63. total mass = 45 g + 29 g + 370 mℓ =
45 g + 29 g + 370 g = 444 g

65. a) V= lwh,   l = 16 m, w = 12 m, h = 12 m

$V = (16)(12)(12) = 2304 \text{ m}^3$

b) 2304 m^3= 2304 kℓ

c) 2304 kℓ = 2304 t

67. Yes, $78^\circ\, F = \dfrac{5}{9}(78 - 32) \approx 25.6^\circ\, C$, not $20^\circ\, C$

69. 4.2 kg = $(4.2 \text{ kg})\left(\dfrac{1\text{ t}}{1000\text{ kg}}\right) = 0.0042$ t

71. 17.4 t = $(17.4\text{ t})\left(\dfrac{1000\text{ kg}}{1\text{ t}}\right) = 17{,}400$ kg =
17,400,000 g

73. 1.2 ℓ = 1200 mℓ a) 1200 g   b) 1200 cm^3

75. a) V = lwh   l = 1 yd = 3 ft   w = 15 in = 1.25 ft
h = 1.5 ft
$V = (3)(1.25)(1.5) = 5.625$ cubic feet

b) $(5.625 \text{ ft}^3)\left(62.5\,\dfrac{\text{lbs}}{\text{ft}^3}\right) = 351.6$ lb

c) $(351.6 \text{ lb})\left(\dfrac{1\text{ gal}}{8.3\text{ lb}}\right) = 42.4$ gal

77. a) $-62.11^\circ\, C$ $F = \dfrac{9}{5}(-62.11) + 32 = -111.798 + 32 = -79.8^\circ\, F$

b) $2.5^\circ\, C$ $F = \dfrac{9}{5}(2.5) + 32 = 4.5 + 32 = 36.5^\circ\, F$

c) $918{,}000{,}000^\circ\, F$ $C = \dfrac{5}{9}(918{,}000{,}000 - 32) = \dfrac{5}{9}(917{,}999{,}968) = 509{,}999{,}982.2 \approx 510{,}000{,}000^\circ\, C$

Exercise Set 8.4

1. **Dimensional analysis** is a procedure used to convert from one unit of measurement to a different unit of measurement.

3. $\dfrac{60 \text{ seconds}}{1 \text{ minute}}$ or $\dfrac{1 \text{ minute}}{60 \text{ seconds}}$ because 60 seconds = 1 minute

5. $\dfrac{1 \text{ lb}}{0.45 \text{ kg}}$ Since we need to eliminate kilograms, kg must appear in the denominator. Since we need to convert to pounds, lb must appear in the numerator.

7. $\dfrac{3.8 \ \ell}{1 \text{ gal}}$ Since we need to eliminate gallons, gal must appear in the denominator. Since we need to convert to liters, l must appear in the numerator.

9. $62 \text{ in.} = (62 \text{ in.})\left(\dfrac{2.54 \text{ cm}}{1 \text{ in.}}\right) = 157.48 \text{ cm}$

11. $4.2 \text{ ft} = (4.2 \text{ ft})\left(\dfrac{30 \text{ cm}}{1 \text{ ft}}\right)\left(\dfrac{1 \text{ m}}{100 \text{ cm}}\right) = 1.26 \text{ m}$

13. $120 \text{ kg} = (120 \text{ kg})\left(\dfrac{1 \text{ lb}}{0.45 \text{ kg}}\right) = 266.\overline{6} \approx 266.67 \text{ lb}$

15. $39 \text{ mi} = (39 \text{ mi})\left(\dfrac{1.6 \text{ km}}{1 \text{ mi}}\right) = 62.4 \text{ km}$

17. $675 \text{ ha} = (675 \text{ ha})\left(\dfrac{1 \text{ acre}}{0.4 \text{ ha}}\right) = 1687.5 \text{ acres}$

19. $15.6 \ \ell = (15.6 \ \ell)\left(\dfrac{1 \text{ pt}}{0.47 \ \ell}\right) = 33.19148936 \approx 33.19 \text{ pints}$

21. $3.8 \text{ km}^2 = (3.8 \text{ km}^2)\left(\dfrac{1 \text{ mi}^2}{2.6 \text{ km}^2}\right) = 1.4615... \approx 1.46 \text{ mi}^2$

23. $120 \text{ lb} = (120 \text{ lb})\left(\dfrac{0.45 \text{ kg}}{1 \text{ lb}}\right) = 54 \text{ kg}$

25. 28 grams

27. 0.45 kilogram

29. 2.54 centimeters, 1.6 kilometers

31. 0.9 meter

33. $505 \text{ m} = (505 \text{ m})\left(\dfrac{1 \text{ yd}}{0.9 \text{ m}}\right) = 561.\overline{1} \approx 561.11 \text{ yd}$

35. $344 \text{ m} = (344 \text{ m})\left(\dfrac{100 \text{ cm}}{1 \text{ m}}\right)\left(\dfrac{1 \text{ ft}}{30 \text{ cm}}\right) = 1146.\overline{6} \approx 1146.67 \text{ ft}$

37. $60 \text{ km} = (60 \text{ km})\left(\dfrac{1 \text{ mi}}{1.6 \text{ km}}\right) = 37.5 \text{ mph}$

39. $(6 \text{ yd})(9 \text{ yd}) = 54 \text{ yd}^2$

$54 \text{ yd}^2 = (54 \text{ yd}^2)\left(\dfrac{0.8 \text{ m}^2}{1 \text{ yd}^2}\right) = 43.2 \text{ m}^2$

41. $80 \text{ km} = (80 \text{ km})\left(\dfrac{1 \text{ mi}}{1.6 \text{ km}}\right) = 50 \text{ mph}$

43. $6 \text{ g} = (6 \text{ g})\left(\dfrac{1 \text{ oz}}{28 \text{ g}}\right) \approx 0.21 \text{ oz}$

45. $(50 \text{ ft})(30 \text{ ft})(8 \text{ ft}) = 12{,}000 \text{ ft}^3$

$12{,}000 \text{ ft}^3 = (12{,}000 \text{ ft}^3)\left(\dfrac{0.03 \text{ m}^3}{1 \text{ ft}^3}\right) = 360 \text{ m}^3$

47. $1 \text{ kg} = (1 \text{ kg})\left(\dfrac{1 \text{ lb}}{0.45 \text{ kg}}\right) = 2.\overline{2} \text{ lb}$

$\dfrac{\$1.10}{2.\overline{2}} = \$0.495 \text{ per pound}$

49. $34.5 \text{ k}\ell = (34.5 \text{ k}\ell)\left(\dfrac{1000 \text{ }\ell}{1 \text{ k}\ell}\right)\left(\dfrac{1 \text{ gal}}{3.8 \text{ }\ell}\right) = 9078.947368 \approx 9078.95 \text{ gal}$

51. $5.7 \text{ }\ell = (5.7 \text{ }\ell)\left(\dfrac{1 \text{ qt}}{0.95 \text{ }\ell}\right) = 6 \text{ qt}$

53. a) $-282 \text{ ft} = (-282 \text{ ft})\left(\dfrac{30 \text{ cm}}{1 \text{ ft}}\right) = -8460 \text{ cm}$

b) $-8460 \text{ cm} = (-8460 \text{ cm})\left(\dfrac{1 \text{ m}}{100 \text{ cm}}\right) = -84.6 \text{ m}$

55. a) $1 \text{ m}^2 = (1 \text{ m}^2)\left(\dfrac{(3.3)^2 \text{ ft}^2}{1 \text{ m}^2}\right) = 10.89 \text{ ft}^2$

b) $1 \text{ m}^3 = (1 \text{ m}^3)\left(\dfrac{(3.3)^3 \text{ ft}^3}{1 \text{ m}^3}\right) = 35.937 \text{ ft}^3$

57. $56 \text{ lb} = (56 \text{ lb})\left(\dfrac{0.45 \text{ kg}}{1 \text{ lb}}\right)\left(\dfrac{1 \text{ mg}}{1 \text{ kg}}\right) = 25.2 \text{ mg}$

59. $76 \text{ lb} = (76 \text{ lb})\left(\dfrac{0.45 \text{ kg}}{1 \text{ lb}}\right)\left(\dfrac{200 \text{ mg}}{1 \text{ kg}}\right) = 6840 \text{ mg}; \ 6840 \text{ mg} = (6840 \text{ mg})\left(\dfrac{1 \text{ g}}{1000 \text{ mg}}\right) = 6.84 \text{ g}$

61. a) $2 \text{ teaspoons} = (2 \text{ teaspoons})\left(\dfrac{12.5 \text{ mg}}{1 \text{ teaspoon}}\right) = 25 \text{ mg}$

b) $12 \text{ fl oz} = (12 \text{ fl oz})\left(\dfrac{30 \text{ m}\ell}{1 \text{ fl oz}}\right)\left(\dfrac{12.5 \text{ mg}}{5 \text{ m}\ell}\right) = 900 \text{ mg}$

63. a) $964 \text{ ft} = (964 \text{ ft})\left(\dfrac{30 \text{ cm}}{1 \text{ ft}}\right)\left(\dfrac{1 \text{ m}}{100 \text{ cm}}\right) = 289.2 \text{ m}$

b) $85{,}000 \text{ tons} = (85{,}000 \text{ tons})\left(\dfrac{0.9 \text{ tonne}}{1 \text{ ton}}\right) = 76\ 500 \text{ t}$

c) $28 \text{ mi} = (28 \text{ mi})\left(\dfrac{1.6 \text{ km}}{1 \text{ mi}}\right) = 44.8 \text{ kph}$

65. a) $\dfrac{2.00 \text{ €}}{1 \text{ kg}} = \left(\dfrac{2.00 \text{ €}}{1 \text{ kg}}\right)\left(\dfrac{1 \text{ kg}}{2.2 \text{ lb}}\right) \approx \dfrac{0.9 \text{ €}}{1 \text{ lb}}$ b) $\dfrac{0.9 \text{ €}}{1 \text{ lb}} = \left(\dfrac{0.9 \text{ €}}{1 \text{ lb}}\right)\left(\dfrac{\$1.30}{1 \text{ €}}\right) = \dfrac{\$1.17}{1 \text{ lb}}$

67. $(0.2 \text{ mg})\left(\dfrac{1 \text{ grain}}{60 \text{ mg}}\right)\left(\dfrac{1 \text{ m}\ell}{\frac{1}{300} \text{ grain}}\right) = 1.0 \text{ cc, or b)}$

69. a) $(4.0\ \ell)\left(\dfrac{1000 \text{ m}\ell}{1 \text{ l}}\right)\left(\dfrac{1 \text{ cm}^3}{1 \text{ m}\ell}\right) = 4000 \text{ cc}$

b) $\left(4000 \text{ cm}^3\right)\left(\dfrac{1 \text{ in.}^3}{(2.54)^3 \text{ cm}^3}\right) = \dfrac{4000}{16.387064} = 244.094... \approx 244.09 \text{ in.}^3$

71. A kilogram
73. A liter
75. A decimeter
77. wonton
79. 1 kilohurtz
81. 1 megaphone
83. 2 kilomockingbirds
85. 1 decoration

Review Exercises

1. $\dfrac{1}{100}$ of base unit

2. $1000\times$ base unit

3. $\dfrac{1}{1000}$ of base unit

4. $100\times$ base unit

5. 10 times base unit

6. $\dfrac{1}{10}$ of base unit

7. $40 \text{ cg} = 0.040 \text{ g}$

8. $3.2\ \ell = 320 \text{ c}\ell$

9. $0.0016 \text{ cm} = 0.016 \text{ mm}$

10. $1\ 000\ 000 \text{ mg} = 1 \text{ kg}$

11. $4.62 \text{ k}\ell = 4620\ \ell$

12. $192.6 \text{ dag} = 19\ 260 \text{ dg}$

13. $2.67 \text{ k}\ell = 2\ 670\ 000 \text{ m}\ell$
    $14\ 630 \text{ c}\ell = 146\ 300 \text{ m}\ell$
    $3000 \text{ m}\ell,\ 14\ 630 \text{ c}\ell,\ 2.67 \text{ k}\ell$

14. $0.047 \text{ km} = 47 \text{ m}$
    $47\ 000 \text{ cm} = 470 \text{ m}$
    $0.047 \text{ km},\ 47\ 000 \text{ cm},$
    4700 m

15. Centimeters

16. Grams

17. Degrees Celsius

18. Millimeters or centimeters

19. Square meters

20. Milliliters or cubic centimeters

21. Millimeters

22. Kilograms or tonnes

23. Kilometers

24. Meters or centimeters

25. a) and b) Answers will vary.

26. a) and b) Answers will vary.

27. c

28. b   29. c   30. a
31. a   32. b

33. $3600 \text{ kg} = (3600 \text{ kg})\left(\dfrac{1 \text{ t}}{1000 \text{ kg}}\right) = 3.6 \text{ t}$ 34. $4.3 \text{ t} = (4.3 \text{ t})\left(\dfrac{1000 \text{ kg}}{1 \text{ t}}\right)\left(\dfrac{1000 \text{ g}}{1 \text{ kg}}\right) = 4\ 300\ 000 \text{ g}$

35. $24° \text{ C} = \dfrac{9}{5}(24) + 32 = 75.2° \text{ F}$ 36. $68° \text{ F} = \dfrac{5}{9}(68 - 32) = 20° \text{ C}$

37. $-6° \text{ F} = \dfrac{5}{9}(-6 - 32) = -21.\overline{1} \approx -21.1° \text{ C}$ 38. $39° \text{ C} = \dfrac{9}{5}(39) + 32 = 102.2° \text{ F}$

39. $l = 4 \text{ cm}, \ w = 1.6 \text{ cm}; \ A = lw = 4(1.6) = 6.4 \text{ cm}^2$

40. $r = 1.5 \text{ cm}; \ A = \pi r^2 \approx 3.14(1.5)^2 = 7.065 \approx 7.07 \text{ cm}^2$

41. a) $V = lwh = (10)(4)(2) = 80 \text{ m}^3$ b) $(80 \text{ m}^3)\left(\dfrac{1 \text{ kl}}{1 \text{ m}^3}\ell\right)\left(\dfrac{1000 \ \ell}{1 \text{ k}\ell}\right)\left(\dfrac{1 \text{ kg}}{1 \ \ell}\right) = 80\ 000 \text{ kg}$

42. a) $A = lw = (33.7)(26.7) = 899.79 \text{ cm}^2$

   b) $899.79 \text{ cm}^2 = (899.79 \text{ cm}^2)\left(\dfrac{1 \text{ m}^2}{10,000 \text{ cm}^2}\right) = 0.089979 \text{ m}^2$

43. a) $V = lwh = (80)(40)(30) = 96\ 000 \text{ cm}^3$ b) $96\ 000 \text{ cm}^3 = (96\ 000 \text{ cm}^3)\left(\dfrac{1 \text{ m}^3}{(100)^3 \ \text{cm}^3}\right) = 0.096 \text{ m}^3$

   c) $96\ 000 \text{ cm}^3 = (96\ 000 \text{ cm}^3)\left(\dfrac{1 \text{ m}\ell}{1 \text{ cm}^3}\right) = 96\ 000 \text{ m}\ell$ d) $0.096 \text{ m}^3 = (0.096 \text{ m}^3)\left(\dfrac{1 \text{ k}\ell}{1 \text{ m}^3}\right) = 0.096 \text{ k}\ell$

44. Since $1 \text{ km} = 100 \times 1 \text{ dam}, \ 1 \text{ km}^2 = 100^2 \times 1 \text{ dam}^2 = 10\ 000 \text{ dam}^2$.

   Thus, 1 square kilometer is 10,000 times larger than a square dekameter.

45. $(25 \text{ cm})\left(\dfrac{1 \text{ in.}}{2.54 \text{ cm}}\right) = 9.842... \approx 9.84 \text{ in.}$ 46. $(105 \text{ kg})\left(\dfrac{1 \text{ lb}}{0.45 \text{ kg}}\right) = 233.\overline{3} \approx 233.33 \text{ lb}$

47. $(83 \text{ yd})\left(\dfrac{0.9 \text{ m}}{1 \text{ yd}}\right) = 74.7 \text{ m}$ 48. $(100 \text{ m})\left(\dfrac{1 \text{ yd}}{0.9 \text{ m}}\right) = 111.\overline{1} \approx 111.11 \text{ yd}$

49. $(45 \text{ mi})\left(\dfrac{1.6 \text{ km}}{1 \text{ mi}}\right) = 72 \text{ kph}$ 50. $(60 \ \ell)\left(\dfrac{1 \text{ qt}}{0.95 \ \ell}\right) = 63.157... \approx 63.16 \text{ qt}$

51. $(20 \text{ gal})\left(\dfrac{3.8 \ \ell}{1 \text{ gal}}\right) = 76 \ \ell$ 52. $(60 \text{ m}^3)\left(\dfrac{1 \text{ yd}^3}{0.76 \text{ m}^3}\right) = 78.947... \approx 78.95 \text{ yd}^3$

53. $(83 \text{ cm}^2)\left(\dfrac{1 \text{ in.}^2}{6.45 \text{ cm}^2}\right) = 12.86821705 \approx 12.87 \text{ in.}^2$ 54. $(4 \text{ qt})\left(\dfrac{0.95 \ \ell}{1 \text{ qt}}\right) = 3.8 \ \ell$

55. $(15 \text{ yd}^3)\left(\dfrac{0.729 \text{ m}^3}{1 \text{ yd}^3}\right) = 10.9 \text{ m}^3$ 56. $(62 \text{ mi})\left(\dfrac{1.6 \text{ km}}{1 \text{ mi}}\right) = 99.2 \text{ km}$

57. $(27 \text{ cm})\left(\dfrac{1 \text{ ft}}{30 \text{ cm}}\right) = 0.9 \text{ ft}$ 58. $(3.25 \text{ in.})\left(\dfrac{2.54 \text{ cm}}{1 \text{ in.}}\right)\left(\dfrac{10 \text{ mm}}{1 \text{ cm}}\right) = 82.55 \text{ mm}$

59. a) $700(1.5 \text{ kg}) = 1050 \text{ kg}$ b) $1050 \text{ kg} = (1050 \text{ kg})\left(\dfrac{1 \text{ lb}}{0.45 \text{ kg}}\right) = 2333.\overline{3} \approx 2333.33 \text{ lb}$

60. $A = lw = (24)(15) = 360 \text{ ft}^2; 360 \text{ ft}^2 = \left(360 \text{ ft}^2\right)\left(\dfrac{0.09 \text{ m}^2}{1 \text{ ft}^2}\right) = 32.4 \text{ m}^2$

61. a) $(50{,}000 \text{ gal})\left(\dfrac{3.8 \, \ell}{1 \text{ gal}}\right)\left(\dfrac{1 \text{ k}\ell}{1000 \, \ell}\right) = 190 \text{ k}\ell$ b) $(190 \text{ k}\ell)\left(\dfrac{1000 \, \ell}{1 \text{ k}\ell}\right)\left(\dfrac{1 \text{ kg}}{1 \, \ell}\right) = 190\,000 \text{ kg}$

62. a) $65 \text{ mi/hr} = (65 \text{ mi/hr})\left(\dfrac{1.6 \text{ km}}{1 \text{ mi}}\right) = 104 \text{ km/hr}$ b) $104 \text{ km/hr} = (104 \text{ km/hr})\left(\dfrac{1000 \text{ m}}{1 \text{ km}}\right) = 104{,}000 \text{ m/hr}$

63. a) $V = lwh = (90)(70)(40) = 252\,000 \text{ cm}^3; 252\,000 \text{ cm}^3 = \left(252\,000 \text{ cm}^3\right)\left(\dfrac{1 \text{ m}\ell}{1 \text{ cm}^3}\right)\left(\dfrac{1 \, \ell}{1000 \text{ m}\ell}\right) = 252 \, \ell$

   b) $252 \, \ell = (252 \, \ell)\left(\dfrac{1 \text{ kg}}{1 \, \ell}\right) = 252 \text{ kg}$

64. $1 \text{ kg} = (1 \text{ kg})\left(\dfrac{1 \text{ lb}}{0.45 \text{ kg}}\right) = 2.\overline{2} \text{ lb}; \dfrac{\$3.50}{2.\overline{2}} = \$1.575 \approx \$1.58 \text{ per pound}$

Chapter Test

1. $4497 \text{ c}\ell = 0.497 \text{ da}\ell$
2. $273 \text{ hm} = 2{,}730{,}000 \text{ m}$
3. $1 \text{ km} = (1 \text{ km})\left(\dfrac{100 \text{ dam}}{1 \text{ km}}\right) = 100 \text{ dam}$ or 100 times greater
4. $400(6) = 2400 \text{ m}; (2400 \text{ m})\left(\dfrac{1 \text{ km}}{1000 \text{ m}}\right) = 2.4 \text{ km}$ 5. b
6. a 7. c
8. c 9. b
10. $1 \text{ m}^2 = \left(1 \text{ m}^2\right)\left(\dfrac{100^2 \text{ cm}^2}{1 \text{ m}^2}\right) = 10\,000 \text{ cm}^2$ or 10,000 times greater
11. $1 \text{ m}^3 = \left(1 \text{ m}^3\right)\left(\dfrac{1000^3 \text{ mm}^3}{1 \text{ m}^3}\right) = 1\,000\,000\,000 \text{ mm}^3$ or 1,000,000,000 times greater
12. $225 \text{ oz} = (225 \text{ oz})\left(\dfrac{28 \text{ g}}{1 \text{ oz}}\right) = 6300 \text{ g}$
13. $45 \text{ km} = (45 \text{ km})\left(\dfrac{1 \text{ mi}}{1.6 \text{ km}}\right) = 28.125 \text{ mi}$
14. $-15° \text{ F} = \dfrac{5}{9}(-15 - 32) = -26.\overline{1} \approx -26.11° \text{ C}$ 15. $20° \text{ C} = \dfrac{9}{5}(20) + 32 = 68° \text{ F}$
16. a) $300 \text{ kg} = (300 \text{ kg})\left(\dfrac{1000 \text{ g}}{1 \text{ kg}}\right) = 300{,}000 \text{ g}$ b) $300 \text{ kg} = (300 \text{ kg})\left(\dfrac{1 \text{ lb}}{0.45 \text{ kg}}\right) = 666.\overline{6} \text{ lb} \approx 670 \text{ lb}$
17. a) $V = lwh = 20(20)(8) = 3200 \text{ m}^3$ b) $3200 \text{ m}^3 = \left(3200 \text{ m}^3\right)\left(\dfrac{1000 \text{ k}\ell}{1 \text{ m}^3}\right) = 3200 \, k\ell$

   c) $3200 \text{ k}\ell = (3200 \text{ k}\ell)\left(\dfrac{1000 \, \ell}{1 \text{ k}\ell}\right)\left(\dfrac{1 \text{ kg}}{1 \, \ell}\right) = 3\,200\,000 \text{ kg}$

18. Total surface area: $2lh + 2wh = 2(20)(6) + 2(15)(6) = 420 \text{ m}^2$

    Liters needed for first coat: $\left(420 \text{ m}^2\right)\left(\dfrac{1 \ \ell}{10 \text{ m}^2}\right) = 42 \ \ell$

    Liters needed for second coat: $\left(420 \text{ m}^2\right)\left(\dfrac{1 \ \ell}{15 \text{ m}^2}\right) = 28 \ \ell$

    Total liters needed: $42 + 28 = 70 \ \ell$

    Total cost: $\left(70 \ \ell\right)\left(\dfrac{\$3.50}{1 \ \ell}\right) = \$245$

CHAPTER NINE

GEOMETRY

Exercise Set 9.1

1. a) Undefined terms, definitions, postulates (axioms), and theorems

   b) First, Euclid introduced **undefined terms**. Second, he introduced certain **definitions**. Third, he stated primitive propositions called **postulates (axioms)** about the undefined terms and definitions. Fourth, he proved, using deductive reasoning, other propositions called **theorems**.

3. Two lines that do not lie in the same plane and do not intersect are called **skew lines**.

5. Two angles in the same plane are **adjacent angles** when they have a common vertex and a common side but no common interior points.

7. Two angles the sum of whose measure is 180° are called **supplementary angles**.

9. An angle whose measure is greater than 90° but less than 180° is an **obtuse angle**.

11. An angle whose measure is less than 90° is an **acute angle**.

13. Ray, $\overrightarrow{BA}$

15. Half line, $\overset{\circ}{\overrightarrow{BA}}$

17. Ray, $\overrightarrow{AB}$

19. Half open line segment, $\overset{\circ}{\overline{AB}}$

21. $\overrightarrow{BG}$

23. $\overset{\circ}{\overrightarrow{BD}}$

25. $\{B, F\}$

27. $\{C\}$

29. $\measuredangle CFG$

31. $\overline{BC}$

33. $\{B\}$

35. $\overrightarrow{BC}$

37. $\measuredangle ABE$

39. $\overline{BF}$

41. $\overset{\circ\circ}{\overline{AC}}$

43. $\overset{\circ}{\overrightarrow{BE}}$

45. Obtuse

47. Straight

49. Right

51. None of these

53. $90° - 26° = 64°$

55. $90° - 32\frac{3}{4}° = 57\frac{1}{4}°$

57. $90° - 64.7° = 25.3°$

59. $180° - 89° = 91°$

61. $180° - 20.5° = 159.5°$

63. $180° - 43\frac{5}{7}° = 136\frac{2}{7}°$

65. b

67. f

69. a

71. Let x = measure of $\measuredangle\,2$

      $x + 8$ = measure of $\measuredangle\,1$

$$x + x + 8 = 90$$
$$2x + 8 = 90$$
$$2x = 82$$
$$x = \frac{81}{2} = 41°,\, m\measuredangle 2$$
$$x + 8 = 41 + 8 = 49°,\, m\measuredangle 1$$

73. Let x = measure of $\measuredangle\,1$

    $180 - x$ = measure of $\measuredangle\,2$

$$x - (180 - x) = 88$$
$$x - 180 + x = 88$$
$$2x - 180 = 88$$
$$2x = 268$$
$$x = \frac{268}{2} = 134°,\, m\measuredangle 1$$
$$180 - x = 180 - 134 = 46°,\, m\measuredangle 2$$

75. $m\measuredangle\,1 + 125° = 180°$

$m\measuredangle\,1 = 55°$

$m\measuredangle\,2 = m\measuredangle\,1$ (vertical angles)

$m\measuredangle\,3 = 125°$ (vertical angles)

$m\measuredangle\,5 = m\measuredangle\,2$ (alternate interior angles)

$m\measuredangle\,4 = m\measuredangle\,3$ (alternate interior angles)

$m\measuredangle\,7 = m\measuredangle\,4$ (vertical angles)

$m\measuredangle\,6 = m\measuredangle\,5$ (vertical angles)

Measures of angles 3, 4, and 7 are each 125°.

Measures of angles 1, 2, 5, and 6 are each 55°.

77. $m\measuredangle\,3 + 120° = 180°$

$m\measuredangle\,3 = 60°$

$m\measuredangle\,4 = 120°$ (vertical angles)

$m\measuredangle\,7 = m\measuredangle\,3$ (vertical angles)

$m\measuredangle\,6 = m\measuredangle\,3$ (alternate interior angles)

$m\measuredangle\,1 = m\measuredangle\,6$ (vertical angles)

$m\measuredangle\,5 = m\measuredangle\,4$ (alternate exterior angles)

$m\measuredangle\,2 = m\measuredangle\,5$ (vertical angles)

Measures of angles 2, 4, and 5 are each 120°.

Measures of angles 1, 3, 6, and 7 are each 60°.

79.
$$x + 2x + 12 = 90$$
$$3x + 12 = 90$$
$$3x = 78$$
$$x = \frac{78}{3} = 26°,\, m\measuredangle\,2$$
$$90 - x = 90 - 26 = 64°,\, m\measuredangle\,1$$

81.
$$x + 2x - 9 = 90$$
$$3x - 9 = 90$$
$$3x = 99$$
$$x = \frac{99}{3} = 33°,\, m\measuredangle\,1$$
$$90 - x = 90 - 33 = 57°,\, m\measuredangle\,2$$

83.
$$x + 3x - 4 = 180$$
$$4x - 4 = 180$$
$$4x = 184$$
$$x = \frac{184}{4} = 46°,\, m\measuredangle\,2$$
$$180 - x = 180 - 46 = 134°,\, m\measuredangle\,1$$

85.
$$x + 5x + 6 = 180$$
$$6x + 6 = 180$$
$$6x = 174$$
$$x = \frac{174}{6} = 29°,\, m\measuredangle\,1$$
$$180 - x = 180 - 29 = 151°,\, m\measuredangle\,2$$

87. a) An infinite number of lines can be drawn through a given point.

    b) An infinite number of planes can be drawn through a given point.

89. An infinite number of planes can be drawn through a given line.

For Exercises 91 - 97, the answers given are one of many possible answers.

91. Plane ABG and plane JCD

93. $\overrightarrow{BG}$ and $\overrightarrow{DG}$

95. Plane $AGB \cap$ plane $ABC \cap$ plane $BCD = \{B\}$

97. $\overleftrightarrow{BC} \cap$ plane $ABG = \{B\}$

99. Always true. If any two lines are parallel to a third line, then they must be parallel to each other.
101. Sometimes true. Vertical angles are only complementary when each is equal to 45°.
103. Sometimes true. Alternate interior angles are only complementary when each is equal to 45°.
105. Answers will vary.
107. No. Line *l* and line *n* may be parallel or skew.

109.

$$m\angle 1 + m\angle 2 = 180°$$
$$m\angle 3 + m\angle 4 = 180°$$
$$180° + 180° = 360°$$

Exercise Set 9.2

1. A **polygon** is a closed figure in a plane determined by three or more straight line segments.
3. The different types of triangles are acute, obtuse, right, isosceles, equilateral, and scalene. Descriptions will vary.
5. If the corresponding sides of two similar figures are the same length, the figures are **congruent figures**.

7. a) Triangle
   b) Regular

9. a) Octagon
   b) Regular

11. a) Rhombus
    b) Not regular

13. a) Octagon
    b) Not regular

15. a) Isosceles
    b) Acute

17. a) Isosceles
    b) Right

19. a) Scalene
    b) Acute

21. a) Scalene
    b) Right

23. Rectangle

25. Square

27. Rhombus

29. The measures of the other two angles of the triangle are 138° and 180° − 155° (supplementary angles). Therefore, the measure of angle *x* is 180° − 138° − (180° − 155°) = 17°.

31. The measure of one angle of the triangle is 27° (by vertical angles). The measure of another angle of the triangle is 180° - 57° = 123°. The measure of the third angle of the triangle is 180° - 27° - 123° = 30°. The measure of angle *x* is 180° - 30° = 150° (The 30° angle and angle *x* form a straight angle.).

33.

Angle	Measure	Reason
1	90°	∡ 1 and ∡ 7 are vertical angles
2	50°	∡ 2 and ∡ 4 are corresponding angles
3	130°	∡ 3 and ∡ 4 form a straight angle
4	50°	Vertical angle with the given 50° angle
5	50°	∡ 2 and ∡ 5 are vertical angles
6	40°	Vertical angle with the given 40° angle
7	90°	∡ 2, ∡ 6, and ∡ 7 form a straight angle
8	130°	∡ 3 and ∡ 8 are vertical angles
9	140°	∡ 9 and ∡ 10 form a straight angle
10	40°	∡ 10 and ∡ 12 are vertical angles
11	140°	∡ 9 and ∡ 11 are vertical angles
12	40°	∡ 6 and ∡ 12 are corresponding angles

35. $n = 6$

$(6 - 2) \times 180° = 4 \times 180° = 720°$

39. $n = 20$

$(20 - 2) \times 180° = 18 \times 180° = 3240°$

37. $n = 8$

$(8 - 2) \times 180° = 6 \times 180° = 1080°$

41. a) The sum of the measures of the interior angles of a triangle is 180°. Dividing by 3, the number of angles, each interior angle measures 60°.

b) Each exterior angle measures $180° - 60° = 120°$.

43. a) The sum of the measures of the interior angles of a pentagon is $(5 - 2) \times 180° = 3 \times 180° = 540°$. Dividing by 5, the number of angles, each interior angle measures 108°.

b) Each exterior angle measures $180° - 108° = 72°$.

45. a) The sum of the measures of the interior angles of a decagon is $(10 - 2) \times 180° = 8 \times 180° = 1440°$. Dividing by 10, the number of angles, each interior angle measures 144°.

b) Each exterior angle measures $180° - 144° = 36°$.

47. Let $x = A'C'$

$$\frac{A'C'}{AC} = \frac{A'B'}{AB}$$

$$\frac{x}{5} = \frac{1}{2.5}$$

$$2.5x = 5$$

$$x = 2$$

Let $y = B'C'$

$$\frac{B'C'}{BC} = \frac{A'B'}{AB}$$

$$\frac{y}{4} = \frac{1}{2.5}$$

$$2.5y = 4$$

$$y = \frac{4}{2.5} = 1.6$$

49. Let $x = DC$

$$\frac{DC}{D'C'} = \frac{AB}{A'B'}$$

$$\frac{x}{6} = \frac{4}{10}$$

$$10x = 24$$

$$x = \frac{24}{10} = \frac{12}{5}$$

Let $y = B'C'$

$$\frac{B'C'}{BC} = \frac{A'B'}{AB}$$

$$\frac{y}{3} = \frac{10}{4}$$

$$4y = 30$$

$$y = \frac{30}{4} = \frac{15}{2}$$

51. Let $x = D'C'$

$$\frac{D'C'}{DC} = \frac{A'D'}{AD}$$

$$\frac{x}{16} = \frac{22.5}{18}$$

$$18x = 360$$

$$x = 20$$

Let $y = A'B'$

$$\frac{A'B'}{AB} = \frac{A'D'}{AD}$$

$$\frac{y}{17} = \frac{22.5}{18}$$

$$18y = 382.5$$

$$y = 21.25$$

53. Let $x = BC$

$$\frac{BC}{EC} = \frac{AB}{DE}$$

$$\frac{x}{1} = \frac{3}{1}$$

$$x = 3$$

55. $AD = AC - DC = 5 - \dfrac{5}{3} = \dfrac{15}{3} - \dfrac{5}{3} = \dfrac{10}{3}$

57. $AC = A'C' = 28$

59. $B'C' = BC = 30$

61. $m\angle ACB = m\angle A'C'B' = 28°$

63. $AD = A'D' = 6$

65. $A'B' = AB = 8$

67. $m\angle A'D'C' = m\angle ADC = 70°$

69. $180° - 125° = 55°$

71. $180° - 90° - 55° = 35°$

73. Let x = height of silo

$$\frac{x}{6} = \frac{105}{9}$$

$$9x = 630$$

$$x = 70 \text{ ft}$$

75. a) $\dfrac{44 \text{ mi}}{0.875 \text{ in.}} = \dfrac{\text{SP-A}}{2.25 \text{ in.}}$

$$\text{SP-A} = \frac{(44)(2.25)}{0.875} \text{ mi} = 113.14 \text{ mi}$$

b) $\dfrac{44 \text{ mi}}{0.875 \text{ in.}} = \dfrac{\text{SP-R}}{1.5 \text{ in.}}$

$$\text{SP-R} = \frac{(44)(1.5)}{0.875} \text{ mi} = 75.43 \text{ mi}$$

77.

$$\frac{DE}{D'E'} = 3$$

$$\frac{12}{D'E'} = 3$$

$$3D'E' = 12$$

$$\overline{D'E'} = 4$$

$$\frac{EF}{E'F'} = 3$$

$$\frac{15}{E'F'} = 3$$

$$3E'F' = 15$$

$$\overline{E'F'} = 5$$

$$\frac{DF}{D'F'} = 3$$

$$\frac{9}{D'F'} = 3$$

$$3D'F' = 9$$

$$\overline{D'F'} = 3$$

79. a) $m\angle HMF = m\angle TMB$, $m\angle HFM = m\angle TBM$, $m\angle MHF = m\angle MTB$

Let x = height of the wall

$$\frac{x}{20} = \frac{5.5}{2.5}$$

$$2.5x = 110$$

$$x = \frac{110}{2.5} = 44 \text{ ft}$$

Exercise Set 9.3

Throughout this section, on exercises involving π, we used the π key on a scientific calculator to determine the answer. If you use 3.14 for π, your answers may vary slightly.

1. a) The **perimeter** of a two-dimensional figure is the sum of the lengths of the sides of the figure.
   b) The **area** of a two-dimensional figure is the region within the boundaries of the figure.
   c)

   $A = lw = 6(2) = 12$ square units

   $P = 2l + 2w = 2(6) + 2(2) = 12 + 4 = 16$ units

3. a) To determine the number of square feet, multiply the number of square yards by $3 \times 3 = 9$.
   b) To determine the number of square yards, divide the number of square feet by $3 \times 3 = 9$.

5. $A = \dfrac{1}{2}bh = \dfrac{1}{2}(5)(7) = 17.5$ in.^{2}

7. $A = \dfrac{1}{2}bh = \dfrac{1}{2}(7)(5) = 17.5$ cm^2

9. $A = lw = 21(10) = 210$ ft^2

   $P = 2l + 2w = 2(21) + 2(10) = 62$ ft

11. $3\text{ m} = 3(100) = 300$ cm

    $A = bh = 300(20) = 6000$ cm^2

    $P = 2l + 2w = 2(300) + 2(27) = 654$ cm

13. $2\text{ ft} = 2(12) = 24$ in.

    $A = \dfrac{1}{2}h(b_1 + b_2) = \dfrac{1}{2}(24)(5 + 19) = 288$ in.^{2}

    $P = s_1 + s_2 + b_1 + b_2 = 25 + 25 + 5 + 19 = 74$ in.

15. $A = \pi r^2 = \pi(4)^2 = 16\pi \approx 50.27$ in.^{2}

    $C = 2\pi r = 2\pi(4) = 8\pi \approx 25.13$ in.

17. $r = \dfrac{13}{2} = 6.5$ ft

    $A = \pi r^2 = \pi(6.5)^2 = 42.25\pi \approx 132.73$ ft^2

    $C = \pi d = \pi(13) \approx 40.84$ ft

19. a) $c^2 = 15^2 + 8^2$

    $c^2 = 225 + 64$

    $c^2 = 289$

    $c = \sqrt{289} = 17$ yd

    b) $P = s_1 + s_2 + s_3 = 8 + 15 + 17 = 40$ yd

    c) $A = \dfrac{1}{2}bh = \dfrac{1}{2}(8)(15) = 60$ yd^2

21. a) $b^2 + 5^2 = 13^2$

    $b^2 + 25 = 169$

    $b^2 = 144$

    $c = \sqrt{144} = 12$ km

    b) $P = s_1 + s_2 + s_3 = 5 + 12 + 13 = 30$ km

    c)

    $A = \dfrac{1}{2}bh = \dfrac{1}{2}(5)(12) = 30$ km^2

23. Area of square: $(10)^2 = 100$ m^2

    Area of circle: $\pi(5)^2 = 25\pi = 78.53981634$ m^2

    Shaded area:

    $100 - 78.53981634 = 21.46018366 \approx 21.46$ m^2

25. Use the Pythagorean Theorem to find the length of a side of the shaded square.

$$x^2 = 2^2 + 2^2$$

$$x^2 = 4 + 4$$

$$x^2 = 8$$

$$x = \sqrt{8}$$

Shaded area: $\sqrt{8}\left(\sqrt{8}\right) = 8$ in.^{2}

27. Find area of trapezoid minus area of unshaded triangle.

Trapezoid: $18\left(\dfrac{9+11}{2}\right) = 180$

Triangle: $\dfrac{1}{2}(18)(10) = 90$

Shaded area: $180 - 90 = 90$ yd^2

29. Area of trapezoid:

$$\frac{1}{2}(8)(9+20) = \frac{1}{2}(8)(29) = 116 \text{ in.}^2$$

Area of circle: $\pi(4)^2 = 16\pi = 50.26548246$ in.^{2}

Shaded area:

$116 - 50.26548246 = 65.73451754 \approx 65.73$ in.^{2}

31. Radius of larger circle: $\dfrac{28}{2} = 14$ cm

Area of large circle:

$\pi(14)^2 = 196\pi = 615.7521601$ cm^2

Radius of each smaller circle: $\dfrac{14}{2} = 7$ cm

Area of each smaller circle:

$\pi(7)^2 = 49\pi = 153.93804$ cm^2

Shaded area:

$615.7521601 - 153.93804 - 153.93804$

$= 307.8760801 \approx 307.88$ cm^2

33. $\dfrac{1}{x} = \dfrac{9}{207}$

$9x = 207$

$x = \dfrac{207}{9} = 23$ yd^2

35. $\dfrac{1}{14.7} = \dfrac{9}{x}$

$x = 14.7(9) = 132.3$ ft^2

37. $\dfrac{1}{23.4} = \dfrac{10,000}{x}$

$x = 23.4(10,000) = 234,000$ cm^2

39. $\dfrac{1}{x} = \dfrac{10,000}{8625}$

$10,000x = 8625$

$x = \dfrac{8625}{10,000} = 0.8625$ m^2

41. Area of living/dining room: $25(22) = 550$ ft^2

   a) $550(10.86) = \$5973$ b) $550(13.86) = \$7623$

43. Area of kitchen: $12(14) = 168$ ft^2

Area of first floor bathroom: $6(10) = 60$ ft^2

Area of second floor bathroom: $8(14) = 112$ ft^2

Area of kitchen and both bathrooms: 340 ft^2

Cost: $340(\$8.50) = \$2890$

45. Area of bedroom 1: $10(14) = 140$ ft^2

Area of bedroom 2: $10(20) = 200$ ft^2

Area of bedroom 3: $10(14) = 140$ ft^2

Total area: $140 + 200 + 140 = 480$ ft^2

Cost: $480(\$6.06) = \$2908.80$

47. Area of entire lawn if all grass:

$400(300) = 120,000$ ft^2

Area of house: $\frac{1}{2}(50)(100+150) = 6250$ ft^2

Area of goldfish pond:

$\pi(20)^2 = 400\pi = 1256.637061$ ft^2

Area of privacy hedge: $200(20) = 4000$ ft^2

Area of garage: $70(30) = 2100$ ft^2

Area of driveway: $40(25) = 1000$ ft^2

Area of lawn:

$120,000 - 6250 - 1256.637061 - 4000 - 2100 - 1000$

$= 105,393.3629$ ft$^2 = \dfrac{105,393.3629}{9}$

$= 11,710.37366$ yd^2

Cost:

$11,710.37366(\$0.02) = \$234.2074732 \approx \$234.21$

49. a) Perimeter $= 2(94) + 2(50) = 288$ ft

b) Area $= (94)(50) = 4700$; 4700 tiles

51. Let $a =$ height on the wall that the ladder reaches

$a^2 + 20^2 = 29^2$

$a^2 + 400 = 841$

$a^2 = 441$

$a = \sqrt{441} = 21$ ft

53. Let d be the distance.

$d^2 = 37^2 + 310^2$

$d^2 = 97494$

$d = \sqrt{97494} \approx 312$ ft

55. a) $A = s^2$

b) $A = (2s)^2 = 4s^2$

c) The area of the square in part b) is four times larger than the area of the square in part a).

57. $s = \dfrac{1}{2}(a+b+c) = \dfrac{1}{2}(8+6+10) = 12$

$A = \sqrt{12(12-8)(12-6)(12-10)}$

$= \sqrt{12(4)(6)(2)} = \sqrt{576} = 24$ cm^2

Exercise Set 9.4

In this section, we use the π key on the calculator to determine answers in calculations involving π. If you use 3.14 for π, your answers may vary slightly.

1. a) **Volume** is a measure of the capacity of a figure.
   b) **Surface area** is the sum of the areas of the surfaces of a three-dimensional solid.
3. A **polyhedron** is a closed surface formed by the union of polygonal regions.
   A **regular polyhedron** is one whose faces are all regular polygons of the same size and shape.

5. A **prism** and a **pyramid** are both polyhedrons, but a prism has a top and a bottom base while a pyramid only has one base.

7. a) $V = lwh = (1)(4)(2) = 8\,\text{m}^3$

    b) $SA = 2lw + 2wh + 2lh$

    $SA = 2(1)(4) + 2(4)(2) + 2(1)(2) = 28\,\text{m}^2$

9. a) $V = s^3;\ V = 5^3 = 125\,\text{ft}^3$

    b) $SA = 6s^2;\ SA = 6\left(5^2\right) = 150\,\text{ft}^2$

11. a) $V = \pi r^2 h = \pi\left(2^2\right)(12) = 48\pi$

    $V \approx 150.80\,\text{in.}^3$

    b) $SA = 2\pi r^2 + 2\pi rh$

    $SA = 2\pi\left(2^2\right) + 2\pi(2)(12) = 56\pi$

    $SA \approx 175.93\,\text{in}^2$

13. a) $V = \dfrac{1}{3}\pi r^2 h = \dfrac{1}{3}\pi\left(3^2\right)(14) = 42\pi$

    $V \approx 131.95\,\text{cm}^3$

    b)

    $SA = \pi r^2 + \pi r\sqrt{r^2 + h^2}$

    $SA = \pi\left(3^2 + 3\sqrt{3^2 + 14^2}\right) = \pi\left(9 + 3\sqrt{205}\right)$

    $SA \approx 163.22\,\text{cm}^2$

15. a) $r = \dfrac{9}{2} = 4.5\,\text{cm}$

    $V = \dfrac{4}{3}\pi r^3$

    $V = \dfrac{4}{3}\pi\left(4.5^3\right) = \dfrac{4}{3}\pi(91.125) \approx 381.70\,\text{cm}^3$

    b) $SA = 4\pi r^2$

    $SA = 4\pi\left(4.5^2\right) = 4\pi(20.25) \approx 254.47\,\text{cm}^2$

17. Area of the base: $B = \dfrac{1}{2}bh = \dfrac{1}{2}(8)(8) = 32\,\text{in.}^2$

    $V = Bh = 32(12) = 384\,\text{in.}^3$

19. Area of the base: $B = s^2 = 11^2 = 121\,\text{cm}^2$

    $V = \dfrac{1}{3}Bh = \dfrac{1}{3}(121)(13) = 524.\overline{3} \approx 524.33\,\text{cm}^3$

21. $V =$ vol. of large prism $-$ vol. of small prism

    $V = (6)(6)(12) - (3)(3)(12) = (36 - 9)(12)$

    $V = (27)(12) = 324\,\text{mm}^3$

23. $V = 2(\text{volume of one small trough})$

    depth of trough $= 7$

    area of triangular ends $= \dfrac{1}{2}(4)(7) = 14$

    $V = 2(14)(14) = 392\,\text{ft}^3$

25. $V =$ volume of cylinder $-$ volume of 3 spheres

    $= \pi(3.5)^2(20.8) - 3\left[\dfrac{4}{3}\pi(3.45)^3\right]$

    $= 254.8\pi - 164.2545\pi = 90.5455\pi$

    $V \approx 284.46\,\text{cm}^3$

27. $V =$ volume of rect. solid $-$ volume of pyramid

    $= 3(3)(4) - \dfrac{1}{3}\left(3^2\right)(4) = 36 - 12 = 24\,\text{ft}^3$

29. $79\text{yd}^3 = 9(27) = 243\,\text{ft}^3$

31. $153\,\text{ft}^3 = \dfrac{153}{27} = 5.\overline{6} \approx 5.67\,\text{yd}^3$

33. $3.7\,\text{m}^3 = 3.7(1,000,000) = 3,700,000\,\text{cm}^3$

35. $7,500,000\,\text{cm}^3 = \dfrac{7,500,000}{1,000,000} = 7.5\,\text{m}^3$

37. Tubs: $V = \pi r^2 h = \pi(3)^2(5) = 45\pi$

    $= 141.3716694 \approx 141.37\,\text{in.}^3$

    Boxes: $V = s^3 = (5)^3 = 125\,\text{in.}^3$

39. $SA = 2lw + 2wh + 2lh$

    $SA = 2(142)(125) + 2(125)(10) + 2(10)(142)$

    $SA = 40,840\,\text{mm}^2$

41. $V = 12(4)(3) = 144 \text{ in.}^3$

   $144 \text{ in.}^3 = 144(0.01736) = 2.49984 \approx 2.50 \text{ qt}$

43.   a) $V = 80(50)(30) = 120,000 \text{ cm}^3$

   b) $120,000 \text{ m}\ell$

   c) $120,000 \text{ m}\ell = \dfrac{120,000}{1000} = 120 \ \ell$

45. $r = \dfrac{3.875}{2} = 1.9375 \text{ in.}$

   Volume of each cylinder:

   $\pi r^2 h = \pi(1.9375)^2(3)$

   $= 11.26171875\pi = 35.37973289$

   Total volume:

   $8(35.37973289) = 283.0378631 \approx 283.04 \text{ in.}^3$

47. $5.5 \text{ ft} = 5.5(12) = 66 \text{ in.}$

   $r = \dfrac{2.5}{2} = 1.25 \text{ in.}$

   $V = \pi r^2 h = \pi(1.25)^2(66) = 103.125\pi$

   $= 323.9767424 \approx 323.98 \text{ in.}^3$

   b) $\dfrac{323.98}{1728} = 0.187488426 \approx 0.19 \text{ ft}^3$

49. $V = \dfrac{1}{3}\pi r^2 h = \dfrac{1}{3}\pi\left(\dfrac{3}{2}\right)^2(6) = 4.5\pi$

   $= 14.13716694 \approx 14.14 \text{ in.}^3$

51. $8 - x + 4 = 2$

   $12 - x = 2$

   $-x = -10$

   $x = 10 \text{ edges}$

53. $x - 8 + 4 = 2$

   $x - 4 = 2$

   $x = 6 \text{ vertices}$

55. $11 - x + 5 = 2$

   $16 - x = 2$

   $-x = -14$

   $x = 14 \text{ edges}$

57. $r_E = \dfrac{12,756.3}{2} = 6378.15 \text{ km}$

   $r_M = \dfrac{3474.8}{2} = 1737.4 \text{ km}$

   a) $SA_E = 4\pi(6378.15^2) \approx 5.11 \times 10^8 \text{ km}^2$

   b) $SA_M = 4\pi(1737.4^2) \approx 3.79 \times 10^7 \text{ km}^2$

   c) $\dfrac{SA_E}{SA_M} = \dfrac{5.11 \times 10^8}{3.79 \times 10^7} \approx 13$

   d) $V_E = \dfrac{4}{3}\pi(6378.15^3) \approx 1.09 \times 10^{12} \text{ km}^3$

   e) $V_M = \dfrac{4}{3}\pi(1737.4^3) \approx 2.20 \times 10^{10} \text{ km}^3$

   f) $\dfrac{V_E}{V_M} = \dfrac{1.09 \times 10^{12}}{2.20 \times 10^{10}} \approx 50$

59.   a) – e) Answers will vary.

   f) If we double the length of each edge of a cube, the new volume will be eight times the original volume.

61.   a) Find the volume of each numbered region. Since the length of each side is $a+b$, the sum of the

   volumes of each region will equal $(a+b)^3$.

   b) $V_1 = a(a)(a) = a^3 \qquad V_2 = a(a)(b) = a^2 b \qquad V_3 = a(a)(b) = a^2 b \qquad V_4 = a(b)(b) = ab^2$

   $V_5 = a(a)(b) = a^2 b \qquad V_6 = a(b)(b) = ab^2 \qquad V_7 = b(b)(b) = b^3$

   c) The volume of the piece not shown is ab^2.

Exercise Set 9.5

1. The act of moving a geometric figure from some starting position to some ending position without altering its shape or size is called **rigid motion**. The four main rigid motions studied in this section are reflections, translations, rotations, and glide reflections.

3. A **reflection** is a rigid motion that moves a figure to a new position that is a mirror image of the figure in the starting position.

5. A **rotation** is a rigid motion performed by rotating a figure in the plane about a specific point.

7. A **translation** is a rigid motion that moves a figure by sliding it along a straight line segment in the plane.

9. A **glide reflection** is a rigid motion formed by performing a translation (or glide) followed by a reflection.

11. A geometric figure is said to have **reflective symmetry** if the positions of a figure before and after a reflection are identical (except for vertex labels).

13. A **tessellation** is a pattern consisting of the repeated use of the same geometric figures to entirely cover a plane, leaving no gaps.

15.
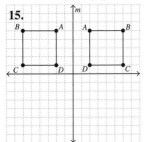

17.
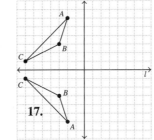

19.
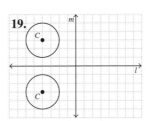

21.
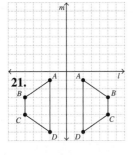

23.
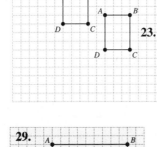

25.
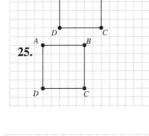

27.
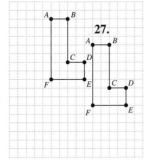

29.
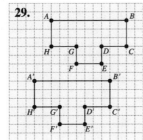

31.
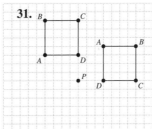

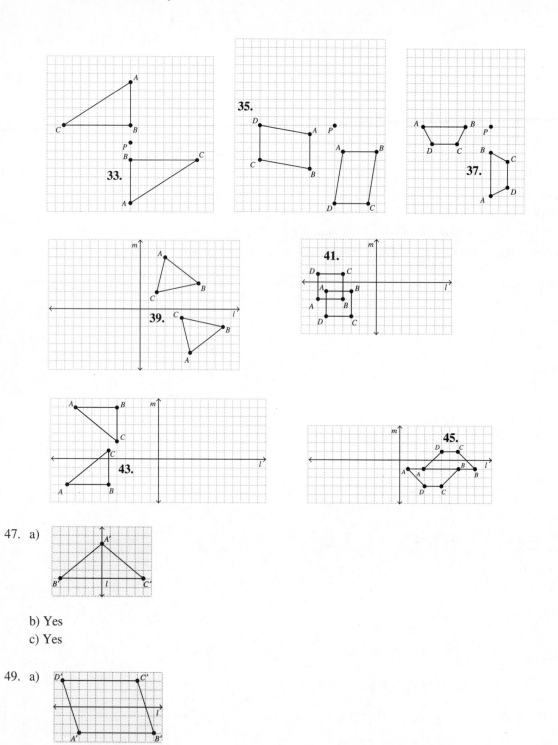

47. a)

b) Yes
c) Yes

49. a)

b) No
c) No

51. a)

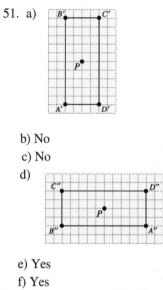

b) No

c) No

d)

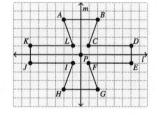

e) Yes

f) Yes

53. a) – c)

d) No. Any 90° rotation will result in the figure being in a different position than the starting position.

55. a) – b)

b) No

c) The order in which the translation and the reflection are performed is important. The figure obtained in part b) is the glide reflection.

57. Answers will vary.

59. a) Answers will vary.

b) A regular pentagon cannot be used as a tessellating shape.

61. Although answers will vary depending on the font, the following capital letters have reflective symmetry about a vertical line drawn through the center of the letter: A, H, I, M, O, T, U, V, W, X, Y.

Exercise Set 9.6

1. Topology is sometimes referred to as "rubber sheet geometry" because it deals with bending and stretching of geometric figures.

3. You can construct a Möbius strip by taking a strip of paper, giving one end a half twist, and taping the ends together.

5. Four

7. A **Jordan curve** is a topological object that can be thought of as a circle twisted out of shape.

9. The number of holes in the object determines the **genus** of an object.

11. 1, 5 – Red; 2, 4 – Green; 3 – Blue

13. 1 – Red; 2, 5 – Yellow; 3, 6 – Blue; 4 – Green

15. 1, 3, 7 – Red; 2, 6, 8 – Blue; 4,5 – Green

17. CA, WA, MT, UT – Red
    OR, WY, AZ – Green
    ID, NM – Blue
    NV, CO – Yellow

19. YT, NU, AB, ON – Red
    NT, QC – Blue
    BC, SK – Green
    MB – Yellow

21. Outside; a straight line from point A to a point clearly outside the curve crosses the curve an even number of times.

23. Inside; a straight line from point F to a point clearly outside the curve crosses the curve an odd number of times.

25. Outside; a straight line from point D to a point clearly outside the curve crosses the curve an even number of times.

27. Outside; a straight line from point B to a point clearly outside the curve crosses the curve an even number of times.

29. 1

31. 5

33. Larger than 5

35. 5

37. 0

39. larger than 5

41. a) - d) Answers will vary.

43. One

45. Two

47. a) No, it has an inside and an outside.
    b) Two   c) Two
    d) Two strips, one inside the other

49. Answers will vary.

51. Answers will vary.

Exercise Set 9.7

1. Benoit Mandelbrot – first to use the word fractal to describe shapes that had several common characteristics, including some form of "self-similarity"

3. Nikolay Ivanovich Lobachevsky - discovered hyperbolic geometry

5. Janos Bolyai - discovered hyperbolic geometry

7. a) Euclidean - Given a line and a point not on the line, one and only one line can be drawn parallel to the given line through the given point.

   b) Hyperbolic - Given a line and a point not on the line, two or more lines can be drawn through the given point parallel to the given line.

   c) Elliptical - Given a line and a point not on the line, no line can be drawn through the given point parallel to the given line.

9. A plane

11. A pseudosphere

13. Spherical - elliptical geometry; flat - Euclidean geometry; saddle-shaped - hyperbolic geometry

15.

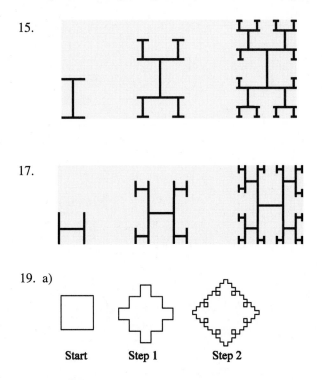

17.

19. a)

Start   Step 1   Step 2

b) Infinite.

c) Finite since it covers a finite or closed area.

Review Exercises

In the Review Exercises and Chapter Test questions, the π key on the calculator is used to determine answers in calculations involving π. If you use 3.14 for π, your answers may vary slightly.

1. $\{B\}$

2. $\overline{AD}$

3. $\triangle BFC$

4. $\overline{BH}$

5. $\{F\}$

6. $\{\ \}$

7. $90° - 23.7° = 66.3°$

8. $180° - 124.7° = 55.3°$

9. Let $x = BC$

$$\frac{BC}{B'C} = \frac{AC}{A'C}$$

$$\frac{x}{3.4} = \frac{12}{4}$$

$$4x = 40.8$$

$$x = \frac{40.8}{4} = 10.2 \text{ in.}$$

10. Let $x = A'B'$

$$\frac{A'B'}{AB} = \frac{A'C}{AC}$$

$$\frac{x}{6} = \frac{4}{12}$$

$$12x = 24$$

$$x = \frac{24}{12} = 2 \text{ in.}$$

11. $m\angle ABC = m\angle A'B'C$

    $m\angle A'B'C = 180° - 88° = 92°$

    Thus, $m\angle ABC = 92°$

    $m\angle BAC = 180° - 30° - 92° = 58°$

12. $m\angle ABC = m\angle A'B'C$

    $m\angle A'B'C = 180° - 88° = 92°$

    Thus, $m\angle ABC = 92°$

13. $m\angle 1 = 50°$

    $m\angle 6 = 180° - 110° = 70°$

    $m\angle 2 = m\angle 1 + m\angle 6 = 120°$

    $m\angle 3 = m\angle 2 = 120°$

    $m\angle 4 = m\angle 6 = 70°$

    $m\angle 5 = 180° - m\angle 4 = 180° - 70° = 110°$

14. $n = 5$

    $(n-2)180° = (5-2)180° = 3(180°) = 540°$

15. a) $A = lw = 10(8) = 80 \text{ cm}^2$

    b) $P = 2l + 2w = 2(10) + 2(8) = 36 \text{ cm}$

16. a) $A = \dfrac{1}{2}h(b_1 + b_2) = \dfrac{1}{2}(2)(4+9) = 13 \text{ in.}^2$

    b) $P = 3.2 + 4 + 3.2 + 9 = 19.4 \text{ in.}$

17. a) $A = bh = 12(7) = 84 \text{ in.}^2$

    b) $P = 2(9) + 2(12) = 42 \text{ in.}$

18. a) $A = \dfrac{1}{2}bh = \dfrac{1}{2}(3)(4) = 6 \text{ km}^2$

    b) $P = 3 + 4 + \sqrt{3^2 + 4^2} = 7 + \sqrt{25} = 12 \text{ km}$

19. a) $A = \pi r^2 = \pi(13)^2 \approx 530.93 \text{ cm}^2$

    b) $C = 2\pi r = 2\pi(13) = 26\pi \approx 81.68 \text{ cm}$

20. $A = $ area of rectangle $- 3$(area of one circle)

    Length of rectangle $= 3$(diameter of circle)

    $\qquad\qquad\qquad = 3(8) = 24$

    Area of rectangle: $(8)(24) = 192$

    Area of circle: $\pi\left(4^2\right) = 16\pi$

    Shaded area: $192 - 3(16\pi) \approx 41.20 \text{ in}^2$

21. Shaded area is the area of an 8 by 8 square minus the four corner squares (each 2 by 2) and minus the area of a circle of diameter 4.

    Shaded area $= (8)(8) - 4(2)(2) - \pi\left(2^2\right)$

    $\qquad\qquad = 48 - 4\pi \approx 35.43 \text{ cm}^2$

22. $A = lw = 14(16) = 224 \text{ ft}^2$

    Cost: $224(\$5.25) = \$1176$

23. a) $V = lwh = 10(3)(4) = 120 \text{ cm}^3$

    b) $SA = 2lw + 2wh + 2lh$

    $SA = 2(10)(3) + 2(3)(4) + 2(10)(4) = 164 \text{ cm}^2$

24. a) $V = \pi r^2 h = \pi\left(6^2\right)(18) = 648\pi \approx 2035.75 \text{ in}^3$

    b) $SA = 2\pi r^2 + 2\pi rh$

    $SA = 2\pi\left(6^2\right) + 2\pi(6)(18) = 288\pi$

    $SA \approx 904.78 \text{ in}^2$

25. a) $r = \dfrac{12}{2} = 6$ mm

$$V = \frac{1}{3}\pi r^2 h = \frac{1}{3}\pi(6^2)(16) = 192\pi$$

$$V \approx 603.19 \text{ mm}^3$$

b) $SA = \pi r^2 + \pi r\sqrt{r^2 + h^2}$

$$SA = \pi\left(6^2 + 6\sqrt{6^2 + 16^2}\right) = \pi\left(36 + 6\sqrt{292}\right)$$

$$SA \approx 435.20 \text{ mm}^2$$

27. $B = \dfrac{1}{2}bh = \dfrac{1}{2}(9)(12) = 54 \text{ m}^2$

$$V = Bh = 54(8) = 432 \text{ m}^3$$

26. a) $V = \dfrac{4}{3}\pi r^3$

$$V = \frac{4}{3}\pi\left(4^3\right) = \frac{4}{3}\pi(64) \approx 268.08 \text{ ft}^3$$

b) $SA = 4\pi r^2$

$$SA = 4\pi\left(4^2\right) = 4\pi(16) \approx 201.06 \text{ ft}^2$$

28. If h represents the height of the triangle which is the base of the pyramid, then

$$\begin{aligned} h^2 + 3^2 &= 5^2 \\ h^2 + 9 &= 25 \\ h^2 &= 16 \\ h &= \sqrt{16} = 4 \text{ ft} \end{aligned}$$

$$B = \frac{1}{2}bh = \frac{1}{2}(6)(4) = 12 \text{ ft}^2$$

$$V = \frac{1}{3}Bh = \frac{1}{3}(12)(7) = 28 \text{ ft}^3$$

29. V = volume of cylinder − volume of cone

$$= \pi(2)^2(9) - \frac{1}{3}\pi(2)^2(9) = 36\pi - 12\pi = 24\pi$$

$$= 75.39822369 \approx 75.40 \text{ cm}^3$$

31.
$$\begin{aligned} h^2 + 1^2 &= 3^2 \\ h^2 + 1 &= 9 \\ h^2 &= 8 \\ h &= \sqrt{8} \end{aligned}$$

$$A = \frac{1}{2}h(b_1 + b_2) = \frac{1}{2}\left(\sqrt{8}\right)(2 + 4) = 8.485281374 \text{ ft}^2$$

a) $V = Bh = 8.485281374(8)$

$$= 67.88225099 \approx 67.88 \text{ ft}^3$$

30. V = vol. of large sphere − vol. of small sphere

$$= \frac{4}{3}\pi(6)^3 - \frac{4}{3}\pi(3)^3 = 288\pi - 36\pi = 252\pi$$

$$= 791.6813487 \approx 791.68 \text{ cm}^3$$

31. b) Weight:

$$67.88(62.4) + 375 = 4610.7 \text{ lb}$$

Yes, it will support the trough filled with water.

c) $(4610.7 - 375) = 4235.7$ lb of water

$$\frac{4235.7}{8.3} = 510.3253 \approx 510.3 \text{ gal}$$

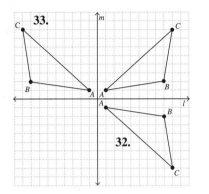

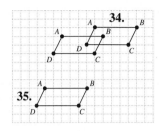

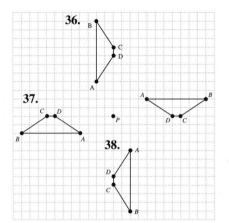

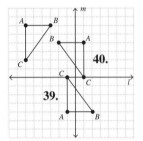

41. Yes   42. No   43. No   44. Yes

45. a) – d) Answers will vary.

46. Saarland, North Rhine-Westphalia, Bremen, Mecklenburg-Western Pomerania, Berlin, Thuringia, Baden-Württemberg, Hamburg – Red

Rhineland-Palatinate, Lower Saxony, Saxony – Green

Schleswig-Holstein, Hesse, Brandenburg – Yellow

Bavaria, Saxony-Anhalt – Blue

47. Outside; a straight line from point A to a point clearly outside the curve crosses the curve an even number of times.

48. Euclidean: Given a line and a point not on the line, one and only one line can be drawn parallel to the given line through the given point.

Elliptical: Given a line and a point not on the line, no line can be drawn through the given point parallel to the given line.

Hyperbolic: Given a line and a point not on the line, two or more lines can be drawn through the given point parallel to the given line.

49.

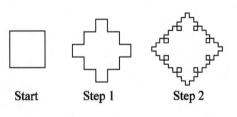

Start   Step 1   Step 2

50.

Chapter Test

1. $\overset{\circ}{\overrightarrow{EF}}$

2. $\triangle BCD$

3. $\{D\}$

4. $\overleftrightarrow{AC}$

5. $90° - 12.4° = 77.6°$

6. $180° - 51.7° = 128.3°$

7. The other two angles of the triangle are $48°$ (by vertical angles) and $180° - 112° = 68°$. Thus, the measure of angle $x = 180° - 48° - 68° = 64°$.

8. $n = 10$

$$(n-2)180° = (10-2)180° = 8(180°) = 1440°$$

9. Let $x = B'C'$

$$\frac{B'C'}{BC} = \frac{A'C'}{AC}$$

$$\frac{x}{7} = \frac{5}{13}$$

$$13x = 35$$

$$x = \frac{35}{13} = 2.692307692 \approx 2.69 \text{ cm}$$

10. a)
$$x^2 + 5^2 = 13^2$$
$$x^2 + 25 = 169$$
$$x^2 = 144$$
$$x = \sqrt{144} = 12 \text{ in.}$$

b) $P = 5 + 13 + 12 = 30$ in.

c) $A = \frac{1}{2}bh = \frac{1}{2}(5)(12) = 30$ in.^{2}

11. $r = \frac{14}{2} = 7$ cm

a) $V = \frac{4}{3}\pi r^3 = \frac{4}{3}\pi(7^3) \approx 1436.76$ cm^3

b) $SA = 4\pi r^2$

$SA = 4\pi(7^2) = 4\pi(49) \approx 615.75$ cm^2

12. Shaded volume $=$

volume of prism $-$ volume of cylinder

Volume of prism: $V = lwh = (6)(4)(3) = 72$ m^3

Volume of cylinder: $V = \pi r^2 h = \pi(1^2)4 = 4\pi$ m^3

Shaded volume $= 72 - 4\pi \approx 59.43$ m^3

13. $B = lw = 4(7) = 28$ ft^2

$$V = \frac{1}{3}Bh = \frac{1}{3}(28)(12) = 112 \text{ ft}^3$$

14.

15.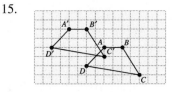

16.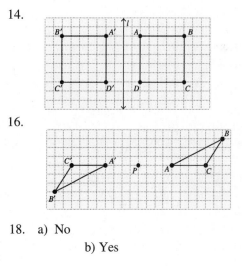

17.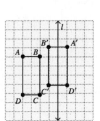

18. a) No

b) Yes

19. A **Möbius strip** is a surface with one side and one edge.

20. a) and b) Answers will vary.

21. **Euclidean:** Given a line and a point not on the line, one and only one line can be drawn parallel to the given line through the given point. **Elliptical:** Given a line and a point not on the line, no line can be drawn through the given point parallel to the given line. **Hyperbolic:** Given a line and a point not on the line, two or more lines can be drawn through the given point parallel to the given line.

CHAPTER TEN

MATHEMATICAL SYSTEMS

Exercise Set 10.1

1. A binary operation is an operation that is performed on two elements, and the result is a single element.

3. Each of these operations can be performed on only two elements at a time and the result is always a single element.   a) $2 + 3 = 5$ b) $5 - 3 = 2$ c) $2 \times 3 = 6$ d) $6 \div 3 = 2$

5. Closure, identity, each element must have a unique inverse, associative property, commutative property.

7. An identity element is an element in a set such that when a binary operation is performed on it and any given element in the set, the result is the given element. The additive identity element is 0, and the multiplicative identity element is 1.   Examples: $5 + 0 = 5$, $5 \times 1 = 5$

9. When a binary operation is performed on two elements in a set and the result is the identity element for the binary operation, then each element is said to be the inverse of the other. The additive inverse of 2 is (-2) since $2 + (-2) = 0$, and the multiplicative inverse of 2 is $(1/2)$ since $2 \times 1/2 = 1$.

11. Yes. For a group, the Commutative property need not apply.

13. d   The Commutative property need not apply.

15. The commutative property of addition stated that $a + b = b + a$, for any elements a, b, and c.
    Example: $3 + 4 = 4 + 3$

17. The associative property of multiplication states that $(a \times b) \times c = a \times (b \times c)$, for any real numbers a, b, and c.   Example: $(3 \times 4) \times 5 = 3 \times (4 \times 5)$

19. $8 \div 4 = 2$, but $4 \div 8 = \frac{1}{2}$

21. $(6 - 3) - 2 = 3 - 2 = 1$, but $6 - (3 - 2) = 6 - 1 = 5$

23. Yes. Satisfies 5 properties needed

25. No. No identity element

27. No. Not closed

29. No. No identity element

31. No. Not all elements have inverses

33. Yes. Satisfies 4 properties needed

35. No. Not closed   ie.: 1/0 is undefined

37. No. Does not satisfy Associative property

39. Yes. Closure: The sum of any two real numbers is a real
        number. The identity element is zero.
          Example: $5 + 0 = 0 + 5 = 5$
    Each element has a unique inverse.
    Example: $6 + (-6) = 0$
    The associative property holds:
    Example: $(2 + 3) + 4 = 2 + (3 + 4)$

41. No; the system is not closed, $\pi + (-\pi) = 0$
   which is not an irrational number.

43. Answers will vary.

Exercise Set 10.2

1. The clock addition table is formed by adding all pairs of integers between 1 and 12 using the 12 hour clock
   to determine the result. Example: If the clock is at 7 and we add 8, then the clock will read 3.
   Thus, $7 + 8 = 3$ in clock arithmetic.

3. a) First add $(2 + 11)$ on the clock, then add that result to 5 on the clock to obtain the final answer.
   b) $(2 + 11) + 5 = 2 \qquad (1) + 5 = 6$

5. a) Add 12 to 2 to get 14, which is larger than 6. Then subtract 6 from 14.
   b) $2 - 6 \qquad 2 + 12 = 14 \qquad 14 - 6 = 8$
   c) Since 12 is the identity element, you can add 12 to any number without changing the answer.

7. Yes. 12

9. Yes. 1 and 11 are inverses, 2 and 10 are inverses, 3 and 9 are inverses, 4 and 8 are inverses, 5 and 7 are
   inverses, 6 is its own inverse, and 12 is its own inverse.

11. Yes. $6 + 9 = 3$ and $9 + 6 = 3$

12. Yes, the five properties are met.
   1) The system is closed. All results are from the set $\{1, 2, 3, 4, 5, 6, 7, 8, 9, 10, 11, 12\}$
   2) The identity element is 12.
   3) Each element has an inverse.
   4) The associative property holds true.
   5) The system is commutative.

13. a) Identity element = 7
    b) Additive inverse of 2 is 5. $2 + 5 = 7$

15. No. Not commutative, Non-symmetrical
    around main diagonal

17. Identity element = C, Row 3 is identical to top
    row and column 3 is identical to left column

19. a) The inverse of A is B because $A \oslash B = C$.
    b) The inverse of B is A because $B \oslash A = C$.
    c) The identity C is its own inverse.

21. $2 + 5 = 7$

23. $8 + 8 = 4$

25. $4 + 12 = 4$

27. $3 + (8 + 9) = 3 + 5 = 8$

29. $(6 + 4) + 8 = 10 + 8 = 6$

31. $(7 + 8) + (9 + 6) = 3 + 3 = 6$

33. $11 - 6 = 5$

35. $6 - 7 = 11$

37. $11 - 8 = 3$

39. $1 - 12 = 1$

41. $5 - 5 = 12$

43. $12 - 12 = 12$

45.

+	1	2	3	4	5	6
1	2	3	4	5	6	1
2	3	4	5	6	1	2
3	4	5	6	1	2	3
4	5	6	1	2	3	4
5	6	1	2	3	4	5
6	1	2	3	4	5	6

55.

+	1	2	3	4	5	6	7
1	2	3	4	5	6	7	1
2	3	4	5	6	7	1	2
3	4	5	6	7	1	2	3
4	5	6	7	1	2	3	4
5	6	7	1	2	3	4	5
6	7	1	2	3	4	5	6
7	1	2	3	4	5	6	7

47. $4 + 5 = 3$

49. $5 - 2 = 3$

51. $2 - 6 = 2$

53. $4 - 6 = 4$

55. See above.

57. $4 + 7 = 4$

59. $7 + 6 = 6$

61. $3 - 6 = 4$

63. $(4 - 5) - 6 = 6 - 6 = 7$

65. Yes. Satisfies 5 required properties

67. a) $\{0, 2, 4, 6\}$

b) £

c) Yes. All elements in the table are in the original set.

d) Identity element is 0.

e) Yes; $0 £ 0 = 0$, $2 £ 6 = 0$, $4 £ 4 = 0$, $6 £ 2 = 0$

f) $(2 £ 4) £ 4 = 6 £ 4 = 2$ and $2 £ (4 £ 4) = 2 £ 0 = 2$

g) Yes; $2 £ 6 = 0 = 6 £ 2$

h) Yes, system satisfies five properties needed.

69. a) $\{*, 5, L\}$

b)

c) Yes. All elements in the table are in the original set.

d) Identity element is L.

e) Yes; $*$ $5 = L$, 5 $* = L$, L $L = L$

f) $(*$ $5)$ $5 = L$ $5 = 5$ and $*$ $(5$ $5) = *$ $* = 5$

g) Yes; L $* = *$ and $*$ $L = *$

h) Yes, system satisfies five properties needed.

71. a) $\{f, r, o, m\}$ b)

c) The system is closed. All elements in the table are elements of the set.

d) $(r$ $o)$ $f = m$ $f = m$

e) $(f$ $r)$ $m) = r$ $m = f$

f) Identity element is f.

g) Inverse of r is m since m $r = f = r$ m.

h) Inverse of m is r since r $m = f = m$ r.

73. Not closed, since $\Gamma \ddagger \Gamma = \Pi$, which is not in the original set of elements.

75. No inverses for $\odot$ and $*$

$(* \boxdot *) \boxdot T = \odot \boxdot T = *$

$* \boxdot (* \boxdot T) = * \boxdot T = P$

Not associative since $* \neq P$

77. No identity element and therefore no inverses.

$(d \Leftrightarrow e) \Leftrightarrow d = d \Leftrightarrow d = e$

$d \Leftrightarrow (e \Leftrightarrow d) = d \Leftrightarrow e = d$

Not associative since $e \neq d$

$e \Leftrightarrow d = e$ $d \Leftrightarrow e = d$

Not commutative since $e \neq d$

79. a)

+	E	O
E	E	O
O	O	E

b) The system is closed, the identity element is E, each element is its own inverse, and the system is commutative since the table is symmetric about the main diagonal. Since the system has fewer than 6 elements and satisfies the above properties, it is a commutative group.

83. a) All elements in the table are in the set {1, 2, 3, 4, 5, 6} so the system is closed. The identity is 6. 5 and 1 are inverses of each other, and 2, 3, 4, and 6 are their own inverses. Thus, if the associative property is assumed, the system is a group.

b) $4 \infty 5 = 2$, but $5 \infty 4 = 3$

85. a)

*	R	S	T	U	V	I
R	V	T	U	S	I	R
S	U	I	V	R	T	S
T	S	R	I	V	U	T
U	T	V	R	I	S	U
V	I	U	S	T	R	V
I	R	S	T	U	V	I

b) Is a group: Is closed; identity = I; each element has a unique inverse;
an example of associativity:
$R * (T * V) = R * U = S$
$(R * T) * V = U * V = S$

c) $R * S = T$ $S * R = U$

Not Commutative since $T \neq U$

81. Student activity - Answers will vary.

83. Examples of associativity
$(2 \infty 3) \infty 4 = 5 \infty 4 = 3$ and
$2 \infty (3 \infty 4) = 2 \infty 5 = 3$
$(1 \infty 3) \infty 5 = 4 \infty 5 = 2$ and
$1 \infty (3 \infty 5) = 1 \infty 4 = 2$

87.

+	0	1	2	3	4
0	0	1	2	3	4
1	1	2	3	4	0
2	2	3	4	0	1
3	3	4	0	1	2
4	4	0	1	2	3

89. 1) Add number in left column to number in top row
2) Divide this sum by 4
3) Replace remainder in table

91. a) Matrix addition is commutative because addition of real numbers is commutative.

b) For example:
$$\begin{bmatrix} 1 & 4 \\ 3 & 7 \end{bmatrix}\begin{bmatrix} 5 & 0 \\ 1 & 2 \end{bmatrix} = \begin{bmatrix} 9 & 8 \\ 22 & 14 \end{bmatrix} \text{ but } \begin{bmatrix} 5 & 0 \\ 1 & 2 \end{bmatrix}\begin{bmatrix} 1 & 4 \\ 3 & 7 \end{bmatrix} = \begin{bmatrix} 5 & 20 \\ 7 & 18 \end{bmatrix}$$

Exercise Set 10.3

1. A modulo m system consists of m elements, 0 through m – 1, and a binary operation.

3. In a modulo 3 system there will be 3 modulo classes. When a number is divided by 3 the remainder will be 0, 1, or 2.

0	1	2
0	1	2
3	4	5
6	7	8
.	.	.

5. In a modulo 16 system there will be 16 modulo classes. When a number is divided by 16 the remainder will be a number 0 – 15.

7. $29 \equiv 4 \pmod 5$: 4, 9, and 204 have the same remainder, 4, when divided by 5.

9. Today is Thursday = Day 4, so $17 \equiv 0 \pmod 7$ is Sunday

11. $365 \div 7 = 52$ remainder 1. So 365 days is 1 day later than today (Thursday), or Friday.

13. 3 years, 34 days = $(3)(365 + 34)$ days = 1129 days
$1129 \div 7 = 161$ remainder 2. So 3 years, 34 days is 2 days later than today (Thursday), or Saturday.

15. 400 days / 7 = 57 remainder 1. So 400 days is one day after today, or Friday.

17. $9 + 16 \equiv 1 \pmod{12}$, so 16 months after October is February.

19. Since 2 years = 24 months and $24 \equiv 0 \pmod{12}$, 2 years 7 months is the same as 7 months;
$9 + 7 \equiv 4$
$\pmod{12}$ so 7 months after October is May.

21. $83 \div 12 = 6$ remainder 11. So 83 months is 11 months after October, which is September.

23. $105 \div 12 = 8$ remainder 9. So 105 months is 9 months after October, which is July.

25. $4 + 7 = 11$ $11 \equiv 1 \pmod 5$

27. $2 + 5 + 5 = 12$ $12 \equiv 2 \pmod 5$

29. $5 - 12 \equiv (10 + 5) - 12 \pmod 5$
$\equiv 15 - 12 \pmod 5$
$\equiv 3 \pmod 5$

31. $8 \bullet 9 = 72$ $72 \equiv 2 \pmod 5$

33. $4 - 8 \equiv (5 + 4) - 8 \pmod 5$
$\equiv 9 - 8 \pmod 5$
$\equiv 1 \pmod 5$

35. $(15 \bullet 4) - 8 = 60 - 8 = 52$ $52 \equiv 2 \pmod 5$

37. $10 \pmod 5 \equiv 0$

39. $48 \pmod{12} \equiv 0$

41. $41 \pmod 9 \equiv 5$

43. $30 \pmod 7 \equiv 2$

45. $-1 \pmod 7 \equiv 6$

47. $-13 \pmod{11} \equiv 9$

49. $205 \pmod{10} \equiv 5$

51. $2 + 3 = 5 \equiv 5 \pmod 6$

53. $1 + 3 \equiv 4 \pmod 5$

55. $4 - 5 \equiv 5 \pmod 6$

57. $5 \bullet 5 \equiv 7 \pmod 9$

59. $2 \bullet \{\ \} \equiv 1 \pmod 6$
No solution

61. $4 \bullet \{\ \} \equiv 4 \pmod{10}$
$\{1, 6\}$

63. $5 - 8 \equiv 9 \pmod{12}$

65. $3 \bullet 0 \equiv 0 \pmod{10}$

67. a) 2020, 2024, 2028, 2032, 2036
b) 3004
c) 2552, 2556, 2560, 2564, 2568, 2572

69.   am/pm am/pm am/pm r am am r r

$\uparrow$

today

a) $28 \equiv 4$ (mod 8); 4 days after today she rests.

b) $60 \equiv 4$ (mod 8); 4 days after today she rests.

c) $127 \equiv 7$ (mod 8); 7 days after today she has morning and afternoon practice.

d) $82 \equiv 2$ (mod 8); 2 days after today she has a morning practice, so she does not have an off day.

73. The waiter's schedule in a mod 14 system is given in the following table:

Day: 0 1 2 3 4 5 6 7 8 9 10 11 12 13

shift: d d d d d e e e d d d e e

$\uparrow$

today

a) $20 \equiv 6$ (mod 14); 6 days from today is an evening shift.

b) $52 \equiv 10$ (mod 14); 10 days from today is the day shift.

c) $365 \equiv 1$ (mod 14); 1 day from today is the day shift.

71. The manager's schedule is repeated every seven weeks. If this is week two of her schedule, then this is her second weekend that she works, or week 1 in a mod 7 system. Her schedule in mod 7 on any given weekend is shown in the following table:

Weekend (mod 7):

Work/off 0   1   2   3   4   5   6

w   w   w   w   w   w   o

a) If this is weekend 1, then in 5 more weeks $(1 + 5 = 6)$ she will have the weekend off.

b) $25 \cong 7 = 3$, remainder 4. Thus $25 \cong$ (mod 7) and 4 weeks from weekend 1 will be weekend 5. She will not have off.

c) $50 \cong 7 = 7$, remainder 1. One week from weekend 1 will be weekend 2. It will be 4 more weeks before she has off. Thus, in 54 weeks she will have the weekend off.

75. a)

+	0	1	2
0	0	1	2
1	1	2	0
2	2	0	1

b) Yes. All the numbers in the table are from the set $\{0, 1, 2, 3\}$.

c) The identity element is 0.

d) Yes.   element + inverse = identity

$0 + 0 = 0$ $1 + 2 = 0$ $2 + 1 = 0$

e) $(1 + 2) + 2 = 2$ $1 + (2 + 2) = 1 + 1 = 2$

Associative since $2 = 2$.

f) Yes, the table is symmetric about the main diagonal. $1 + 2 = 0 = 2 + 1$

g) Yes. All five properties are satisfied.

h)   Yes. The modulo system behaves the same no matter how many elements are in the system.

77. a)

×	0	1	2	3
0	0	0	0	0
1	0	1	2	3
2	0	2	0	2
3	0	3	2	1

    b) Yes. All the elements in the table are from the set $\{0, 1, 2, 3\}$.

    c) Yes. The identity element is 1.

    d) elem. $\times$ inverse = identity

       $0 \times$ none $= 1$ $1 \times 1 = 1$ $2 \times$ none $= 1$

       $3 \times 3 = 1$ Elements 0 and 2 do not have inverses.

    e) $(1 \times 3) \times 0 = 3 \times 0 = 0$

       $1 \times (3 \times 0) = 1 \times 0 = 0$ Yes, Associative

    f) Yes. $2 \times 3 = 2 = 3 \times 2$

    g) No. Not all elements have inverses.

For the operation of division in modular systems, we define $n \div d = n \bullet i$, where i is the multiplicative inverse of d.

79. $? \div 5 \equiv 5 \pmod 9$

    Since $5 \cdot 2 \equiv 1 \pmod 9$, 2 is the inverse of 5.

    $? \div 5 \pmod 9 \equiv ? \cdot 2 \pmod 9 \equiv 5 \pmod 9$

    Since $7 \cdot 2 \equiv 5 \pmod 9$, $? = 7$

81. $? \div ? \equiv 1 \pmod 4$ $0 \div 0$ is undefined.

    $1 \div 1 \equiv 1 \pmod 4$ $2 \div 2 \equiv 1 \pmod 4$

    $3 \div 3 \equiv 1 \pmod 4$ $? = \{1, 2, 3\}$

83. $5k \equiv x \pmod 5$ $5(1) \equiv 0 \pmod 5$

    $5(2) = 10 \equiv 0 \pmod 5$ $x = 0$

85. $4k - 2 \equiv x \pmod 4$ $4(0) - 2 = -2 \equiv 2 \pmod 4$

    $4(1) - 2 = 2 \equiv 2 \pmod 4$ $4(2) - 2 = 6 \equiv 2 \pmod 4$

    $x = 2$

87. a) $365 \equiv 1 \pmod 7$ so his birthday will be one day later in the week next year and will fall on Tuesday.

    b) $366 \equiv 2 \pmod 7$ so if next year is a leap year his birthday will again fall Wednesday.

89. (365 days)(24 hrs./day)(60 min./hr.) = 525,600 hrs.

    $(525,600)/(4) = 131,400$ rolls $131400 \cong 0 \pmod 4$

Review Exercises

1. A set of elements and at least one binary operation.

2. A binary operation is an operation that can be performed on two and only two elements of a set. The result is a single element.

3. No. Example: $2 - 3 = -1$, but -1 is not a whole number.

4. Yes. The difference of any two real numbers is always a real number.

5. $8 + 7 = 15 \equiv 3 \pmod{12}$ 6. $5 + 12 = 17 \equiv 5 \pmod{12}$ 7. $4 - 9 = -3 \equiv 9 \pmod{12}$

8. $4 + 7 + 9 = 20 \equiv 8 \pmod{12}$ 9. $7 - 4 + 6 = 9 \equiv 9 \pmod{12}$ 10. $2 - 8 - 7 = -13 \equiv 11 \pmod{12}$

11. a) The system is closed. If the binary operation is $\boxdot$ then for any elements a and b in the set, a $\boxdot$ b is a member of the set.

    b) There exists an identity element in the set. For any element a in the set, if a $\boxdot$ i = i $\boxdot$ a = a, then i is called the identity element.

    c) Every element in the set has a unique inverse. For any element a in the set, there exists an element b such that a $\boxdot$ b = b $\boxdot$ a = i. Then b is the inverse of a, and a is the inverse of b.

    d) The set is associative under the operation For elements a, b, and c in the set, (a $\boxdot$ b) $\boxdot$ c = a $\boxdot$ (b $\boxdot$ c).

12. An Abelian group is a group in which the operation has the commutative property.

13. Yes

14. The set of integers with the operation of multiplication does not form a group since not all elements have an inverse. Example: $4 \cdot \dfrac{1}{4} = 1$, but $\dfrac{1}{4}$ is not an integer. Only 1 and –1 have inverses.

15. No, there is no identity element.

16. The set of rational numbers with the operation of multiplication does not form a group since zero does not have an inverse. $0 \bullet \underline{?} = 1$

17. There is no identity element. Therefore the system does not form a group.

18. Not every element has an inverse.
    Not Associative   Example: (P ? P) ? 4 = L ? 4 = #   P ? (P ? 4) = P ? L = 4;   # $\neq$ 4

19. Not Associative   Example: (! $\square$ p) $\square$? = p $\square$? = ! and ! $\square$ (p $\square$?) = ! $\square$! = ?;   ! $\neq$?

20. a) { ☺ , ●, ♀, ♂ }

b) ⌂

c) Yes. All the elements in the table are from the set { ☺ , ●, ♀, ♂ }.

d) The identity element is ☺ .

e) Yes.

☺ ⌂ ☺ = ☺   ● ⌂ ♂ = ☺

♀ ⌂ ♀ = ☺   ♂ ⌂ ● = ☺

Every element has an inverse.

f) Yes, Associative

(☺ ⌂ ♀) ⌂ ♂ = ♀ ⌂ ♂ = ●

☺ ⌂ (♀ ⌂ ♂) = ☺ ⌂ ● = ●

g) Yes. The elements in the table are symmetric around the diagonal.

♂ ⌂ ♀ = ● = ♀ ⌂ ♂

h) Yes, all five properties are satisfied.

21. $15 \div 5 = 3$, remainder 0 $15 \equiv 0 \pmod 5$

22. $26 \div 7 = 3$, remainder 5 $26 \equiv 5 \pmod 7$

23. $31 \div 6 = 5$, remainder 1 $31 \equiv 1 \pmod 6$

24. $59 \div 8 = 7$, remainder 3 $59 \equiv 3 \pmod 8$

25. $71 \div 12 = 5$, remainder 11 $71 \equiv 11 \pmod{12}$

26. $54 \div 4 = 13$, remainder 2 $54 \equiv 2 \pmod 4$

27. $52 \div 12 = 4$, remainder 4 $52 \equiv 4 \pmod{12}$

28. $54 \div 14 = 3$, remainder 12 $54 \equiv 12 \pmod{14}$

29. $97 \div 11 = 8$, remainder 9 $97 \equiv 9 \pmod{11}$

30. $31 \div 20 = 1$, remainder 1 $31 \equiv 11 \pmod{20}$

31. $3 + 7 = 10 \equiv 1 \pmod 9$; $? = 1$

32. $? - 4 \equiv 2 \pmod 5$

$0 - 4 \equiv 1 \pmod 5$ $1 - 4 \equiv 2 \pmod 5$

$2 - 4 \equiv 3 \pmod 5$ $3 - 4 \equiv 4 \pmod 5$

$? = 1$

33. $4 \bullet ? \equiv 3 \pmod 7$

$4 \bullet 0 \equiv 0 \pmod 7$ $4 \bullet 1 \equiv 4 \pmod 7$

$4 \bullet 2 = 8 \equiv 1 \pmod 7$ $4 \bullet 3 = 12 \equiv 5 \pmod 7$

$4 \bullet 4 = 16 \equiv 2 \pmod 4$ $4 \bullet 5 = 20 \equiv 6 \pmod 7$

$4 \bullet 6 = 24 \equiv 3 \pmod 7$

$? = 6$

34. $6 - ? \equiv 5 \pmod 7$

$6 - 0 \equiv 6 \pmod 7$ $6 - 1 \equiv 5 \pmod 7$

$6 - 2 \equiv 4 \pmod 7$ $6 - 3 \equiv 3 \pmod 7$

$6 - 4 \equiv 2 \pmod 7$ $6 - 5 \equiv 1 \pmod 7$

Replace ? with 1.

35. $? \bullet 4 \equiv 0 \pmod 8$

$0 \bullet 4 \equiv 0 \pmod 8$ $1 \bullet 4 \equiv 4 \pmod 8$

$2 \bullet 4 = 8 \equiv 0 \pmod 8$ $3 \bullet 4 = 12 \equiv 4 \pmod 8$

$4 \bullet 4 = 16 \equiv 0 \pmod 8$ $5 \bullet 4 = 20 \equiv 4 \pmod 8$

$6 \bullet 4 = 24 \equiv 0 \pmod 8$ $7 \bullet 4 = 28 \equiv 4 \pmod 8$

Replace ? with {0, 2, 4, 6}.

36. $10 \bullet 7 \equiv ? \pmod{11}$

$10 \bullet 7 = 70$; $70 \div 11 = 6$, remainder 4

Thus, $10 \bullet 7 \equiv 4 \pmod{11}$.

Replace ? with 4

37. $3 - 5 \equiv ? \pmod 7$

$3 - 5 = (3+7) - 5 = 5 \equiv 5 \pmod 7$

Replace ? with 5.

38. $? \bullet 7 \equiv 3 \pmod{10}$

$0 \bullet 7 \equiv 0 \pmod{10}$ $1 \bullet 7 \equiv 7 \pmod{10}$

$2 \bullet 7 = 14 \equiv 4 \pmod{10}$ $3 \bullet 7 = 21 \equiv 1 \pmod{10}$

$4 \bullet 7 = 28 \equiv 8 \pmod{10}$ $5 \bullet 7 = 35 \equiv 5 \pmod{10}$

$6 \bullet 7 = 42 \equiv 2 \pmod{10}$ $7 \bullet 7 = 49 \equiv 9 \pmod{10}$

$8 \bullet 7 = 56 \equiv 6 \pmod{10}$ $9 \bullet 7 = 63 \equiv 3 \pmod{10}$

$10 \bullet 7 = 70 \equiv 0 \pmod{10}$

Replace ? with 9.

39. $5 \bullet ? \equiv 3 \pmod 8$

$5 \bullet 0 \equiv 0 \pmod 8$ $5 \bullet 1 \cong 5 \pmod 8$

$5 \bullet 2 = 10 \equiv 2 \pmod 8$ $5 \bullet 3 = 15 \equiv 7 \pmod 8$

$5 \bullet 4 = 20 \equiv 4 \pmod 8$ $5 \bullet 5 = 25 \equiv 1 \pmod 8$

$5 \bullet 6 = 30 \equiv 6 \pmod 8$ $5 \bullet 7 = 35 \equiv 3 \pmod 8$

Replace ? with 7.

40. $7 \bullet ? \equiv 2 \pmod 9$

$7 \bullet 0 \equiv 0 \pmod 9$ $7 \bullet 1 \equiv 7 \pmod 9$

$7 \bullet 2 = 14 \equiv 5 \pmod 9$ $7 \bullet 3 = 21 \equiv 3 \pmod 9$

$7 \bullet 4 = 28 \equiv 1 \pmod 9$ $7 \bullet 5 = 35 \equiv 7 \pmod 9$

$7 \bullet 6 = 42 \equiv 6 \pmod 9$ $7 \bullet 7 = 49 \equiv 4 \pmod 9$

$7 \bullet 8 = 56 \equiv 2 \pmod 9$

Replace ? with 8.

41. a)

+	0	1	2	3	4	5
0	0	1	2	3	4	5
1	1	2	3	4	5	0
2	2	3	4	5	0	1
3	3	4	5	0	1	2
4	4	5	0	1	2	3
5	5	0	1	2	3	4

41. b) Since all the numbers in the table are elements of
{0, 1, 2, 3, 4, 5}, the system has the closure
property.

c) The identity element is 0 and the inverses of
each element are 0 – 0, 1 – 5, 2 – 4, 3 – 3,
4 – 2, 5 – 1

d) The associative property holds as
illustrated by the example:
(2 + 3) + 5 = 4 = 2 +(3 + 5)

e) The commutative property holds since the
elements are symmetric about the main
diagonal

f) The modulo 6 system forms a commutative
group under addition.

42. a)

×	0	1	2	3
0	0	0	0	0
1	0	1	2	3
2	0	2	0	2
3	0	3	2	1

b) The identity element is 1, but because 0 and 2
have no inverses, the modulo 4 system does not
form a group under multiplication.

43. Day (mod 10): 0 1 2 3 4 5 6 7 8 9 10 11 12 13

Work/off : w w w o o o o w w w w o o o

↑

today

a) Since $30 \equiv 2 \pmod{14}$, Linda will be working 30 days after today.

b) Since $45 \equiv 3 \pmod{14}$, Linda will have the day off in 45 days.

Chapter Test

1. A mathematical system consists of a set of elements and at least one binary operation.

2. Closure, identity element, inverses, associative property, and commutative property.

3. Yes.

4.

+	1	2	3	4	5
1	2	3	4	5	1
2	3	4	5	1	2
3	4	5	1	2	3
4	5	1	2	3	4
5	1	2	3	4	5

5. Yes. It is closed since the only elements in the table are from the set $\{1, 2, 3, 4, 5\}$. The identity element is 5. The inverses are $1 \leftrightarrow 4, 2 \leftrightarrow 3, 3 \leftrightarrow 2,$ $4 \leftrightarrow 1,$ and $5 \leftrightarrow 5$. The system is associative. The system is commutative since the table is symmetric about the main diagonal. Thus, all five properties are satisfied.

6. $1 + 3 + 4 = 8 = 3$

7. On a clock with numbers 1 to 5, starting at 2 and moving counterclockwise 5 places ends back at 2. So $2 - 5 = 2$.

8. a) The binary operation is $\square$.

   b) Yes. All elements in the table are from the set $\{W, S, T, R\}$.

   c) The identity element is T, since $T \square x = x = x \square T$, where x is any member of the set $\{W, S, T, R\}$.

   d) The inverse of R is S, since $R \square S = T$ and $S \square R = T$.

   e) $(T \square R) \square W = R \square W = S$

9. The system is not a group. It does not have the closure property since $c * c = d$, and d is not a member of $\{a, b, c\}$.

10. Since all the numbers in the table are elements of $\{1, 2, 3\}$, the system is closed. The commutative property holds since the elements are symmetric about the main diagonal. The identity element is 2 and the inverses are $1 - 3, 2 - 2, 3 - 1$. If it is assumed the associative property holds as illustrated by the example: $(1 \ ? \ 2) \ ? \ 3 = 2 = 1 \ ? \ (2 \ ? \ 3)$, then the system is a commutative group.

11. Since all the numbers in the table are elements of $\{@, \$, \&, \%\}$, the system is closed. The commutative property holds since the elements are symmetric about the main diagonal. The identity element is $\$$ and the inverses are $@ - \&, \$ - \$, \& - @, \% - \%$. It is assumed the associative property holds as illustrated by the example: $(@ \ O \ \$) \ O \ \% = \& = @ \ O \ (\$ \ O \ \%)$, then the system is a commutative group.

12. $27 \div 5 = 5$, remainder 2 $27 \equiv 2 \pmod 5$

13. $91 \div 13 = 7$, remainder 0 $91 \equiv 0 \pmod{13}$

14. $7 + 7 \equiv 6 \pmod 8$

15. $2 - 3 = (5 + 2) - 3 = 4 \equiv 4 \pmod 5$; replace ? with 2.

16. $3 - 5 \equiv 7 \pmod 9$

    $3 - 5 = (3 + 9) - 5 = 12 - 5 \equiv 7 \pmod 9$

    $12 - 5 \equiv 7 \pmod 9$

    Replace ? with 5.

17. $4 \bullet 2 = 8$ and $8 \equiv 2 \pmod 6$

    $4 \bullet 2 \equiv 2 \pmod 6$

    Replace ? with 2.

18. $3 \bullet ? \bullet \equiv 2 \pmod 6$

    $3 \bullet 0 \equiv 0 \pmod 6$ $3 \bullet 1 \equiv 3 \pmod 6$

    $3 \bullet 2 \equiv 0 \pmod 6$ $3 \bullet 3 \equiv 3 \pmod 6$

    $3 \bullet 4 \equiv 0 \pmod 6$ $3 \bullet 5 \equiv 3 \pmod 6$

    There is no solution for ?   The answer is { }.

19. $239 \div 7 = 29$, remainder 7

    $293 \equiv 7 \pmod 8$

20. a)

$\times$	0	1	2	3	4
0	0	0	0	0	0
1	0	1	2	3	4
2	0	2	4	1	3
3	0	3	1	4	2
4	0	4	3	2	1

  b) The system is closed. The identity is 1.
     However, 0 does not have an inverse, so
     the system is <u>not</u> a group.

CHAPTER ELEVEN

CONSUMER MATHEMATICS

Exercise Set 11.1

1. A percent is a ratio of some number to 100.
3. (i) Divide the number by the denominator.
   (ii) Multiply the quotient by 100 (which has the effect of moving the decimal point two places to the right).
   (iii) Add a percent sign.

5. Percent change = $\dfrac{\left(\text{Amount in latest period}\right) - \left(\text{Amount in previous period}\right)}{\text{Amount in previous period}} \times 100$

7. $\dfrac{3}{4} = 0.75 = (0.75)(100)\% = 75.0\%$

9. $\dfrac{11}{20} = 0.55 = (0.55)(100)\% = 55.0\%$

11. $0.007654 = (0.007654)(100)\% = 0.8\%$

13. $3.78 = (3.78)(100)\% = 378\%$

15. $5\% = \dfrac{5}{100} = 0.05$

17. $5.15\% = \dfrac{5.15}{100} = 0.0515$

19. $\dfrac{1}{4}\% = 0.25\% = \dfrac{0.25}{100} = 0.0025$

21. $\dfrac{1}{5}\% = 0.2\% = \dfrac{0.2}{100} = 0.002$

23. $1\% = \dfrac{1}{100} = 0.01$

25. $\dfrac{5}{20} = \dfrac{25}{100} = 25\%$

27. $8(.4125) = 3.3 \qquad 8.0 - 3.3 = 4.7 \text{ g}$

29. $(2,604,262)(.08) = \$208,341$

31. $(2,604,262)(.25) = \$651,066$

33. $(5000)(0.31) = \$1550$

35. $(5000)(0.42) = \$2100$

37. $\dfrac{231\text{million}}{559 \text{ million}} = 0.413 \qquad 0.413 = 41.3\%$

39. $\dfrac{56 \text{ million}}{559 \text{ million}} = 0.100 \qquad 0.100 = 10.0\%$

41. $\dfrac{636,677 - 642,200}{642,200} = -0.009 = 0.9\% \text{ decrease}$

43. a) $\dfrac{0.654 - 0.329}{0.329} = 0.988 = 98.8\% \text{ increase}$

   b) $\dfrac{2.125 - 0.654}{0.645} = 2.249 = 224.9\% \text{ increase}$

   c) $\dfrac{5.225 - 2.125}{2.125} = 1.459 = 145.9\% \text{ increase}$

   d) $\dfrac{8.283 - 5.225}{5.225} = 0.585 = 58.5\% \text{ increase}$

45. a) $\dfrac{11,500 - 6450}{6450} = 0.783 = 78.3\% \text{ increase}$

   b) $\dfrac{8340 - 11,500}{11,500} = 0.275 = 27.5\% \text{ decrease}$

   c) $\dfrac{10,720 - 8340}{8340} = 0.285 = 28.5\% \text{ increase}$

   d) $\dfrac{10,720 - 6450}{6450} = 0.662 = 66.2\% \text{ increase}$

47. (.15)(45) = $ 6.75   49. 24/96 = .25; (.25)(100) = 25%   51. .05x = 75;   x = 75/.05 = 300

53. a) tax = 6% of $43.50 = (0.06)(43.50) = $2.61
    b) total bill before tip = $43.50 + $2.61 = $46.11
    c) tip = 15% of 46.11 = 0.15(46.11) = $6.92
    d) total cost = 46.11 + 6.92 = $53.03

55. $1.50(x) = 18 \qquad x = \dfrac{18}{1.50} = 12$

    12 students got an A on the 2nd test.

57. Mr. Browns' increase was 0.07(36,500) = $2,555
    His new salary = $36,500 + $2,555 = $39,055

59. Percent change = $\left(\dfrac{407-430}{430}\right)(100) =$

    $\left(\dfrac{-23}{430}\right)(100) = -5.3\%$

    There was a 5.3% decrease in the # of units sold.

61. $\dfrac{37.0-32.9}{32.9} = 0.125 = 12.5\%$ increase

63. Percent decrease from regular price =

    $\left(\dfrac{\$439-539.62}{539.62}\right)(100) = \left(\dfrac{-100.62}{539.62}\right)(100) =$

    - 18.6 %

    The sale price is 18.6% lower than the regular price.

65. (0.18)(sale price) = $675

    sale price = $\dfrac{675}{0.18} = \$3,750$

67. $1000 increased by 10% is $1000 + 0.10($1000) = $1000 + $100 = $1,100.
    $1,100 decreased by 10% is $1,100 – 0.10($1,100) = $1,100 – $110 = $990.
    Therefore if he sells the car at the reduced price he will lose $10.

69. The profit must be 0.40($5901.79) = $2,360.72. With a 40% profit, the total revenue must be
    $5901.79 + $2360.72 = $8,262.51; the revenue from the first sale would be 100 × $9.00 = $900, and the
    revenue from the second sale would be 150 × $12.50 = $1,875.00. The revenue needed for the 250 ties
    would be $8,262.51 – $900.00 – $1,875.00 = $5,487.51. Thus, the selling price should be
    $\dfrac{\$5,487.51}{250} = \$21.95.$

Exercise Set 11.2

1. Interest is the money the borrower pays for the use of the lender's money.

3. Security is anything of value pledged by the borrower that the lender may sell or keep if the borrower does not
   repay the loan.

5. i = interest, p = principal, r = interest rate, t = time
   The rate and time must be expressed for the same period of time, i.e. days, months or years.

7. The United States Rule states that if a partial payment is made on a loan, interest is computed on the principal
   from the first day of the loan (or previous partial payment) up to the date of the partial payment. For each
   partial payment, the partial payment is used to pay the interest first, then the remainder of the payment is
   applied to the principle. On the due date of the loan the interest is calculated from the date of the last partial
   payment.

9. $i = prt = (250)(.03)25) = \$15.00$

11. $i = (1100)(.0875)(90/360) = \$24.06$

13. $i = \text{prt} = (587)(0.00045)(60) = \$15.85$

15. $i = (2{,}756.78)(0.1015)\left(\dfrac{103}{360}\right) = \$80.06$

17. $i = (1372.11)(.01375)(6) = \$113.20$

19. $(2000)(r)(4) = 400 \qquad r = \dfrac{400}{(2000)(4)}(100) = 5\%$

21. $12.00 = p(0.08)\left(\dfrac{3}{12}\right) = p(0.02) \quad p = \dfrac{12.00}{0.02} = \$600$

23. $124.49 = (957.62)(0.065)(t) = 62.2453t$

$\dfrac{124.49}{62.2453} = t \qquad t = 2$ years

25. $i = (6000)(0.035)(3) = 630$

Repayment $= p + i = 6000 + 630 = \$6630$

27. a) $i = \text{prt} \qquad i = (3500)(0.075)(6/12) = \$131.25$
   b) $A = p + i \qquad A = 3500 + 131.25 = \$3{,}631.25$

29. a) $i = \text{prt} \qquad I = (3650)(0.075)(8/12) = \$182.50$
   b) $3650.00 - 182.50 = \$3467.50$, which is the amount Julie received.
   c) $i = \text{prt} \qquad 182.50 = (3467.50)(r)(8/12) = 2311.67r$

$\dfrac{182.50}{2311.67} = r = 0.0789$ or 7.9%

31. Amt. collected $= (470)(4500/2) = \$1{,}057{,}500$

$i = \text{prt} = (1{,}057{,}500)(0.054)(5/12) = \$23{,}793.75$

33. [Feb 14–Nov 25]: $329 - 45 = 284$ days

35. [02/02–03/17]: $304 - 33 = 271$ days
   Because of Leap Year, $271 + 1 = 272$ days

37. [08/24–05/15]: $(365 - 236) + 135 = 129 + 135 = 264$ days

39. [03/15] for 90 days: $74 + 90 = 164$, which is June 13

41. [11/25] for 120 days: $329 + 120 = 449$;
   $449 - 365 = 84 \qquad 84 - 1$ leap year day = day 83, which is March 24

43. [05/01 to 06/01]: $152 - 121 = 31$ days
   $(3000)(.04)(31/360) = 10.33$
   $1000.00 - 10.33 = 989.67$
   $3000.00 - 989.67 = \$2010.33$

   [06/01 to 07/01]: $182 - 152 = 30$ days
   $(2010.33)(.04)(30/360) = 6.70$
   $2010.33 + 6.70 = \$2017.03$

45. [02/01 to 05/01]: $= 121 - 32 = 89$ days
   $(2400)(.055)(89/360) = \$32.63$
   $1000.00 - 32.63 = \$967.37$
   $2400 - 967.37 = \$1432.63$

   [05/01 to 08/31]: $= 243 - 121 = 122$ days
   $(1432.63)(.055)(122/360) = \$26.70$
   $1432.63 + 26.70 = \$1459.33$

47. [07/15 to 12/27]: $361 - 196 = 165$ days
   $(9000)(.06)(165/360) = \$247.50$
   $4000.00 - 247.50 = \$3752.50$
   $9000.00 - 3752.50 = \$5247.50$

   [12/27 to 02/01]: $(365 - 361) + 32 = 36$ days
   $(5247.50)(.06)(36/360) = \$31.49$
   $5247.50 + 31.49 = \$5278.99$

49. [08/01 to 09/01]: 244 − 213 = 31 days
    (1800)(.15)(31/360) =$ 23.25
    500.00 − 23.25 = $476.75
    1800.00 − 476.75 = $1323.25

    [09/01 to 10/01]: 274 − 244 = 30 days
    (1323.25)(.15)(30/360) = $16.54
    500.00 − 16.54 = $483.46
    1323.25 − 483.46 = $839.79

    [10/01 to 11/01]: 305 − 274 = 31 days
    (839.79)(.15)(31/360) = $10.85
    839.79 + 10.85 = $850.64

53. [03/01 to 05/01]: 121 − 60 = 61 days
    (6500)(.105)(61/360) = $115.65
    1750.00 − 115.65 = $1634.35
    6500.00 − 1634.35 = $4865.65

    [05/01 to 07/01]: 182 − 121 = 61 days
    (4865.65)(.105)(61/360) = $86.57
    2350.00 − 86.57 = $2263.43
    4865.65 − 2263.43 = $2602.22

    Since the loan was for 180 days, the maturity
    date occurs in 180 − 61 − 61 = 58 days
    (2602.22)(.105)(58/360) = $44.02
    2602.22 + 44.02 = $2646.24

57. a) Amt. received = 743.21 − 39.95 = $703.26
    i = prt
    39.95 = (703.26)(r)(5/360)
    39.95 = (9.7675)(r)
    r = 39.95/9.7675 = 4.09 or 409%
    b) 39.95 = (703.26)(r)10/360)
    39.95 = (19.535)(r)
    r = 39.95/19.535 = 2.045 or 204.5%
    c) 39.95 = (703.26)(r)(20/360)
    39.95 = (39.07)r
    r = 39.95/39.07 = 1.023 or 102.3%

51. [03/01 to 08/01]: 213 − 60 = 153 days
    (11600)(.06)(153/360) = $95.80
    2000.00 − 295.80 = $1704.20
    11600.00 − 1704.20 = $9895.80

    [08/01 to 11/15]: 319 − 213 = 106 days
    (9895.80)(.06)(106/360) = $174.83
    4000.00 − 174.83 = $3825.17
    9895.8 − 3825.17 = $6070.63

    [11/15 to 12/01]: 335 − 319 = 16 days
    (6070.63)(.06)(16/360) = $16.19
    6070.63 + 16.19 = $6086.82

55. a) May 5 is day 125   125 + 182 = 307
    day 307 is Nov. 3
    b) i = (1000)(0.0434)(182/360) = $21.94
    Amt. paid = 1000 − 21.94 = $978.06
    c) interest = $21.94
    d) $r = \dfrac{i}{pt} = \dfrac{21.94}{978.06\left(\dfrac{182}{360}\right)} = 0.0444$ or 4.44%

59. a) $\dfrac{93337}{100000} = 0.93337$
    1.00000 − 0.93337 = .06663 or 6.663 %
    b) $\dfrac{100000}{93337} = 1.071386$
    1.071386 − 1.000000 = .071386 or 7.139 %
    c) 100000 − 93337 = $6663.00
    d) (6663)(.05)(1) = 33.15
    6663.00 + 33.15 = $6696.15

Exercise Set 11.3

1. An investment is the use of money or capital for income or profit.

3. For a variable investment neither the principal nor the interest is guaranteed.

5. The effective annual yield on an investment is the simple interest rate that gives the same amount of interest as a compound rate over the same period of time.

7. $n = 1,\ r = 0.04,\ t = 5,\ p = 1000$

    a) $A = 1000\left(1 + \dfrac{0.04}{1}\right)^{(1)(5)} = \$1216.65$

    b) $i = 1216.56 - 1000 = \$216.65$

9. $n = 4,\ r = 0.03,\ t = 4,\ p = 2000$

    a) $A = 2000\left(1 + \dfrac{0.03}{4}\right)^{(4)(4)} = \$2253.98$

    b) $i = 2253.98 - 2000 = \$253.98$

11. $n = 12,\ r = 0.055,\ t = 3,\ p = 7000$

    a) $A = 7000\left(1 + \dfrac{0.055}{1}\right)^{(12)(3)} = \$8252.64$

    b) $i = 8252.64 - 7000 = \$1252.64$

13. $n = 360,\ r = 0.04,\ t = 2,\ p = 8000$

    a) $A = 8000\left(1 + \dfrac{0.04}{360}\right)^{(360)(2)} = \$8666.26$

    b) $i = 8666.26 - 8000 = \$666.26$

15. $n = 1,\ r = 0.05,\ t = 10,\ A = 50,000$

    $p = \dfrac{50,000}{\left(1 + \dfrac{0.05}{1}\right)^{(1)(10)}} = \$30,695.66$

17. $n = 4,\ r = 0.04,\ t = 4,\ A = 100,000$

    $p = \dfrac{100,000}{\left(1 + \dfrac{0.04}{4}\right)^{(4)(4)}} = \$85,282.13$

19. $A = 50,000\left(1 + \dfrac{0.04}{4}\right)^{(4)(2)} = \$54,142.84$

21. $A = 1500\left(1 + \dfrac{0.039}{12}\right)^{12 \cdot 2.5} = \$1653.36$

23. $p = 800 + 150 + 300 + 1000 = \$2250$

    $A = 2250\left(1 + \dfrac{0.02}{360}\right)^{360 \cdot 2} = \$2,341.82$

25. a) $A = 2000\left(1 + \dfrac{0.05}{2}\right)^{2 \cdot 15} = \$4,195.14$

    b) $A = 2000\left(1 + \dfrac{0.05}{4}\right)^{2 \cdot 15} = \$4,214.36$

27. $A = 3000\left(1 + \dfrac{0.0175}{4}\right)^{4 \cdot 2} = \$3106.62$

29. $A = 6000\left(1 + \dfrac{0.08}{4}\right)^{12} = \$7,609.45$

31. $A = 1\left(1 + \dfrac{0.035}{12}\right)^{12 \cdot 1} \approx 1.0356$

    $i = A - 1 = 1.0356 - 1 = 0.0356$

    APY = 3.56%

33. $A = 1\left(1 + \dfrac{0.024}{12}\right)^{12 \cdot 1} \approx 1.0243$

    $i = A - 1 = 1.0243 - 1 = 0.0243$

    Yes, APY = 2.43 %, not 2.6 %

35. The effective rate of the 4.75% account is:

    $A = 1\left(1 + \dfrac{0.0475}{12}\right)^{12} \approx 1.0485$

    APY = 1.0485 − 1.00 = 0.0485 or 4.85%

    Therefore the 5% simple interest account pays more interest.

37. a) $925,000 - 370,000 = \$555,000$

    b) $\dfrac{555,000}{\left(1 + \dfrac{0.075}{12}\right)^{(12)(30)}} = \$58,907.60$

    c) surcharge $= \dfrac{58,907.60}{598} = \$98.51\,8$

39. Present value $= \dfrac{275,000}{\left(1+\dfrac{0.0515}{12}\right)^{(12)(5)}} = \$212,687.10$

41. Present value $= \dfrac{50,000}{\left(1+\dfrac{0.08}{4}\right)^{72}} = \$12,015.94$

45. $p = 1.35,\ r = 0.025,\ t = 5,\ n = 1$

$A = 1.35(1 + 0.025)^5 = \$1.53$

43. a) $A = 1000\left(1+\dfrac{0.02}{2}\right)^4 = \$1,040.60$

$i = \$1040.60 - \$1000 = \$40.60$

b) $A = 1000\left(1+\dfrac{0.04}{2}\right)^4 = \$1,082.43$

$i = \$1082.43 - \$1000 = \$82.43$

c) $A = 1000\left(1+\dfrac{0.08}{2}\right)^4 = \$1,169.86$

$i = \$1169.86 - \$1000 = \$169.86$

d) No predictable outcome.

47. a) $72/3 = 24$ years   b) $72/6 = 12$ years
c) $72/8 = 9$ years   d) $72/12 = 6$ years
e) $72/r = 22$ $72 = 22r$ $r = 72/22 = 0.0327$
$r = 3.27\%$

49. $i = prt = (100,000)(0.05)(4) = \$20,000$

$A = 100,000\left(1+\dfrac{0.05}{360}\right)^{360\cdot4} = \$122,138.58$

interest earned: $\$122,138.58 - 100,000 = \$22,138.58$

Select investing at the compounded daily interest because the compound interest is greater by $\$2,138.58$.

Exercise Set 11.4

1. An open-end installment loan is a loan on which you can make different payment amounts each month. A fixed installment loan is one in which you pay a fixed amount each month for a set number of months.

3. The APR is the true rate of interest charged on a loan.

5. The total installment price is the sum of all the monthly payments and the down payment, if any.

7. The unpaid balance method and the average daily balance method.

9. a) Amount financed $= 10,000 - 0.15(10,000) = \$8500.00$
   From Table 11.2 the finance charge per $100 at 7.5 % for 60 payments is 20.23.

   Total finance charge $= (20.23)\left(\dfrac{8500}{100}\right) = \$1719.55$

   b) Total amount due after down payment $= 8500.00 + 1719.55 = \$10,219.55$

   Monthly payment $= \dfrac{10,219.55}{60} = \$170.33$

11. a) Amount financed $= 9900 - 0.10(9900) = \$8910$
    From Table 11.2, the finance charge per $100 financed at 9.5% for 48 months is $20.59.

    Total finance charge is $(20.59)\left(\dfrac{8910}{100}\right) = \$1834.57$

    b) Total amount due $= 8910 + 1834.57 = \$10,744.57$

    Monthly payment $= \dfrac{10,744.57}{48} = \$223.85$

13. a) Down payment = 0.20(3200) = $640; Cheryl borrowed 3200 − 640 = $2560
Total installment price = (60 • 53.14) = $3188.40
Finance charge = 3188.40 − 2560 = $628.40

b) $\left(\dfrac{\text{finance charge}}{\text{amt. financed}}\right)(100) = \left(\dfrac{628.40}{2560}\right)(100) = 24.55$

From Table 11.2 for 60 payments, the value of 24.55 corresponds with an APR of 9.0 %.

15. a) Total installment price = (224)(48) = $10752.00
Finance charge = 10752.00 − 9000.00 = $1752.00

b) $\left(\dfrac{\text{finance charge}}{\text{amt. financed}}\right)(100) = \left(\dfrac{1752.00}{9000}\right)(100) = 19.47$

From Table 11.2, for 48 payments, the value of $19.47 is closest to $19.45 which corresponds with an APR of 9.0 %.

17. Down payment = $0.00   Amount financed = $12,000.00   Monthly payments = $232.00

a) Installment price = (60)(232) = 13,920
Finance charge = $13,920.00 − $12,000.00 = $1920.00

$\left(\dfrac{\text{finance charge}}{\text{amt. financed}}\right)(100) = \left(\dfrac{1920.00}{12000}\right)(100) = 16.00$

From Table 11.2, for 6 payments, the value of $16.00 corresponds to an APR of 6.0 %.

b) From Table 11.2, the monthly payment per $100 for the remaining 36 months at 6.0% APR is $9.52.

$$u = \frac{nPV}{100+V} = \frac{(36)(232)(9.52)}{(100+9.52)} = \frac{79511.04}{109.52} = 725.9956 = \$726.00$$

c) (232)(23) = 5336;   5336 + 726 = 6062;   13920 − 6062 = $7858.00

19. a) Amount financed = 32000 − 10000 = $22000
From Table 11.2, the finance charge per 100 financed at 8 % for 36 payments is 12.81.

$\text{Total finance charge} = (12.81)\left(\dfrac{22000}{100}\right) = 2818.20$

b) Total amt. due = 22000 + 2818.20 = $24,818.20

$\text{Monthly payment} = \dfrac{24818.20}{36} = \$689.39$

c) From Table 11.2, the monthly payment per $100 for the remaining 12 months at 8.0% APR is $4.39.

$$u = \frac{nPV}{100+V} = \frac{(12)(689.89)(4.39)}{100+4.39} = \frac{36317.07}{104.39} = \$347.90$$

d) (23)(689.39) = 15855.97;   15855.97 − 347.90 = 16203.87;   24818.20 − 16203.87 = $8614.17

21. a) (0.01)(2600) = $26

b) Principal on which interest is paid during April:

2600 − 500 = $2100

Interest for April:

(2100)(0.00039698)(30) = $25.01

1% of outstanding principal:

(0.01)(2100) = $21.00

Minimum monthly payment:

25.01 + 21.00 = 46.01 which rounds up to $47.

23. a) Total charges: $677 + 452 + 139 + 141 = \$1409$

    $(0.015)(1409) = \$21.14$ which rounds up to $22

    b) Principal on which interest is paid during December:

    $1409 - 300 = \$1109$

    Interest for December:

    $(1109)(0.0005163)(31) = \$17.75$

    1.5% of outstanding principal:

    $(0.015)(1109) = \$16.64$

    Minimum monthly payment:

    $17.75 + 16.64 = 34.39$ which rounds up to $35.

25. a) Finance charge = $(1097.86)(0.018)(1) = \$19.76$

    b) Bal. due May 5 = $(1097.86 + 19.76 + 425.79) - 800 = \$743.41$

27. a) Finance charge = $(124.78)(0.0125)(1) = \$1.56$

    b) old balance + finance charge – payment + art supplies + flowers + music CD = new balance

    $124.78 + 1.56 - 100.00 + 25.64 + 67.23 + 13.90 = \$133.11$

29. a)

Date	Balance Due	Number of Days	(Balance)(Days)
May 12	$378.50	1	$(378.50)(\ 1) = \$\ \ 378.50$
May 13	$508.29	2	$(508.29)(\ 2) = \ 1{,}016.58$
May 15	$458.29	17	$(458.29)(17) = \ 7{,}790.93$
June 01	$594.14	7	$(594.14)(\ 7) = \ 4{,}158.98$
June 08	$631.77	4	$(631.77)(\ 4) = \ 2{,}527.08$
		31	sum = $15,872.07

Average daily balance $= \dfrac{15872.07}{31} =$ $512

   b) Finance charge = prt =

   $(512.00)(0.013)(1) = \$6.66$

   c) Balance due = $631.77 + 6.66 =$

   $638.43

31. a)

Date	Balance Due	Number of Days	(Balance)(Days)
Feb. 03	$124.78	5	$(124.78)(5) = \$623.90$
Feb. 08	$150.42	4	$(150.42)(4) = \ \ 601.68$
Feb. 12	$ 50.42	2	$(50.42)(2) = \ \ \ 100.84$
Feb. 14	$117.65	11	$(117.65)(11) = \ 1294.15$
Feb. 25	$131.55	6	$(131.55)(6) = \ \ \ 789.30$
Mar 3		28	sum = $3,409.87

Average daily balance $= \dfrac{3409.87}{28} =$ $121.78

   b) Finance charge = prt =

   $(121.78)(0.0125)(1) = \$1.52$

   c) Balance due = $131.55 + 1.52 =$

   $133.07

   d) The finance charge using the avg. daily balance method is $0.04 less than the finance charge using the unpaid balance method.

33. a) $i = (875)(0.0004273)(32) = \$11.96$ b) $A = 875 + 11.96 = \$886.96$

35. $1000.00   5 %   6 payments
  a) State National Bank (SNB): $(1000)(.05)(.5) = \$25.00$
  b) Consumers Credit Union (CCU): $(1000)(x)(1) = 35.60$ $(86.30)(12) = 1035.60$
     $1035.60 - 1000.00 = \$35.60$

  c) $\left(\dfrac{25}{1000}\right)(100) = 2.50$ In Table 11.2, $2.49 is the closest value to $2.50, which corresponds to an
     APR of 8.5 %.

  d) $\left(\dfrac{35.60}{1000}\right)(100) = 3.56$ In Table 11.2, $3.56 corresponds to an APR of 6.5 %.

37. Let p = amount Ken borrowed   Since i = prt we have:
       p + 2500 = purchase price $13,662 - p = (p)(.06)(3) = 13,662 - p = .18p$
     Installment price: $2500 + (379.50)(36) = \$16,162$ $13,662 = .18p + p$ $p = 11,577.97$
     Interest = Installment price – purchase price   purchase price = 11,577.97 + 2500 =
     $i = 16,162 - (p + 2500) = 16,162 - p - 2500 = 13,662 - p$ $14,077.97

39. $35,000   15 % down payment   60 month fixed loan   APR = 8.5 %
       $(35000)(.15) = 5250$ $35000 - 5250 = 29750$
  a) From Table 11.2, 60 payments at an APR of 8.5 % yields a finance charge of $23.10 per $100.

     $\left(\dfrac{29750}{100}\right)(23.10) = \$6872.25$

  b) $29750.00 + 6872.25 = 36622.25$ $\dfrac{36622.25}{60} = \$610.37$

  c) In Table 11.2, 36 payments at an APR of 8.5 % yields a finance charge of $13.64 per $100.

     $u = \dfrac{(36)(610.37)(13.64)}{100+13.64} = \dfrac{299716.08}{113.64} = \$2637.42$

  d) $u = \dfrac{f \cdot k(k+1)}{n(n+1)} = \dfrac{(6872.25)(36)(37)}{60(61)} = \dfrac{9153837}{3660} = \$2501.05$

41. $m = \dfrac{4500\left(\dfrac{0.07}{12}\right)}{1-\left(1+\dfrac{0.07}{12}\right)^{-(12)(2)}} = \$201.48$

Exercise Set 11.5

1. A mortgage is a long-term loan in which the property is pledged as security for payment of the difference between the down payment and the sale price.

3. The major difference between these two types of loans is that the interest rate for a conventional loan is fixed for the duration of the loan, whereas the interest rate for a variable-rate loan may change every period, as specified in the loan agreement.

5. A buyer's adjusted monthly income is found by subtracting any fixed monthly payment with more than 10 months remaining from the gross monthly income.

7. An amortization schedule is a list of the payment number, interest, principal, and balance remaining on the loan.

9. Equity is the difference between the appraised value of your home and the loan balance.

11. a) Down payment = 20% of $350,000
    $(0.20)(350000) = \$70,000$
    b) amt. of mortgage = 350000 − 70,000 = 28,000
    Table 11.4 yields $8.71 per $1000 of mortgage

    Monthly payment = $\left(\dfrac{28,000}{1000}\right)(8.71)$
    $= \$2438.80$

13. a) Down payment = 15% of $1,750,000
    $(0.15)(1750000) = \$262,500$
    b) amt. of mortgage =
    1750000 − 262500 = 1487500
    Table 11.4 yields $6.44 per $1000 of mortgage
    Monthly payment = $\left(\dfrac{1,487,500}{1000}\right)(6.44)$
    $= \$9579.50$

15. a) Down payment = 20% of $195,000
    $(0.20)(195000) = \$39,000$
    b) amt. of mortgage = 195000 − 39000 = $156,000
    c) $(156000)(.02) = \$3120.00$

17. $3200 = monthly income
    3200 − 335 = 2865 = adjusted income
    a) $(0.28)(2865) = \$802.20$
    b) Table 11.4 yields $7.91 per $1000 of mortgage
    $\left(\dfrac{150000}{1000}\right)(7.91) = \$1186.50$

    1186.50 + 225.00 = $1411.50
    c) No; $1411.50 > $802.20

19. a) $(490.24)(30)(12) = \$176,486.40$
    176486.40 + 11250.00 = $187,736.40
    b) 187736.40 − 75000 = $112,736.40
    c) i = prt = $(63750)(.085)(1/12) = 451.56$
    490.24 − 451.56 = $38.68

21. a) down payment = $(0.15)(113500) = \$17,025$
    b) amount of loan = 113500 − 17025
    b) amount of mortgage = 113500 − 17,025
    $= \$96,475$
    cost of three points = $(0.03)(96475) = \$2894.25$
    c) 4750 − 420 = $4330.00 adjusted monthly income
    $(0.28)(4330.00) = \$1212.40 = 28\%$ of adjusted income
    d) At a rate of 10% for 20 years, Table 11.4 yields 9.66.

    mortgage payment = $\left(\dfrac{96475}{1000}\right)(9.66) = \$931.95$

e) Insurance + taxes = $\dfrac{1200+320}{12} = \$126.67$ /mo.
    931.95 + 126.67 = $1058.62 total mo. payment
f) Since $1,212.40 is greater than $1058.62, the Rosens qualify.
g) interest on first payment = i = prt =
    $(96475)(0.10)(1/12) = \$803.96$
    amount applied to principal = 931.95 − 803.96
    $= \$127.99$

23. <u>Bank A</u>   Down payment = (0.10)(105000) = $10,500

amount of mortgage   105000 – 10500 = $94,500
At a rate of 10% for 30 years, Table 11.4 yields $8.70 .

monthly mortgage payment =

$$\left(\frac{94500}{1000}\right)(8.70) = \$822.15$$

cost of three points = (0.03)(94500) = $2835
Total cost of the house =
   10500 + 2835 + (822.15)(12)(30) = $309,309

<u>Bank B</u>   Down payment = (0.20)(105000) = $21,000

amount of mortgage   105000 – 21000 = $84,000
At a rate of 11.5% for 25 years, Table 11.4 yields $10.16.

monthly mortgage payment =

$$\left(\frac{84000}{1000}\right)(10.16) = \$853.44$$

cost of the house = 21000 + (853.44)(12)(25) = $277,032

The Riveras should select Bank B.

25.   Down payment: (0.20)(450000) = $90,000
Amount of mortgage: 450000 – 90000 = $360,000
1 point: (0.01)(360000) = $3600

a) Monthly payment: $(13.22)\left(\dfrac{360000}{1000}\right) = \$4759.20$

Total payments: $90000 + 3600 + (10)(12)(4759.20) = \$664,704$

b) Monthly payment: $(9.65)\left(\dfrac{360000}{1000}\right) = \$3474.00$

Total payments: $90000 + 3600 + (20)(12)(3474.00) = \$927,360$

c) Monthly payment: $(8.78)\left(\dfrac{360000}{1000}\right) = \$3160.80$

Total payments: $90000 + 3600 + (30)(12)(3160.80) = \$1,231,488$

27. a) Amount of mortgage = 105000 - 5000 = $100000   Initial monthly payment = $\left(\dfrac{100000}{1000}\right)(8.05) = \$805.00$

b)

Payment #	Interest	Principal	Balance
1	$750.00	$55.00	$99,945.00
2	749.59	55.41	99,889.59
3	749.17	55.83	99,833.76

c) effective interest rate = 6.13% + 3.25% = 9.38%.   The new rate is 9.38%.

d)

Payment #	Interest	Principal	Balance
4	$780.37	$24.63	$99,809.13
5	780.17	24.83	99,784.30
6	779.98	25.02	99,759.28

e)   New rate = 6.21% + 3.25% = 9.46%

29. a) The variable rate mortgage would be the cheapest.

b) To find the payments for the variable rate mortgage, add the 6 monthly rates and multiply by 12, giving $52,336.80. The payments over 6 years for the fixed rate mortgage are 6 times 12 times 90 times the table value 8.41 for a 30-year mortgage at 9.5%. This yields $54,496.80. The variable rate saves $2160.

Exercise Set 11.6

1. An annuity is an account into which, or out of which, a sequence of scheduled payments is made.

3. a) An annuity into which payments are made at regular intervals, with interest compounded at the end of each interval and with a fixed interest rate for each compounding period, is called an ordinary annuity.

   b) A sinking fund is a type of annuity in which the goal is to save a specific amount of money in a specified amount of time.

5. An immediate annuity is an annuity that is established with a lump sum of money for the purpose of providing the investor with regular, usually monthly, payments for the rest of the investor's life.

7. With a 410k plan, the money invested is not subject to taxes; however, when money is withdrawn, it is subject to income taxes. With a Roth 401k plan, the investor has already paid taxes on the money invested, and money withdrawn is not subject to income taxes.

9. $A = \dfrac{3000\left[\left(1+\dfrac{0.05}{1}\right)^{(1)(30)}-1\right]}{\dfrac{0.05}{1}} = \$199{,}316.54$

11. $A = \dfrac{400\left[\left(1+\dfrac{0.08}{4}\right)^{(4)(35)}-1\right]}{\dfrac{0.08}{4}} = \$299{,}929.32$

13. $p = \dfrac{20{,}000\left(\dfrac{0.04}{2}\right)}{\left(1+\dfrac{0.04}{2}\right)^{(2)(15)}-1} = \$493.00$

15. $p = \dfrac{250{,}000\left(\dfrac{0.06}{12}\right)}{\left(1+\dfrac{0.06}{12}\right)^{(12)(35)}-1} = \$175.48$

17. $A = \dfrac{50\left[\left(1+\dfrac{0.03}{12}\right)^{(12)(35)}-1\right]}{\dfrac{0.03}{12}} = \$37{,}078.18$

19. $A = \dfrac{2000\left[\left(1+\dfrac{0.07}{2}\right)^{(2)(10)}-1\right]}{\dfrac{0.07}{2}} = \$56{,}559.36$

21. $p = \dfrac{25{,}000\left(\dfrac{0.09}{12}\right)}{\left(1+\dfrac{0.09}{12}\right)^{(12)(5)}-1} = \$331.46$

23. $p = \dfrac{1000000\left(\dfrac{0.08}{4}\right)}{\left(1+\dfrac{0.08}{4}\right)^{(4)(20)}-1} = \$5160.71$

25. a) $A = \dfrac{100\left[\left(1+\dfrac{0.12}{12}\right)^{(12)(10)}-1\right]}{\dfrac{0.12}{12}} = \$23{,}003.87$

   b) $A = 23{,}003.87\left(1+\dfrac{0.12}{12}\right)^{(12)(30)} = \$826{,}980.88$

   c) $A = \dfrac{100\left[\left(1+\dfrac{0.12}{12}\right)^{(12)(30)}-1\right]}{\dfrac{0.12}{12}} = \$349{,}496.41$

   d) $(100)(12)(10) = \$12{,}000$

   e) $(100)(12)(30) = \$36{,}000$

   f) Alberto

Review Exercises

1. $3/4 = 0.75 = 75.0\%$

2. $5/6 \approx 0.833 = 83.3\%$

3. $5/8 = 0.625 = 62.5\%$

4. $0.041 = 4.1\%$

5. $0.0098 = 0.98\% \approx 1.0\%$

6. $3.141 = 314.1\%$

7. 9% $\dfrac{9}{100} = .09$

8. 14.1% $\dfrac{14.1}{100} = 0.141$

9. 123% $\dfrac{123}{100} = 1.23$

10. $\dfrac{1}{4}\% = 0.25\%$ $\dfrac{.25}{100} = .0025$

11. $\dfrac{5}{6} = 0.8\bar{3}\%$ $\dfrac{0.8\bar{3}}{100} = 0.008\bar{3}$

12. 0.00045% $\dfrac{0.00045}{100} = 0.0000045$

13. $\dfrac{71,500 - 60,790}{60,790} \approx 0.176 = 17.6\%$

14. $\dfrac{51,300 - 46,200}{46,200} \approx 0.1104 = 11.0\%$

15. $(x\%)(80) = 15$ $x\% = 15/80 = .1875$
    $0.1875 = 18.75\%$
    15 is 18.75% of 80.

16. $0.55x = 44$ $x = 44/0.55 = 80$
    44 is 5% of 80.

17. $(0.17)(540) = x$ $91.8 = x$
    17% of 540 is 91.8.

18. Tip $= 15\%$ of $\$42.79 = (0.15)(42.79) = \$6.42$

19. $0.20(x) = 8$ $x = 8/0.20 = 40$
    The original number was 40 people.

20. $\dfrac{(95 - 75)}{75} = \dfrac{20}{75} = .2\bar{6}$ $(.267)(100) = 26.7$
    The increase was 26.7%.

21. $i = (6000)(.02)(30/360) = \$10.00$

22. $37.50 = (2700)(r)(100/360)$
    $37.50 = \left(\dfrac{270000}{360}\right)(r)$ $r = 0.05$ or 5%

23. $114.75 = (p)(0.085)(3)$
    $114.75 = (p)(0.255)$ $\$450 = p$

24. $316.25 = (5500)(0.115)(t)$
    $316.25 \qquad = (632.50)(t)$ $t = 0.5$ yrs. or 6 mos.

25. $i = (5300)(.0575)(3) = 914.25$
    Total amount due at maturity $= 5300 + 914.25 = \$6214.25$

26. a) $i = (3000)(0.081)(240/360) = \$162$
    She paid $3000 + 162 = \$3,162$

27. a) $i = (6000)(0.115)(24/120 = \$1380.00$
    b) amount received: $6000.00 - 1380.00 = 4,620.00$
    c) $i = prt$ $1380 = (4620)(r)(24/12) = 9240r$
       $r = (1380)(9240) = .1494$ $(.1494)(100) = 14.9\%$

28. a) $5\dfrac{1}{2}\% + 2\% = 7\dfrac{1}{2}\%$
    b) $i = (800)(0.75)(6/12) = \$30$
       $A = \$800 + \$30 = \$830.00$
    c) x = amount of money in the account
       85% of x = 800 $0.85x = 800$ $x = \$941.18$

29. a) $A = 5000\left(1 + \dfrac{0.06}{1}\right)^{(1)(5)} = \$6691.13$ $\$6691.13 - \$5000 = \$1691.13$

b) $A = 5000\left(1 + \dfrac{0.06}{2}\right)^{(2)(5)} = \$6719.58$ $\$6719.58 - \$5000 = \$1719.58$

c) $A = 5000\left(1 + \dfrac{0.06}{4}\right)^{(4)(5)} = \$6734.28$ $\$6734.28 - \$5000 = \$1734.28$

d) $A = 5000\left(1 + \dfrac{0.06}{12}\right)^{(12)(5)} = \$6744.25$ $\$6744.25 - \$5000 = \$1744.25$

e) $A = 5000\left(1 + \dfrac{0.06}{360}\right)^{(360)(5)} = \$6749.13$ $\$6749.13 - \$5000 = \$1749.13$

30. $A = p\left(1 + \dfrac{r}{n}\right)^{nt}$

$A = 2500\left(1 + \dfrac{0.0475}{4}\right)^{4 \bullet 15} = \$5,076.35$

31. Let $p = 1.00$. Then $A = 1\left(1 + \dfrac{0.56}{360}\right)^{360} = 1.05759$

$i = 1.05759 - 1.00 = 0.05759$
The effective annual yield is 5.76% .

32. $p\left(1 + \dfrac{0.055}{4}\right)^{80} = 40000$ $p = \dfrac{40000}{(1.01375)^{80}} = 13415.00$ You need to invest \$13,415.00

33. 48 mo.   \$176.14/mo.   \$7500   24 payments

a) $(176.14)(48) = 8454.72$ $8454.72 - 7500 = \$954.72$ $\left(\dfrac{954.72}{7500}\right)(100) = \$12.73/\$100$

From Table 11.2, \$12.73 indicates an APR of 6.0%

b) $n = 24,\ p = 176.14,\ v = 6.37$ $u = \dfrac{(24)(176.14)(6.37)}{100 + 6.37} = \dfrac{26928.28}{106.37} = \$253.16$

c) $(176.14)(48) = 8454.72$ $(176.14)(23) = 4051.22$ $8454.72 - 4051.22 = \$4403.50$
   $4403.50 - 253.16 = \$4150.34$

34. Amount borrowed: \$4000 − \$500 = \$3500

a) Total of payments: $(24)(155.91) = \$3741.84$

Finance charge: $3741.84 - 3500 = \$241.84$

$\left(\dfrac{241.84}{3500}\right)(100) = \$6.91/\$100$

From Table 11.2, \$6.91 for a 24-month loan indicates an APR of 6.5%

b) $n = 12,\ p = 155.91,\ v = 6.37$ $u = \dfrac{(12)(155.91)(3.56)}{100 + 3.56} = \$64.32$

c) $(155.91)(11) = 1715.01$ $3741.84 - 1715.01 = \$2026.83$
   $2026.83 - 64.32 = \$1962.51$

35. 24 mo. $111.73/mo.   Down payment = $860   24 payments

   a) $3420 - 860 = \$2560.00$ $(111.73)(24) = 2681.52$ $2681.52 - 2560.00 = \$121.52$

$$\left(\frac{121.52}{2560}\right)(100) = \$4.75/\$100$$

   From Table 11.2, $4.75 indicates an APR of 4.5%

   b) $n = 12$, $p = 111.73$, $v = 2.45$ $u = \dfrac{(12)(111.73)(2.45)}{100 + 2.45} = \dfrac{3284.86}{102.45} = \$32.06$

   c) $(111.73)(11) = 1229.03$ $2681.52 - 1229.03 = 1452.49$ $1452.49 - 32.06 = \$1420.43$

36. Balance = $485.75 as of June 01 $i = 1.3\%$

   June 04: $485.75 - 375.00) = \$110.75$ June 08: $110.75 + 370.00 = \$480.75$

   June 21: $480.75 + 175.80 = \$656.55$ June 28: $656.55 + 184.75 = \$841.30$

   a) $(485.75)(.013)(1) = \$6.31$ b) $841.30 + 6.31 = \$847.61$

   c) $(485.75)(3) + (110.75)(4) + (480.75)(13) + (656.55)(7) + (841.30)(3) = \$15269.75$

     $15269.75/30 = \$508.99$

   d) $(508.99)(.013)(1) = \$6.62$ e) $841.30 + 6.62 = \$847.92$

37. a) $i = (185.72)(0.14)(1) = \$2.60$

   b) Aug. 01: $185.72
     Aug. 05: $185.72 + 2.60 = \$188.32$
     Aug. 08: $188.32 + 85.75 = \$274.07$
     Aug. 10: $274.07 - 75.00 = \$199.07$
     Aug. 15: $199.07 + 72.85 = \$271.92$
     Aug. 21: $271.92 + 275.00 = \$546.92$
     As of Sep 5, the new account balance
     is $546.92.

   d) $i = (\$382.68)(0.014)(1) = \$5.36$

   c)

Date	Balance	# of Days	(Balance)(Days)
Aug. 05	185.72	3	$(185.72)(3) =$ 557.16
Aug. 08	271.47	2	$(271.47)(2) =$ 542.94
Aug. 10	196.47	5	$(196.47)(5) =$ 982.35
Aug. 15	269.32	6	$(269.32)(6) =$ 1615.92
Aug. 21	544.32	15	$(544.32)(15) = 8164.80$
Sep 05		31	sum = $11,863.17

   avg. daily balance = $\dfrac{11,863.17}{31} = \$382.68$

   e) $544.32 + 5.36 = \$549.68$ is the amount due on Sep 5.

38. a) $(\$72,000)(0.20) = \$14,400$

   b) $72,000 - 14,400 = \$57,600$

   c) $(1100.26)(60) = \$66,015.60$;
     $66,015.60 - 57,600 = \$8415.60$

   d) $\left(\dfrac{\text{finance charge}}{\text{amount financed}}\right)(100)$

     $= \left(\dfrac{8415.60}{57,600}\right)(100) = 14.61;$

     Using Table 11.2, where no. of
     payments is 60, APR = 5.5%

39. a) $275 - 50 = \$225$; $(19.62)(12) = \$235.44$;
     $235.44 - 225 = \$10.44$ interest paid

   b) $\left(\dfrac{\text{finance charge}}{\text{amount financed}}\right)(100)$

     $= \left(\dfrac{10.44}{225}\right)(100) = 4.64;$

     Using Table 11.2, where no. of payments is
     12, $4.64 is closest to $4.66, which yields an
     APR of 8.5%

40. a) down payment = (0.25)(135700) = \$33,925
    b) gross monthly income = 64000/12 = \$5,333.33
        adjusted monthly income:
        5333.33 − 528.00 = \$4,805.33
        28% of adjusted monthly income:
        (0.28)(4805.33) = \$1,345.49
    c) $\left(\dfrac{101,775}{1000}\right)(8.41) = \$855.93$
    d) total monthly payment:
        855.93 + 316.67 = \$1172.60
    e) Yes, \$1345.49 is greater than \$1172.60.

41. a) down payment = (0.15)(89900) = \$13,485
    b) amount of mortgage = 89,900 − 13,485 =
        \$76,415
        At 11.5% for 30 years, Table 11.4 yields 9.90.
        monthly mortgage payment:
        $\left(\dfrac{76415}{1000}\right)(9.90) = \$756.51$
    c) i = prt = (76415)(0.115)(1/12) = \$732.31
        amount applied to principal:
        756.51 − 732.51 = \$24.20
    d) total cost of house: 13485 + (756.51)(12)(30) =
        \$285,828.60
    e) total interest paid: 285,828.60 − 89900 =
        \$195,928.60

42. a) amount of mortgage: 105,000 − 26,250 = \$78,750   First payment = $\left(\dfrac{78,750}{1000}\right)(6.99) = \$550.46$

    b) 5.00% + 3.00% = 8.00%   c) 4.75% + 3.00% = 7.75%

43. $A = \dfrac{250\left[\left(1+\dfrac{0.09}{12}\right)^{(12)(10)} - 1\right]}{\dfrac{0.09}{12}} = \$48,378.57$

44. $p = \dfrac{100,000\left(\dfrac{0.07}{4}\right)}{\left(1+\dfrac{0.07}{4}\right)^{(4)(15)} - 1} = \$955.34$

Chapter Test

1. a) i = (2000)(0.04)(1/2) = \$40.00   b) 288 = (1200)(0.08)(t); 288 = 96t; t = 3 years

2. i = prt = (5700)(0.0575)(30/12) = \$819.38

3. Total amount paid to the bank
    5700 + 819.38 = \$6519.38

4. Partial payment on Sept. 15 (45 days)
    i = (5400)(0.125)(45/360) = \$84.375
    \$3000.00 - 84.375 = \$2,915.625
    5400.00 − 2915.625 = \$2484.375

    i = (2484.375)(0.125)(45/360) = \$38.82
    2484.38 + 38.82 = \$2523.20

5. 84.38 + 38.82 = \$123.20

6. a) $A = 7500\left(1+\dfrac{0.03}{4}\right)^{8} = \$7961.99$
       interest = 7961.99 − 7500.00 = \$461.99

   b) $A = 2500\left(1+\dfrac{0.065}{12}\right)^{36} = \$3036.68$
       interest = 3036.68 − 2500.00 = \$536.68

7. (2350)(.15) = 352.50
    2350 − 352.50 = \$1997.50

8. (90.79)(24) = 2178.96
    2178.96 − 1997.50 = \$181.46

9. $\left(\dfrac{181.46}{1997.50}\right)(100) = \$9.08 \; /\$100$ In Table 11.2, \$9.08 is closest to \$9.09 which yields an APR of 8.5% .

10. \$7500   36 mo.   \$223.10 per mo. $(223.10)(36) = 8031.60$ $8031.60 - 7500.00 = 531.60$

   a) $\left(\dfrac{181.46}{1997.50}\right)(100) = \$9.08$ In Table 11.2, \$9.08 yields an APR of 4.5% .

   b) $u = \dfrac{n \cdot P \cdot V}{100 + V} = \dfrac{(12)(223.10)(2.45)}{100 + 2.45} = \dfrac{6559.14}{102.45} = \$64.02$

   c) $(223.10)(23) = 5131.30$ $8031.60 - 5131.30 = \$2900.30$ $2900.30 - 64.02 = \$2836.28$

11. Mar. 23: \$878.25
   Mar. 26: $878.25 + 95.89 = \$974.14$
   Mar. 30: $974.14 + 68.76 = \$1042.90$
   Apr. 03: $1042.90 - 450.00 = \$592.90$
   Apr. 15: $592.90 + 90.52 = \$683.42$
   Apr. 22: $683.42 + 450.85 = \$1134.27$

   a) $i = (878.25)(.014)(1) = \$12.30$
   b) $1134.27 + 12.30 = \$1146.57$

   c)

Date	Balance	# of Days	Balance-Days
Mar. 23	878.25	3	$(878.25)(3) = 2634.75$
Mar. 26	974.14	4	$(974.14)(4) = 3896.56$
Mar. 30	1042.90	4	$(1042.90)(4) = 4171.60$
Apr. 03	592.90	12	$(592.90)(12) = 7114.80$
Apr. 15	683.42	7	$(683.42)(7) = 4783.94$
Apr. 22	1134.27	1	$(1134.27)(1) = 1134.27$
Apr 23		31	sum = \$23,735.92

   avg. daily balance $= \dfrac{23,735.92}{31} = \$765.67$

   d) $(765.67)(.014)(1) = \$10.72$
   e) $1134.27 + 10.72 = \$1144.99$

12. down payment $= (0.15)(144500) = \$21,675.00$

13. amount of loan $= 144500 - 21675 = \$122,825$
   points $= (0.02)(122825) = \$2456.50$

14. gross monthly income $= 86500 \div 12 = \$7208.33$
   $7,208.33 - 605.00 = \$6,603.33$ adj. mo. income
   maximum monthly payment $= (0.28)(6603.33) = \$1,848.93$

15. At 10.5% interest for 30 years, Table 11.4 yields \$9.15.

   monthly payments $= \left(\dfrac{122825}{1000}\right)(9.15) = \$1,123.85$

16. $1123.85 + 304.17 = \$1428.02$ total mo. payment

17. Yes.

18. a) Total cost of the house:
   $21675 + 2456.50 + (1123.85)(12)(30)$
   $= \$428,717.50$
   b) interest $= 428,717.50 - 144,500 = \$284,217.50$

19. $A = \dfrac{500\left[\left(1 + \dfrac{0.06}{12}\right)^{(12)(20)} - 1\right]}{\dfrac{0.06}{12}} = \$231,020.45$

20. $p = \dfrac{15,000\left(\dfrac{0.06}{12}\right)}{\left(1 + \dfrac{0.06}{12}\right)^{(12)(4)} - 1} = \$277.28$

CHAPTER TWELVE

PROBABILITY

1. An experiment is a controlled operation that yields a set of results.

3. Empirical probability is the relative frequency of occurrence of an event. It is determined by actual observation of an experiment.

$$P(E) = \frac{\text{number of times the event occurred}}{\text{number of times the experiment was performed}}$$

5. Relative frequency over the long run can accurately be predicted, not individual events or totals.

7. Not necessarily, but it does mean that if a coin was flipped many times, about one-half of the tosses would land heads up.

9. a) Roll a die 100 times and determine the number of times that a 5 occurs out of 100.
    b) Answers will vary (AWV).   c) AWV

11. AWV   13. AWV

15. Of 30 birds: 14 finches   10 cardinals   6 blue jays
    a) P(f) = 14/30 = 7/15   b) P(c) = 10/30 = 1/3   c) P(bj) = 6/30 = 1/5

17. Of 105 animals: 45 are dogs.   40 are cats   15 are birds   5 are rabbits

    a) P(dog) = 45/105 = 3/7   b) P(cat) = 8/21   c) P(rabbit) = 5/105 = 1/21

19. a) $\dfrac{4737}{129,098} \approx 0.0367$ b) $\dfrac{21,922}{129,098} \approx 0.1698$ c) $\dfrac{11,418}{129,098} \approx 0.0884$

21. a) $\dfrac{29}{186}$

    b) $\dfrac{98}{186} = \dfrac{1}{2}$

    c) $\dfrac{12}{186} = \dfrac{2}{31}$

23. Of 80 votes: 22 for Allison   18 for Emily   20 for Kimberly   14 for Johanna   6 for others
    a) P(A) = 22/80 = 11/40   b) P(E) = 18/80 = 9/40   c) P(K) = 20/80 = 1/4   d) P(J) = 14/80 = 7/40
    e) P(others) = 6/80 = 3/40

25. a) P(bulls-eye) = $\dfrac{6}{20} = \dfrac{3}{10}$

    b) P(not bulls-eye) = $\dfrac{14}{20} = \dfrac{7}{10}$

    c) P(at least 20 pts.) = $\dfrac{14}{20} = \dfrac{7}{10}$

    d) P(does not score) = $\dfrac{2}{20} = \dfrac{1}{10}$

29. a) P(white flowers) = $\dfrac{224}{929} = 0.24$ b) P(purple flowers) = $\dfrac{705}{929} = 0.76$

31. Answers will vary.

27. a) P(affecting circular) = $\dfrac{0}{150} = 0$

    b) P(affecting elliptical) = $\dfrac{50}{250} = 0.2$

    c) P(affecting irregular) = $\dfrac{100}{100} = 1$

Exercise Set 12.2

1. If each outcome of an experiment has the same chance of occurring as any other outcome, they are said to be equally likely outcomes.

3. P(A) + P(not A) = 1

5. P(event will not occur) = $1 - 0.7 = 0.3$

7. P(event will occur) = $1 - \dfrac{5}{12} = \dfrac{12}{12} - \dfrac{5}{12} = \dfrac{7}{12}$

9. None of the possible outcomes is the event in question.

11. All probabilities are between 0 and 1.

13. a) P(correct) = 1/5   b) P(correct) = 1/4

15. P(you win) = $\dfrac{\text{one choice}}{40 \text{ possible choices}} = \dfrac{1}{40}$

17. P(5) = $\dfrac{4}{52} = \dfrac{1}{13}$

19. P(not 5) = $\dfrac{48}{52} = \dfrac{12}{13}$

21. P(black) = $\dfrac{13+13}{52} = \dfrac{26}{52} = \dfrac{1}{2}$

23. P(red or black) = $\dfrac{26+26}{52} = \dfrac{52}{52} = \dfrac{1}{1} = 1$

25. P(>4 and <9) = P(5,6,7,8) = $\dfrac{16}{52} = \dfrac{4}{13}$

27. a) P(red) = $\dfrac{2}{4} = \dfrac{1}{2}$ b) P(green) = $\dfrac{1}{4}$

    c) P(yellow) = $\dfrac{1}{4}$ d) P(blue) = 0

29. a) P(red) = $\dfrac{2}{6} = \dfrac{1}{3}$ b) P(green) = $\dfrac{1}{6}$

    c) P(yellow) = $\dfrac{2}{6} = \dfrac{1}{3}$ d) P(blue) = $\dfrac{1}{6}$

Of 100 cans: 30 are cola (c)   40 are orange (o)   10 are ginger ale (ga)   20 are root beer (rb)

31. $P(rb) = \dfrac{20}{100} = \dfrac{1}{5}$

33. $P(c, rb, o) = \dfrac{90}{100} = \dfrac{9}{10}$

35. $P(600) = \dfrac{1}{12}$

37. $P(lose/bankrupt) = \dfrac{2}{12} = \dfrac{1}{6}$

Of 30 basketballs: 10 are Wilson (w)   15 are Spalding (s)   5 are other (o)

39. $P(s) = \dfrac{15}{30} = \dfrac{1}{2}$

41. $P(not\ w) = \dfrac{20}{30} = \dfrac{2}{3}$

For a traffic light: 25 seconds on red (r)   5 seconds on yellow (y)   55 seconds on green (g)

43. $P(g) = \dfrac{55}{85} = \dfrac{11}{17}$

45. $P(not\ r) = \dfrac{60}{85} = \dfrac{12}{17}$

Of 9 letters: 1 = T   4 = E   2 = N   2 = S

47. $P(S) = \dfrac{2}{9}$

49. $P(consonant) = \dfrac{5}{9}$

51. $P(W) = 0$

53. $\dfrac{3}{51}$

55. $\dfrac{27}{51}$

57. $P(15) =$

$\dfrac{1}{26}$

59. $P(\geq 22) =$

$\dfrac{5}{26}$

61. $P(car) = \dfrac{85}{130} = \dfrac{17}{26}$

63. $P(GM) = \dfrac{83}{130}$

65. $P(GM\ car) = \dfrac{55}{130} = \dfrac{11}{26}$

67. $P(Jif) = \dfrac{12}{38} = \dfrac{6}{19}$

69. $P(chunky) = \dfrac{15}{38}$

71. $P(Peter\ Pan\ smooth) = \dfrac{10}{38} = \dfrac{5}{19}$

73. $P(red) = \dfrac{2}{18} + \dfrac{1}{12} + \dfrac{1}{6} = \dfrac{4}{36} + \dfrac{3}{36} + \dfrac{6}{36} = \dfrac{13}{36}$

75. $P(yellow) = \dfrac{1}{6} + \dfrac{1}{12} + \dfrac{1}{12} = \dfrac{2}{12} + \dfrac{2}{12} = \dfrac{4}{12} = \dfrac{1}{3}$

77. $P(yellow\ or\ green) = \dfrac{1}{3} + \dfrac{11}{36} = \dfrac{23}{36}$

79. a) $P(CC) = 0$ b) $P(CC) = 1$

81. a) $P(R/R) = \dfrac{2}{4} \cdot \dfrac{2}{4} = \dfrac{4}{16} = \dfrac{1}{4}$

83. $4 \cdot 7 + 1 = 29$

 b) $P(G/G) = \dfrac{2}{4} \cdot \dfrac{2}{4} = \dfrac{4}{16} = \dfrac{1}{4}$

 c) $P(R/G) = \dfrac{2}{4} \cdot \dfrac{2}{4} = \dfrac{4}{16} = \dfrac{1}{4}$

Exercise Set 12.3

1. a) The odds against an event are found by dividing the probability that the event will not occur by the probability that the event will occur.

   b) The odds in favor of an event are found by dividing the probability that the event will occur by the probability that the event will not occur.

3. Odds against are more commonly used.

5. 7 : 2 or 7 to 2

7. a) P(event occurs) = $\dfrac{1}{1+1} = \dfrac{1}{2}$

   b) P(event fails to occur) = $\dfrac{1}{1+1} = \dfrac{1}{2}$

9. a) P(tie goes well) = $\dfrac{8}{15}$

   b) P(tie does not go well) = $\dfrac{7}{15}$

   c) odds against tie going well =

   $\dfrac{\text{P(tie does not go well)}}{\text{P(tie goes well)}} = \dfrac{7/15}{8/15} = \dfrac{7}{8}$ or 7 : 8

   d) odds in favor of it going well are 8 : 7

11. Against: 6 : 1; in favor: 1 : 6

13. Since there is only one 4, the odds against a 4 are 5 : 1.

15. odds against rolling less than 3 =

   $\dfrac{\text{P(3 or greater)}}{\text{P(less than 3)}} =$

   $\dfrac{4/6}{2/6} = \dfrac{4}{6} \cdot \dfrac{6}{2} = \dfrac{4}{2} = \dfrac{2}{1}$ or 2:1

17. odds against a queen =

   $\dfrac{\text{P(failure to pick a queen)}}{\text{P(pick a queen)}} =$

   $\dfrac{48/52}{4/52} = \dfrac{48}{52} \cdot \dfrac{52}{4} = \dfrac{48}{4} = \dfrac{12}{1}$ or 12:1

   Therefore, odds in favor of picking a queen are 1:12.

19. odds against a picture card =

   $\dfrac{\text{P(failure to pick a picture)}}{\text{P(pick a picture)}} = \dfrac{40/52}{12/52} = \dfrac{40}{12} = \dfrac{10}{3}$

   or 10:3

   Therefore, odds in favor of picking a picture card are 3:10.

21. odds against red =

   $\dfrac{\text{P(not red)}}{\text{P(red)}} = \dfrac{1/2}{1/2} = \dfrac{1}{2} \cdot \dfrac{2}{1} = \dfrac{2}{2} = \dfrac{1}{1}$ or 1:1

23. odds against red = $\dfrac{\text{P(not red)}}{\text{P(red)}} = \dfrac{5/8}{3/8} = \dfrac{5}{8} \cdot \dfrac{8}{3} = \dfrac{5}{3}$

   or 5:3

25. a) odds against selecting female =
$$\frac{\text{P(failure to select female)}}{\text{P(select female)}} = \frac{16/30}{14/30} = \frac{16}{14} = \frac{8}{7}$$
    or 8 : 7 .

    b) odds against selecting male =
$$\frac{\text{P(failure to select male)}}{\text{P(select male)}} = \frac{14/30}{16/30} = \frac{14}{16} = \frac{7}{8}$$
    or 7 : 8 .

29. odds in favor of even are $\dfrac{\text{P(even)}}{\text{P(not even)}} =$

$$\frac{7/15}{8/15} = \frac{7}{15} \cdot \frac{15}{8} = \frac{7}{8} \quad \text{or } 7{:}8$$

33. a) $\dfrac{3}{10}$

    b) 7 : 3

37. a) P(Wendy wins) = $\dfrac{7}{7+4} = \dfrac{7}{11}$

    b) P(Wendy loses) = $\dfrac{4}{7+4} = \dfrac{4}{11}$

41. P(G) = $\dfrac{15}{75} = \dfrac{1}{5}$

45. Odds against G are 4:1

49. $\dfrac{66}{34} = \dfrac{33}{17}$ or 33 : 17

53. If P(selling out) = $0.9 = \dfrac{9}{10}$, then

    P(do not sell your car this week) = $1 - \dfrac{9}{10} = \dfrac{1}{10}$.

    The odds against selling out = $\dfrac{1/10}{9/10} = \dfrac{1}{9}$ or 1:9.

27. odds against a stripe = $\dfrac{\text{P(not a stripe)}}{\text{P(stripe)}} =$

$$\frac{8/15}{7/15} = \frac{8}{15} \cdot \frac{15}{7} = \frac{8}{7} \quad \text{or } 8{:}7$$

31. odds against a ball with 9 or greater are
$$\frac{\text{P(less than 9)}}{\text{P(9 or greater)}} = \frac{8/15}{7/15} = \frac{8}{15} \cdot \frac{15}{7} = \frac{8}{7} \quad \text{or } 8{:}7$$

35. The odds against testing negative =
$$\frac{\text{P(test positive)}}{\text{P(test negative)}} = \frac{4/76}{72/76} = \frac{4}{72} = \frac{1}{18} \quad \text{or } 1:18$$

39. Odds against   4 : 11   P(promoted) =
$$\frac{11}{4+11} = \frac{11}{15}$$

43. Odds in favor of G = $\dfrac{\text{P(G)}}{\text{P(not G)}} = \dfrac{1/5}{4/5} = \dfrac{1}{4}$ or 1:4

47. P(A+) = $\dfrac{34}{100} = 0.34$

51. P(O or O-) = $\dfrac{43}{100} = \dfrac{43}{43+57}$ or 43 : 57

55. If P(all parts are present) = $\dfrac{7}{8}$, then the odds in favor of all parts being present are 7 : 1 .

57. a) $P(\text{birth defect}) = \dfrac{1}{33}$

   b) Number without birth defect: $33 - 1 = 32$
   Odds against birth defect: $32 : 1$

59. $P(\#\ 1\ \text{wins}) = \dfrac{2}{9}$ $P(\#\ 2\ \text{wins}) = \dfrac{1}{3}$

   $P(\#\ 3\ \text{wins}) = \dfrac{1}{16}$ $P(\#\ 4\ \text{wins}) = \dfrac{5}{12}$

   $P(\#\ 5\ \text{wins}) = \dfrac{1}{2}$

61. If multiple births are 3% of births, then single births are 97% of births, and the odds against a multiple birth are $97 : 3$.

Exercise Set 12.4

1. Expected value is used to determine the average gain or loss of an experiment over the long run.

3. The fair price is the amount charged for the game to be fair and result in an expected value of 0.

5. To obtain fair price, add the cost to play to the expected value.

7. $0.50. Since you would lose $1.00 on average for each game you played, the fair price of the game should be $1.00 less. Then the expected value would be 0, and the game would be fair.

9. $3(-\$0.40) = -\$1.20$

11. $E = P_1A_1 + P_2A_2 = 0.30(14{,}000) + 0.70(8400) = 10{,}080$ people

13. $E = P_1A_1 + P_2A_2 = 0.50(78) + 0.50(62) = 39 + 31 = 70$ points

15. $E = P_1A_1 + P_2A_2 = 0.40(1.2\text{ M}) + 0.60(1.6\text{ M}) = .48\text{ M} + .96\text{ M} = 1.44\text{ M}$ viewers

17. a) $E = P_1A_1 + P_2A_2 = (.60)(10000) + (.10)(0) + (.30)(-7200) = 6000 + 0 + -2160 = \$3840$

19. $E = P_1A_1 + P_2A_2 + P_3A_3 = P(\$1\text{ off})(\$1) + P(\$2\text{ off})(\$2) + P(\$5\text{ off})(\$5)$
    $E = (1/10)(1) + (2/10)(2) + (1/10)(5) = 7/10 + 4/10 + 5/10 = 16/10 = \$1.60$

21. a) $(3/6)(5) + (2/6)(2) + (1/6)(-15) \approx -\$0.17$

   b) Gabriel's expectation is the negative of Alyssa's, or $0.17.

23. a) $(1/5)(5) + (0)(0) + (4/5)(-1) = 1 - 4/5 = 1/5$
   Yes, positive expectations $= 1/5$

   b) $(1/4)(5) + (0)(0) + (3/4)(-1) = 5/4 - 3/4 = 1/2$
   Yes, positive expectations $= 1/2$

25. a) $\left(\dfrac{1}{500}\right)(398) + \left(\dfrac{499}{500}\right)(-2) = \dfrac{398 - 998}{500} =$

   $\dfrac{-600}{500} = \dfrac{-300}{250} = -\$1.20$

   b) Fair price $= -1.20 + 2.00 = \$.80$

27. a) $\left(\dfrac{1}{2000}\right)(997) + \left(\dfrac{2}{2000}\right)(497) + \left(\dfrac{1997}{2000}\right)(-3) = \dfrac{997 + 994 - 5991}{2000} = \dfrac{-4000}{2000} = -\$2.00$

   b) Fair price $= -2.00 + 3.00 = \$1.00$

29. $\frac{1}{2}(1)+\frac{1}{2}(5)=\frac{1}{2}+\frac{5}{2}=3=\$3.00$

31. $\frac{1}{2}(10)+\frac{1}{4}(-5)+\frac{1}{4}(-20)=5-1.25-5=-\$1.25$

33. $(500)\left(\frac{2}{5}\right)+(1000)\left(\frac{3}{5}\right)=\$800$

35. $(100+200+300+400+1000)\left(\frac{1}{5}\right)=\$400$

37. a) $(8)\left(\frac{1}{2}\right)+(1)\left(\frac{1}{2}\right)=\$5.50$

   b) Fair price = \$5.50 + 2.00 = \$7.50

39. a) $(8)\left(\frac{1}{4}\right)+(3)\left(\frac{1}{4}\right)+(-1)\left(\frac{1}{2}\right)=\$2.25$

   b) Fair price = 2.25 + 2.00 = \$4.25

41. a) $(5-10)\left(\frac{2}{4}\right)+(20-10)\left(\frac{2}{4}\right)=\$2.50$

   b) Fair price = 2.50 + 10.00 = \$12.50

43. a)
$(0-10)\left(\frac{1}{4}\right)+(1-10)\left(\frac{1}{4}\right)+(5-10)\left(\frac{1}{4}\right)+(10-10)\left(\frac{1}{4}\right)=-\$6.00$

   b) Fair price = −6.00 + 10.00 = \$4.50

45. $E = P_1A_1 + P_2A_2 + P_3A_3 + P_4A_4 + P_5A_5 =$
   $0.17(1) + 0.10(2) + 0.02(3) + 0.08(4) + 0.63(0) =$
   0.75 base

47. $E = P_1A_1 + P_2A_2 + P_3A_3$
   $= \frac{3}{10}(4)+\frac{5}{10}(3)+\frac{2}{10}(1) = 1.2 + 1.5 + 0.2$
   $= 2.9$ points

49. $E = (0.65)(75)+(0.35)(20) = 55.75$
   Expected number of new employees is 56.

51. $E = (0.75)(10,000)+(0.1)(0)+(0.15)(-2,000)$
   $= \$7200$

53. $E = P(1)(1) + P(2)(2) + P(3)(3) + P(4)(4) + P(5)(5)$
   $+ P(6)(6)$
   $= \frac{1}{6}(1)+\frac{1}{6}(2)+\frac{1}{6}(3)+\frac{1}{6}(4)+\frac{1}{6}(5)+\frac{1}{6}(6)$
   $= \frac{21}{6} = 3.5$ points

55. $E = P_1A_1 + P_2A_2 + P_3A_3$
   $= \frac{200}{365}(110) + \frac{100}{365}(160) + \frac{65}{365}(210)$
   $= 60.27 + 43.84 + 37.40 = 141.51$ calls/day

57. a) $P(1) = \frac{1}{2}+\frac{1}{16} = \frac{8}{16}+\frac{1}{16} = \frac{9}{16}$, $P(10) = \frac{1}{4} = \frac{4}{16}$,
   $P(\$20) = \frac{1}{8} = \frac{2}{16}$, $P(\$100) = \frac{1}{16}$
   b) $E = P_1A_1 + P_2A_2 + P_3A_3 + P_4A_4$
   $= \frac{9}{16}(\$1)+\frac{4}{16}(\$10)+\frac{2}{16}(\$20)+\frac{1}{16}(\$100)$
   $= \frac{9}{16}+\frac{40}{16}+\frac{40}{16}+\frac{100}{16} = \frac{189}{16} = \$11.81$
   c) fair price = expected value − cost to play =
   \$11.81 − 0 = \$11.81

59. $E = P(\text{insured lives})(\text{cost}) + P(\text{insured dies})(\text{cost} -$
   \$40,000)
   $= 0.97(\text{cost}) + 0.03(\text{cost} - 40,000)$
   $= 0.97(\text{cost}) + 0.03(\text{cost}) - 1200$
   $= 1.00(\text{cost}) - 1200$
   Thus, in order for the company to make a profit,
   the cost must exceed \$1,200

61. E = P(win)(amount won) + P(lose)(amount lost)

$$= \left(\frac{1}{38}\right)(35) + \left(\frac{37}{38}\right)(-1) = \frac{35}{38} - \frac{37}{38} = -\frac{2}{38}$$

$$= -\$0.053$$

63. a) $E = \frac{1}{12}(100) + \frac{1}{12}(200) + \frac{1}{12}(300) + \frac{1}{12}(400) + \frac{1}{12}(500) + \frac{1}{12}(600) + \frac{1}{12}(700) + \frac{1}{12}(800) + \frac{1}{12}(900)$

$\frac{1}{12}(1000) = \left(\frac{5500}{12}\right) = \$458.3\overline{3}$

b) $E = \frac{1}{12}(5500) + \frac{1}{12}(-1800) = \frac{3700}{12} = \$308.33$

Exercise Set 12.5

1. If a first experiment can be performed in *M* distinct ways and a second experiment can be performed in *N* distinct ways, then the two experiments in that specific order can be performed in *M·N* distinct ways.

3. (2)(7) = 14 ways. Using the counting principle.

5. The first selection is made. Then the second selection is made before the first selection is returned to the group of items being selected from.

7. a) (50)(50) = 2500   b) (50)(49) = 2450

9. a) (7(7)(7) = 343   b) (7)(6)(5) = 210

11. a) (2)(2) = 4 points

b)

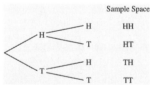

Sample Space

H	HH
T	HT
H	TH
T	TT

c) P(no heads) = 1/4

d) P(exactly one head) = 2/4 = 1/2

e) P(two heads) = 1/4

13. a) (3)(3) = 9 points

b)

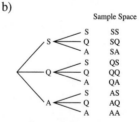

Sample Space

S	SS
Q	SQ
A	SA
S	QS
Q	QQ
A	QA
S	AS
Q	AQ
A	AA

c) P(two apples) = 1/9

d) P(sun and then question mark) = 1/9

e) P(at least one apple) = 5/9

15. a) (4)(3) = 12 points

b)

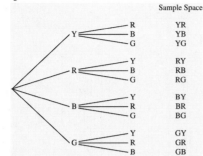

Sample Space

c) P(exactly one red) = 6/12 = ½

d) P(at least one is not red) = 12/12 = 1

e) P(no green) = 6/12 = 1/2

17. a) (3)(3) = 9 points

b)

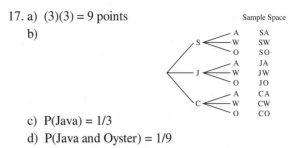

Sample Space

c) P(Java) = 1/3

d) P(Java and Oyster) = 1/9

e) P(paint other than Java) = 2/3

19. a) (6)(6) = 36 points

b)

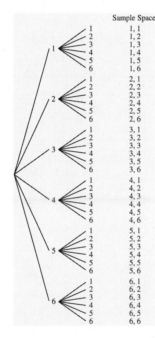

Sample Space

c) P(double) = 6/36 = 1/6

d) P(sum of 7) = 6/36 = 1/6

e) P(sum of 2) = 1/36

f) No; the P(sum of 2) < P(sum of 7)

21. a) (3)(2)(1) = 6 points

b)

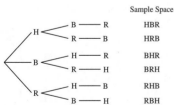

Sample Space

c) P(BB – 1st) = 2/6 = 1/3

d) P(Home Depot – 1st / RL - last) = 1/6

e) P(BB, RL, Home Depot) = 1/6

23. a) (3)(2)(3) = 18 points
    b)

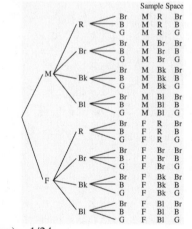

    c) P(two bedroom) = 6/18 = 1/3
    d) P(two bedroom, fireplace) = 2/18 = 1/9
    e) P(no balcony) = 12/18 = 2/3

25. a) (3)(3)(3) = 27
    b)

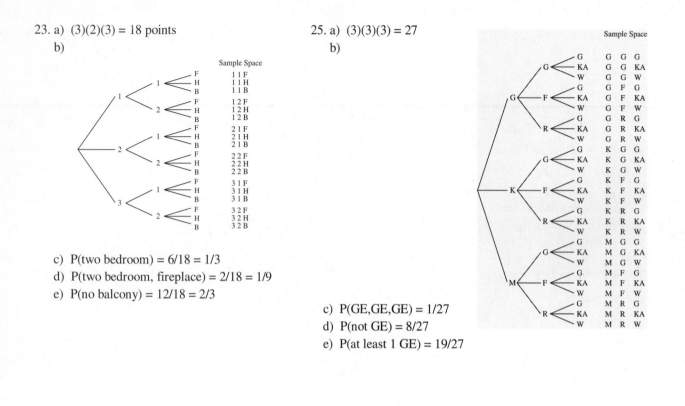

    c) P(GE,GE,GE) = 1/27
    d) P(not GE) = 8/27
    e) P(at least 1 GE) = 19/27

27. a) (2)(4)(3) = 24 sample points
    b)

    c) P(M, black, blue) = 1/24
    d) P(F, blonde) = 3/24 = 1/8

29. a) P(white) = 1/3
    b) P(red) = 2/3
    c) No; P(white) < P(red)
    d)

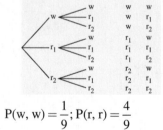

$$P(w, w) = \frac{1}{9}; P(r, r) = \frac{4}{9}$$

31. 1 red, 1 blue, and 1 brown

Exercise Set 12.6

1. a) A occurs, B occurs, or both occur   b) "and" means both events, A and B, must occur.

3. a) Two events are mutually exclusive if it is impossible for both events to occur simultaneously.
   b) $P(A \text{ or } B) = P(A) + P(B)$

5. We assume that event A has already occurred.

7. a) No, it is possible for both to like classical music.
   b) No, if the mother likes classical music, the daughter might be more likely to like classical music.

9. If the events are mutually exclusive, the events cannot happen simultaneously and thus $P(A \text{ and } B) = 0$.

11. $P(A \text{ and } B) = 0.3$
    $P(A \text{ or } B) = P(A) + P(B) - P(A \text{ and } B)$
    $= 0.6 + 0.4 - 0.3 = 1.0 - 0.3 = 0.7$

13. $P(B) = P(A \text{ or } B) + P(A \text{ and } B) - P(A)$
    $= 0.7 + 0.3 - 0.6 = 0.4$

15. $P(M \text{ and } E) = 0.55$
    $P(M \text{ or } E) = P(M) + P(E) - P(M \text{ and } E)$
    $= 0.7 + 0.6 - 0.55 = 1.3 - 0.55 = 0.75$

17. $P(2 \text{ or } 5) = 1/6 + 1/6 = 2/6 = 1/3$

19. $P(\text{greater than 4 or less than 2}) = P(5, 6. \text{ or } 1) = 2/6 + 1/6 = 3/6 = 1/2$

21. Since these events are mutually exclusive,
    $P(\text{ace or } 2) = P(\text{ace}) + P(2) =$
    $= \dfrac{4}{52} + \dfrac{4}{52} = \dfrac{8}{52} = \dfrac{2}{13}$

23. Since it is possible to obtain a card that is a picture card and a red card, these events are not mutually exclusive.
    $P(\text{picture or red})$
    $= P(\text{pict.}) + P(\text{red}) - P(\text{pict. \& red})$
    $= \dfrac{12}{52} + \dfrac{26}{52} - \dfrac{6}{52} = \dfrac{32}{52} = \dfrac{8}{13}$

25. Since it is possible to obtain a card less than 8 that is a club, these events are not mutually exclusive.
    $P(<8 \text{ or club}) = \dfrac{28}{52} + \dfrac{13}{52} - \dfrac{7}{52} = \dfrac{34}{52} = \dfrac{17}{26}$

27. a) $P(\text{monkey and monkey}) =$
    $\dfrac{5}{20} \cdot \dfrac{5}{20} = \dfrac{1}{4} \cdot \dfrac{1}{4} = \dfrac{1}{16}$
    b) $P(\text{monkey and monkey}) =$
    $\dfrac{5}{20} \cdot \dfrac{4}{19} = \dfrac{1}{4} \cdot \dfrac{4}{19} = \dfrac{1}{19}$

29. a) $P(\text{lion and bird}) = \dfrac{5}{20} \cdot \dfrac{5}{20} = \dfrac{1}{4} \cdot \dfrac{1}{4} = \dfrac{1}{16}$
    b) $P(\text{lion and bird}) = \dfrac{5}{20} \cdot \dfrac{5}{19} = \dfrac{1}{4} \cdot \dfrac{5}{19} = \dfrac{5}{76}$

31. a) $P(\text{red bird and monkey}) =$
    $\dfrac{3}{20} \cdot \dfrac{5}{20} = \dfrac{3}{20} \cdot \dfrac{1}{4} = \dfrac{3}{80}$
    b) $P(\text{red bird and monkey}) =$
    $\dfrac{3}{20} \cdot \dfrac{5}{19} = \dfrac{15}{380} = \dfrac{3}{76}$

33. a) $P(\text{odd and odd}) = \dfrac{12}{20} \cdot \dfrac{12}{20} = \dfrac{3}{5} \cdot \dfrac{3}{5} = \dfrac{9}{25}$
    b) $P(\text{odd and odd}) = \dfrac{12}{20} \cdot \dfrac{11}{19} = \dfrac{3}{5} \cdot \dfrac{11}{19} = \dfrac{33}{95}$

35. $P(\text{frog or even}) = \dfrac{5}{20} + \dfrac{8}{20} - \dfrac{2}{20} = \dfrac{11}{20}$

37. $P(\text{lion or a } 5) = \dfrac{5}{20} + \dfrac{4}{20} - \dfrac{1}{20} = \dfrac{8}{20} = \dfrac{2}{5}$

39. $P(2 \text{ reds}) = \dfrac{1}{2} \cdot \dfrac{1}{2} = \dfrac{1}{4}$

41. $P(\text{red and green}) = \dfrac{1}{4} \cdot \dfrac{1}{2} = \dfrac{1}{8}$

43. $P(2 \text{ yellows}) = P(\text{red and red}) = \dfrac{3}{8} \cdot \dfrac{3}{8} = \dfrac{9}{64}$

45. $P(2 \text{ reds}) = \dfrac{1}{2} \cdot \dfrac{1}{4} = \dfrac{1}{8}$

47. $P(\text{both not red}) = \dfrac{1}{2} \cdot \dfrac{3}{4} = \dfrac{3}{8}$

49. $P(\text{green or red}) = 4/7$

51. $P(\text{both red}) = (3/7)(3/7) = 9/49$

53. $P(\text{all red}) = (3/7)(2/6)(1/5) = 1/35$

55. $P(R, B, B) = (3/7)(2/6)(1/5) = 1/35$

57. $P(3 \text{ girls}) = P(1^{\text{st}} \text{ girl}) \bullet P(2^{\text{nd}} \text{ girl}) \bullet P(3^{\text{rd}} \text{ girl})$

$= \dfrac{1}{2} \bullet \dfrac{1}{2} \bullet \dfrac{1}{2} = \dfrac{1}{8}$

59. $P(G,G,B) = P(1^{\text{st}} \text{ girl}) \bullet P(2^{\text{nd}} \text{ girl}) \bullet$
$P(3^{\text{rd}} \text{ boy})$

$= \dfrac{1}{2} \bullet \dfrac{1}{2} \bullet \dfrac{1}{2} = \dfrac{1}{8}$

61. a) $P(5 \text{ boys}) = P(b) \bullet P(b) \bullet P(b) \bullet P(b) \bullet P(b)$

$= \dfrac{1}{2} \bullet \dfrac{1}{2} \bullet \dfrac{1}{2} \bullet \dfrac{1}{2} \bullet \dfrac{1}{2} = \dfrac{1}{32}$

b) $P(\text{next child is a boy}) = \dfrac{1}{2}$

63. a) $P(\text{Titleist/Pinnacle}) = \dfrac{4}{7} \bullet \dfrac{1}{7} = \dfrac{4}{49}$

b) $P(\text{Titleist/Pinnacle}) = \dfrac{4}{7} \bullet \dfrac{1}{6} = \dfrac{4}{42} = \dfrac{2}{21}$

65. a) $P(\text{at least 1 Top Flite}) =$
$\dfrac{2}{7} \bullet \dfrac{5}{7} + \dfrac{5}{7} \bullet \dfrac{2}{7} + \dfrac{2}{7} \bullet \dfrac{2}{7} = \dfrac{24}{49}$

b) $P(\text{at least 1 Top Flite}) =$
$\dfrac{2}{7} \bullet \dfrac{5}{6} + \dfrac{5}{7} \bullet \dfrac{2}{6} + \dfrac{2}{7} \bullet \dfrac{1}{6} = \dfrac{11}{21}$

67. $P(\text{neither had trad. ins.}) = \dfrac{31}{50} \bullet \dfrac{30}{49} = \dfrac{93}{245}$

69. $P(\text{at least one trad.}) = 1 - P(\text{neither trad.}) =$
$1 - \dfrac{93}{245} = \dfrac{152}{245}$ (see Exercise 67)

71. $P(\text{all recommended}) = \dfrac{23}{40} \bullet \dfrac{22}{39} \bullet \dfrac{21}{38} = \dfrac{1771}{9880}$

73. $P(\text{no/no/not sure}) = \dfrac{7}{40} \bullet \dfrac{6}{39} \bullet \dfrac{10}{38} = \dfrac{7}{988}$

75. The probability that any individual reacts favorably is 70/100 or 0.7.
$P(\text{Mrs. Rivera reacts favorably}) = 0.7$

77. $P(\text{all 3 react favorably}) = 0.7 \bullet 0.7 \bullet 0.7 = 0.343$

79. Since each question has four possible answers of which only one is correct, the probability of guessing correctly on any given question is 1/4.
$P(\text{correct answer on any one question}) = ¼$

81. $P(\text{only the } 3^{\text{rd}} \text{ and } 4^{\text{th}} \text{ questions correct}) =$
$\left(\dfrac{3}{4}\right)\left(\dfrac{3}{4}\right)\left(\dfrac{1}{4}\right)\left(\dfrac{1}{4}\right)\left(\dfrac{3}{4}\right) = \dfrac{27}{1024}$

83. $P(\text{none of the 5 questions correct}) =$
$\left(\dfrac{3}{4}\right)\left(\dfrac{3}{4}\right)\left(\dfrac{3}{4}\right)\left(\dfrac{3}{4}\right)\left(\dfrac{3}{4}\right) = \dfrac{243}{1024}$

85. $P(\text{orange on } 1^{\text{st}} \text{ reel}) = 5/22$

87. $P(\text{no bar/no bar/no bar}) =$
$\left(\dfrac{20}{22}\right)\left(\dfrac{20}{22}\right)\left(\dfrac{21}{22}\right) = \dfrac{1050}{1331}$

89. $P(\text{blue/blue}) = \left(\dfrac{2}{8}\right)\left(\dfrac{2}{12}\right) = \dfrac{4}{96} = \dfrac{1}{24}$

91. $P(\text{not red on outer and not red on inner}) =$
$\dfrac{8}{12} \bullet \dfrac{5}{8} = \dfrac{5}{12}$

93. $P(\text{no hit/no hit}) = (0.6)(0.6) = 0.36$

95. $P(\text{both hit}) = (0.4)(0.9) = 0.36$

97. a) No; The probability of the 2nd depends on the outcome of the first.

  b) P(one afflicted) = .001

  c) P(both afflicted) = (.001)(.04) = .00004

  d) P(afflicted/not afflicted) = (0.001)(.96) = (.00096)

  e) P(not afflicted/afflicted) = (.999)(.001) = .000999

  f) P(not affl/not affl) = (.999)(.999) = .998001

105. $P(\text{no diamonds}) = \left(\dfrac{39}{52}\right)\left(\dfrac{38}{51}\right) = \dfrac{1482}{2652} = .56$

The game favors the dealer since the probability of no diamonds is greater than 1/2.

99. P(audit this year) = .036

101. P(audit/no audit) = (.036)(.964) = .034704

103. P(2 - same color) = P(2 r) + P(2 b) + P(2 y)

$$= \left(\frac{5}{10}\right)\left(\frac{4}{9}\right) + \left(\frac{3}{10}\right)\left(\frac{2}{9}\right) + \left(\frac{2}{10}\right)\left(\frac{1}{9}\right)$$

$$= \left(\frac{20}{90}\right) + \left(\frac{6}{90}\right) + \left(\frac{2}{90}\right) = \frac{28}{90} = \frac{14}{45}$$

107. P(2/2) = (2/6)(2/6) = 4/36 = 1/9

109. P(even or < 3) = 2/6 + 3/6 – 2/6 = 3/6 = 1/2

Exercise Set 12.7

1. The probability of E_2 given that E_1 has occurred.

3. $P(E_2 \mid E_1) = \dfrac{n(E_1 \text{ and } E_2)}{n(E_1)} = \dfrac{4}{12} = \dfrac{1}{3}$

5. P(5 | orange) = 1/3

7. P(even | not orange) = 2/3

9. P(red | orange) = 2/3

11. P(circle | odd) = 3/4

13. P(red | even) = 2/3

15. P(circle or square | < 4) = 2/3

17. P(4 | purple) = 1/5

19. P(purple | odd) = 2/6 = 1/3

21. P(> 4 | purple) = 3/5

23. P(gold | > 5) = 1/7

25. P(1 and 1) = (1/4)(1/4) = 1/16

27. P(5 | at least a 5) = 1/7

29. P(sum = 6) = 5/36

31. P(6 | 3) = 1/6

33. P(> 7 | 2nd die = 5) = 4/6 = 2/3

35. P(C4) = 3/9 =1/3

37. P(at least $20B | C4) = 1/3

39. P(C5 | at least $25B) = 1/2

41. $P(\text{car}) = \dfrac{1462}{2461} = 0.5941$

43. $P(\text{E-Z} \mid \text{car}) = \dfrac{527}{1462} = 0.3605$

45. $P(\text{car} \mid \text{E-Z}) = \dfrac{527}{843} = 0.6251$

47. $P(\text{agg}) = \dfrac{350}{650} = \dfrac{7}{13}$

49. $P(\text{no sale} \mid \text{pass}) = \dfrac{80}{300} = \dfrac{4}{15}$

51. $P(\text{sale} \mid \text{pass}) = \dfrac{220}{300} = \dfrac{11}{15}$

53. $P(\text{male}) = \dfrac{145}{295} = \dfrac{29}{59}$

55. $P(\text{15-64} \mid \text{female}) = \dfrac{99}{150} = \dfrac{33}{50}$

57. $P(\text{female} \mid \text{0-14}) = \dfrac{30}{61}$

59. $P(\text{good}) = \dfrac{300}{330} = \dfrac{10}{11}$

61. $P(\text{defective} \mid \text{20 watts}) = \dfrac{15}{95} = \dfrac{3}{19}$

63. $P(\text{good} \mid \text{50 or 100 watts}) = \dfrac{220}{235} = \dfrac{44}{47}$

65. $P(\text{ABC or NBC}) = \dfrac{110}{270} = \dfrac{11}{27}$

67. $P(\text{ABC or NBC} \mid \text{man}) = \dfrac{50}{145} = \dfrac{10}{29}$

69. $P(\text{ABC,NBC,or CBS} \mid \text{man}) = \dfrac{55}{145} = \dfrac{11}{29}$

71. P(large company stock) = 93/200

73. $P(\text{blend} \mid \text{medium co. stock}) = 15/52$

75. a) n(A) = 140   b) n(B) = 120

    c) P(A) = 140/200 = 7/10

    d) P(B) = 120/200 = 6/10 = 3/5

    e) $P(A \mid B) = \dfrac{n(B \text{ and } A)}{n(B)} = \dfrac{80}{120} = \dfrac{2}{3}$

    f) $P(B \mid A) = \dfrac{n(A \text{ and } B)}{n(B)} = \dfrac{80}{140} = \dfrac{4}{7}$

    g) $P(A) \bullet P(B) = \left(\dfrac{7}{10}\right)\left(\dfrac{3}{5}\right) = \dfrac{21}{50}$

      $P(A \mid B) \quad P(A) \bullet P(B) \qquad \dfrac{2}{3} \neq \dfrac{21}{50}$

    A and B are not independent events.

77. a) $P(A \mid B) = \dfrac{P(A \text{ and } B)}{P(B)} = \dfrac{0.15}{0.5} = 0.3$

    b) $P(B \mid A) = \dfrac{P(A \text{ and } B)}{P(A)} = \dfrac{0.15}{0.3} = 0.5$

    c) Yes, $P(A) = P(A \mid B)$ and $P(B) = P(B \mid A)$.

79. $P(+ \mid \text{orange circle}) = \frac{1}{2}$

81. P(green + | +) = 1/3

83. P(orange circle w/green + | +) = 0

Exercise Set 12.8

1. If a first experiment can be performed in M distinct ways and a second experiment can be performed in N distinct ways, then the two experiments in that specific order can be performed in $M \cdot N$ distinct ways.

3. a) $n! = n(n-1)(n-2) \cdots 3 \cdot 2 \cdot 1$

    b) $\dfrac{n!}{n_1! n_2! \cdots n_r!}$

5. The number of permutations of n items taken r at a time.

7. $_n P_r = \dfrac{n!}{(n-r)!}$

9. $4! = 24$

11. $_7 P_2 = \dfrac{7!}{5!} = 7 \bullet 6 = 42$

13. $0! = 1$

15. $_8 P_0 = \dfrac{8!}{8!} = 1$

17. $_9 P_4 = \dfrac{9!}{5!} = 9 \bullet 8 \bullet 7 \bullet 6 = 3024$

19. $_8 P_3 = \dfrac{8!}{5!} = 8 \bullet 7 \bullet 6 = 336$

21. (10)(10)(10)(10) = 10000

23. a) (26)(25)(24)(10)(9) = 1,404,000

    b) (26)(26)(26)(10)(10) = 1,757,600

25. a) $5^5 = 3125$

    b) $\dfrac{1}{3125} = 0.00032$

27. (34)(36)(36)(36)(36) = 57,106,944

29. $(7)(4)(3)(4) = 336$

31. a) $6! = 720$ b) $5! = 120$
    c) $4! = 24$ d) $5! \bullet 5 = 600$

33. $_{10}P_3 = \dfrac{10!}{(10-3)!} = \dfrac{10!}{7!} = \dfrac{10 \cdot 9 \cdot 8 \cdot 7!}{7!} = 720$

35. a) There are 12 individuals and they can be arranged in $12! = 479,001,600$ ways
    b) $10! = 3,628,800$ different ways
    c) $5! \cdot 5! = 14,400$ different ways

37. $(26)(25)(24)(10)(9)(8)(7) = 78,624,000$

39. $(26)(10)(9)(8)(7) = 131,040$

41. $(10)(10)(10)(26)(26) = 676,000$

43. $(5)(4)(8)(26)(25) = 104,000$

45. a) $(8)(10)(10)(10)(10)(10)(10) = 8,000,000$
    b) $(8)(10)(10)(8,000,000) = 6,400,000,000$
    c) $(8)(10)(10)(8)(10^{10}) = (64)(10^{12})$
       $= 64,000,000,000,000$

47. $(15)(14)(13)(12)(11)(10) = 3,603,600$

49. $_{7}P_7 = \dfrac{7!}{0!} = \dfrac{7!}{1} = 7! = 5,040$

51. $(5)(4)(7)(2) = 280$ systems

53. $_{9}P_9 = \dfrac{9!}{0!} = 9! = 362,880$

55. $\dfrac{10!}{3!2!} = 302,400$

57. $\dfrac{7!}{2!2!2!} = 630$
    (There are 2 6's, 2 7's, and 2 8's.)

59. The order of the flags is important. Thus, it is a permutation problem.
    $_{9}P_5 = \dfrac{9!}{(9-5)!} = \dfrac{9!}{4!} = (9)(8)(7)(6)(5) = 15,120$

61. a) Since the pitcher must bat last, there is only one possibility for the last position. $\underline{\;\;\;\;\;\;\;\;1}$
    There are 8 possible batters left for the 1st position. Once the 1st batter has been selected, there are
    7 batters left for the 2nd position, 6 for the third, etc. $\underline{(8)}\;\underline{(7)}\;\underline{(6)}\;\underline{(5)}\;\underline{(4)}\;\underline{(3)}\;\underline{(2)}\;\underline{(1)}\;\underline{(1)} = 40,320$
    b) $9! = (9)(8)(7)(6)(5)(4)(3)(2)(1) = 362,880$

63. a) $5^5 = 3125$ different keys
    b) $400,000 \div 3,125 = 128$ cars
    c) $\dfrac{1}{3125} = 0.00032$

65. $_{7}P_5 = \dfrac{7!}{2!} = \dfrac{7 \cdot 6 \cdot 5 \cdot 4 \cdot 3 \cdot 2!}{2!} = 2,520$ different
    letter permutations; $2500 \times \dfrac{1}{12} = 210$ minutes
    or $3\dfrac{1}{2}$ hours

67. No, Ex. $_{3}P_2 \neq {_3}P_{(3\text{-}2)}$
    $\dfrac{3!}{1!} \neq \dfrac{3!}{2!}$ because $6 \neq 3$

69. $(25)(24) = 600$ tickets

71. a) $\dfrac{7!}{2!} = 2520$ b) SCROOGE

Exercise Set 12.9

1. The selection of a certain number of items without regard to their order.

3. $_nC_r = \dfrac{n!}{(n-r)! \cdot r!}$

5. If the order of the items is important then it is a permutation problem. If order is not important then it is a combination problem.

7. $_4C_2 = \dfrac{4!}{(4-2)!2!} = \dfrac{(4)(3)(2)(1)}{(2)(1)(2)(1)} = 6$

9. a) $_6C_4 = \dfrac{6!}{2!4!} = \dfrac{(6)(5)}{(2)(1)} = 15$

   b) $_6P_4 = \dfrac{6!}{(6-4)!} = \dfrac{6!}{2!} = (6)(5)(4)(3) = 360$

11. a) $_8C_0 = \dfrac{8!}{8!0!} = 1$

    b) $_8P_0 = \dfrac{8!}{(8-0)!} = \dfrac{8!}{8!} = 1$

13. a) $_{10}C_3 = \dfrac{10!}{7!3!} = \dfrac{(10)(9)(8)(7!)}{(7!)(3)(2)(1)} = 120$

    b) $_{10}P_3 = \dfrac{10!}{(10-3)!} = \dfrac{(10)(9)(8)(7!)}{7!} = 720$

15. $\dfrac{_5C_3}{_5P_3} = \dfrac{\frac{5!}{2!3!}}{\frac{5!}{2!}} = \left(\dfrac{5!}{2!3!}\right)\left(\dfrac{2!}{5!}\right) = \dfrac{1}{3!} = \dfrac{1}{6}$

17. $\dfrac{_9C_4}{_9C_2} = \dfrac{\frac{9!}{5!4!}}{\frac{9!}{7!2!}} = \left(\dfrac{9!}{5!4!}\right)\left(\dfrac{7!2!}{9!}\right) = \dfrac{(7)(6)}{(4)(3)} = \dfrac{7}{2}$

19. $\dfrac{_9P_5}{_{10}C_4} = \dfrac{\frac{9!}{4!}}{\frac{10!}{6!4!}} = \dfrac{(9)(8)(7)(6)(5)}{\frac{(10)(9)(8)(7)}{(4)(3)(2)(1)}} = \dfrac{144}{2} = 72$

21. $_8C_2 = \dfrac{8!}{6!2!} = \dfrac{(8)(7)(6!)}{(2)(1)(6!)} = 28$ ways

23. $_5C_4 = \dfrac{5!}{1!4!} = 5$

25. $_8C_3 = \dfrac{8!}{5!3!} = \dfrac{(8)(7)(6)}{(3)(2)(1)} = 56$

27. $_7C_4 = \dfrac{7!}{3!4!} = \dfrac{(7)(6)(5)}{(3)(2)(1)} = 35$

29. $_{12}C_8 = \dfrac{12!}{4!8!} = \dfrac{(12)(11)(10)(9)}{(4)(3)(2)(1)} = 495$

31. $_8C_4 = \dfrac{8!}{4!4!} = \dfrac{(8)(7)(6)(5)}{(4)(3)(2)(1)} = 70$

33. $_8C_2 = \dfrac{8!}{6!2!} = \dfrac{(8)(7)}{(2)(1)} = 28$ tickets

35. $_{12}C_3 \bullet {_8C_2} =$

$\left(\dfrac{12!}{9!3!}\right)\left(\dfrac{8!}{6!2!}\right)$

$= \left(\dfrac{(12)(11)(10)}{(3)(2)(1)}\right)\left(\dfrac{(8)(7)}{(2)(1)}\right) = 6160$

37. Red: $_{10}C_4 = \dfrac{10!}{6!4!} = \dfrac{(10)(9)(8)(7)}{(4)(3)(2)(1)} = 210$

   White. $_8C_2 = \dfrac{8!}{6!2!} = \dfrac{(8)(7)}{(2)(1)} = 28$

   $(210)(28) = 5880$ different choices

39. Regular: $_{10}C_5 =$

   $\dfrac{10!}{5!5!} = \dfrac{(10)(9)(8)(7)(6)}{(5)(4)(3)(2)(1)} = 252$

   Diet: $_7C_3 =$

   $\dfrac{7!}{3!4!} = \dfrac{(7)(6)(5)}{(3)(2)(1)} = 35$

   $(252)(35) = 8820$ ways to select the sodas

41. $_8C_4 \bullet {_5C_2} =$

$$\left(\frac{8!}{4!4!}\right)\left(\frac{5!}{3!2!}\right) = \frac{(8)(7)(6)(4)}{(4)(3)(2)(1)}\left(\frac{(5)(4)}{(2)(1)}\right) = 700$$

43. $_6C_3 \bullet {_5C_2} \bullet {_4C_2} =$

$$\left(\frac{6!}{3!3!}\right)\left(\frac{5!}{3!2!}\right)\left(\frac{4!}{2!2!}\right) =$$

$$\left(\frac{(6)(5)(4)}{(3)(2)(1)}\right)\left(\frac{(5)(4)}{(2)(1)}\right)\left(\frac{(4)(3)}{(2)(1)}\right) = 1200$$

45. a) $_{10}C_8 = \frac{10!}{2!8!} = \frac{(10)(9)}{(2)(1)} = 45$

b) $_{10}C_9 = \frac{10!}{1!9!} = \frac{(10)(9!)}{(1)(9!)} = 10$

$_{10}C_{10} = \frac{10!}{10!} = 1$

$_{10}C_8 + {_{10}C_9} + {_{10}C_{10}} = 45 + 10 + 1 = 56$

47. a)
```
            1
         1     1
      1     2     1
   1     3     3     1
1     4     6     4     1
```

b)   **1   5   10   10   5   1**

49. a) $4! = 24$ b) $4! = 24$

51. $(15)(14) \cdot {_{13}C_3} = (15)(14)\left(\frac{13!}{10!\,3!}\right)$

$$= (15)(14)\left(\frac{(13)(12)(11)}{(3)(2)(1)}\right) = 60,060$$

Exercise Set 12.10

1. $P(\text{4 red balls}) = \dfrac{\text{no. of 4 red ball comb.}}{\text{no. of 4 ball comb.}} = \dfrac{_8C_4}{_{12}C_4}$

3. $P(\text{3 vowels}) = \dfrac{\text{no. of 3 vowel comb.}}{\text{no. of 3 letter comb.}} = \dfrac{_5C_3}{_{26}C_3}$

5. $P(\text{all 5 yellow Labs}) =$

$\dfrac{\text{no. of 5 yellow Lab comb.}}{\text{no. of 5 puppy comb.}} = \dfrac{_8C_5}{_{15}C_5}$

7. $P(\text{none of the 8 are Nike}) =$

$\dfrac{\text{no. of 8 non-Nike comb.}}{\text{no. of 8 ball comb.}} = \dfrac{_{22}C_8}{_{50}C_8}$

9. $_5C_3 = \frac{5!}{2!3!} = \frac{(5)(4)}{(2)(1)} = 10$

$_9C_3 = \frac{9!}{6!3!} = \frac{(9)(8)(7)}{(3)(2)(1)} = 84$

$P(\text{3 reds}) = \frac{10}{84} = \frac{5}{42}$

11. $_8C_5 = \frac{8!}{3!5!} = \frac{(8)(7)(6)}{(3)(2)(1)} = 56$

$_{14}C_5 = \frac{14!}{5!9!} = \frac{(14)(13)(12)(11)(10)}{(5)(4)(3)(2)(1)} = 2002$

$P(\text{5 men's names}) = \frac{56}{2002} = \frac{4}{143}$

13. $_5C_3 = \dfrac{5!}{2!3!} = \dfrac{(5)(4)}{(2)(1)} = 10$

 $_{10}C_3 = \dfrac{10!}{7!3!} = \dfrac{(10)(9)(8)}{(3)(2)(1)} = 120$

 P(3 greater than 4) $= \dfrac{10}{120} = \dfrac{1}{12}$

15. $_6C_2 = \dfrac{6!}{4!2!} = \dfrac{(6)(5)}{(2)(1)} = 15$

 $_2C_1 = 2$

 $_{11}C_3 = \dfrac{11!}{8!3!} = \dfrac{(11)(10)(9)}{(3)(2)(1)} = 165$

 P(2 from mfg, 1 from acct.) $= \dfrac{(15)(2)}{165} = \dfrac{2}{11}$

17. $_{46}C_6 = \dfrac{46!}{40!6!} = 9,366,819$ $_6C_6 = 1$

 P(win grand prize) $= \dfrac{1}{9,366,819}$

19. $_3C_2 = \dfrac{3!}{1!2!} = 3$ $_5C_2 = \dfrac{5!}{3!2!} = \dfrac{(5)(4)}{(2)(1)} = 10$

 P(no cars) $= \dfrac{3}{10}$

21. P(at least 1 car) $= 1 - $ P(no cars) $=$

 $1 - 1 - \dfrac{3}{10} = \dfrac{7}{10}$

23. $_6C_3 = \dfrac{6!}{3!3!} = \dfrac{(6)(5)(4)}{(3)(2)(1)} = 20$

 $_{25}C_3 = \dfrac{25!}{3!22!} = \dfrac{(25)(24)(23)}{(3)(2)(1)} = 2300$

 P(3 infielders) $= \dfrac{20}{2300} = \dfrac{1}{115}$

25. $_{10}C_2 = \dfrac{10!}{8!2!} = 45$ $_6C_1 = \dfrac{6!}{5!1!} = 6$

 P(2 pitchers and 1 infielder) $= \dfrac{(45)(6)}{2300} = \dfrac{27}{230}$

For problems 27–29, use the fact that $_{25}C_6 = \dfrac{25!}{19!6!} = 177,100$

27. $_{10}C_6 = \dfrac{10!}{4!6!} = 210$

 P(all mid) $= \dfrac{210}{177,100} = 0.0012$

29. $_{10}C_3 = \dfrac{10!}{7!3!} = 120$

 $_{15}C_3 = \dfrac{15!}{12!3!} = 455$

 P(2 mid/4 compact) $= \dfrac{(120)(455)}{117,100} = 0.3083$

For problems 31–33, use the fact that $_{12}C_4 = \dfrac{12!}{8!4!} = 495$

31. $_4C_2 = \dfrac{4!}{2!2!} = 6$ $_3C_2 = \dfrac{3!}{1!2!} = 3$

 P(2 in FL, 2 in TX) $= \dfrac{(6)(3)}{495} = \dfrac{2}{55}$

33. $_4C_1 = 4$
 $_5C_1 = 5$
 $_3C_2 = 3$
 P(1 in FL, 1 in CA, 2 in TX) $=$
 $\dfrac{(4)(5)(3)}{495} = \dfrac{4}{33}$

For problems 35 – 37, use the fact that $_{11}C_5 = \dfrac{11!}{6!5!} = \dfrac{(11)(10)(9)(8)(7)}{(5)(4)(3)(2)(1)} = 462$

35. $_6C_5 = \dfrac{6!}{1!5!} = 6$

   $P(\text{5 women first}) = \dfrac{6}{462} = \dfrac{1}{77}$

37. Any one of the 6 women can sit in any one of the five seats - 30 possibilities.

   $P(\text{exactly 1 woman}) = \dfrac{30}{462} = \dfrac{5}{77}$

39. $_{24}C_3 = \dfrac{24!}{21!3!} = 2024$

   $_{10}C_3 = \dfrac{10!}{7!3!} = 120$

   $P(\text{all 3 are cashiers}) = \dfrac{120}{2024} = \dfrac{15}{253}$

41.   a) $P(\text{royal spade flush}) = \dfrac{_{47}C_2}{_{52}C_7} = \dfrac{1}{123760}$

   b) $P(\text{any royal flush}) = \dfrac{4}{123760} = \dfrac{1}{30,940}$

43. a) $\dfrac{\left((_4C_2)(_4C_2)(_{44}C_1)\right)}{_{52}C_5} = \dfrac{1584}{2598960} = \dfrac{33}{54,145}$

   $P(\text{2 aces/2 8's/other card} \quad \text{ace or 8}) = \dfrac{33}{54,145}$

   b) P(aces of spades and clubs/8's of spades and clubs/9 of diamonds) =

   $\dfrac{1}{_{52}C_5} = \dfrac{1}{2,598,960}$

45. a) $\left(\dfrac{1}{15}\right)\left(\dfrac{1}{14}\right)\left(\dfrac{1}{13}\right)\left(\dfrac{5}{12}\right)\left(\dfrac{4}{11}\right)\left(\dfrac{3}{10}\right)\left(\dfrac{2}{9}\right)\left(\dfrac{1}{8}\right)$

   $= \dfrac{120}{259459200} = \dfrac{1}{2,162,160}$

   b) $P(\text{any 3 of 8 for officers}) = \dfrac{(8)(7)(6)}{2162160} = \dfrac{1}{6435}$

47. Since there are more people than hairs, 2 or more people must have the same number of hairs.

Exercise Set 12.11

1. A probability distribution shows the probability associated with each specific outcome of an experiment. In a probability distribution every possible outcome must be listed and the sum of all the probabilities must be 1.

3. $P(x) = {_nC_x}\,p^x q^{n-x}$

5. $P(3) = {_5C_3}(0.2)^3(0.8)^{5-3}$

   $= \dfrac{5!}{2!3!}(0.008)(0.64) = 0.0512$

7. $P(2) = {_5C_2}(0.4)^2(0.6)^{5-2}$

   $= \dfrac{5!}{2!3!}(0.16)(0.216) = 0.3456$

9. $P(0) = {_6C_0}(0.5)^0(0.5)^{6-0}$

   $= \dfrac{6!}{0!6!}(1)(.0156252) = 0.015625$

11. $p = 0.14, \ q = 1 - p = 1 - 0.14 = 0.86$

   a) $P(x) = {_nC_x}(0.14)^x(0.86)^{n-x}$

   b) $n = 12, \ x = 2, \ p = 0.14, \ q = 0.86$

   $P(2) = {_{12}C_2}(0.14)^2(0.86)^{12-2}$

13. $P(4) = {}_6C_4(0.8)^4(0.2)^{6-4}$

$= \dfrac{6!}{4!2!}(0.4096)(0.04) = 0.24576$

15. $P(5) = {}_8C_5(0.7)^5(0.3)^{8-5}$

$= \dfrac{8!}{3!5!}(0.16807)(0.027) = 0.25412$

17. $P(4) = {}_6C_4(0.92)^4(0.08)^{6-4}$

$= \dfrac{6!}{4!2!}(.7164)(.0064) = 0.06877$

19. $P(4) = {}_5C_4(.8)^4(.2)^{5-4}$

$= \dfrac{5!}{1!4!}(.4096)(.2) = 0.4096$

21. a) $P(\text{all five}) = {}_5C_5(0.25)^5(0.75)^{5-5}$

$= \dfrac{5!}{5!}(0.0009765625)(1) \approx 0.00098$

b) $P(\text{exactly three}) = {}_5C_3(0.25)^3(0.75)^{5-3}$

$= \dfrac{5!}{2!3!}(0.015625)(0.5625) \approx 0.08789$

c) $P(\text{at least } 3) = P(3) + P(4) + P(5)$

$P(4) = {}_5C_4(0.25)^4(0.75)^{5-1}$

$= \dfrac{5!}{1!4!}(0.00390625)(0.75) \approx 0.01465$

$P(\text{at least } 3) = 0.08789 + 0.01465 + 0.00098$

$= 0.10352$

23. a) $P(3) = {}_6C_3\left(\frac{12}{52}\right)^3\left(\frac{40}{52}\right)^3$

$\dfrac{6!}{3!3!}(.01229)(.45517) = 0.11188$

b) $P(2) = {}_6C_2\left(\frac{13}{52}\right)^2\left(\frac{39}{52}\right)^4$

$= \dfrac{6!}{2!4!}(.0625)(.3164) = 0.29663$

25. The probability that the sun would be shining would equal 0 because 72 hours later would occur at midnight.

Review Exercises

1. Relative frequency over the long run can accurately be predicted, not individual events or totals.
2. Roll the die many times then compute the relative frequency of each outcome and compare with the expected probability of 1/6.

3. $P(\text{SUV}) = \dfrac{8}{40} = \dfrac{1}{5}$

4. Answers will vary.

5. $P(\text{watches ABC}) = \dfrac{80}{200} = \dfrac{2}{5}$

6. $P(\text{even}) = \dfrac{5}{10} = \dfrac{1}{2}$

7. $P(\text{odd or} > 5) = \dfrac{5}{10} + \dfrac{4}{10} - \dfrac{2}{10} = \dfrac{7}{10}$

8. $P(> 3 \text{ or} < 6) = \dfrac{6}{10} + \dfrac{6}{10} - \dfrac{2}{10} = \dfrac{10}{10} = 1$

9. $P(\text{even and} > 6) = \dfrac{1}{10}$

10. $P(\text{Jack}) = \dfrac{14}{50} = \dfrac{7}{25}$

11. $P(\text{Muenster}) = \dfrac{11}{50}$

12. $P(\text{Cheddar or Colby}) =$

$\dfrac{18}{50} + \dfrac{14}{50} = \dfrac{32}{50} = \dfrac{16}{25}$

13. $P(\text{not Swiss}) = \dfrac{50-7}{50} = \dfrac{43}{50}$

14. a) 69:31   b) 31:69

15. 5:3

16. P(wins Triple Crown) = $\dfrac{3}{85}$

17. 7:3

18. a) E = P(win \$200)•\$198 + P(win \$100)•\$98
       + P(lose)•(–\$2)
    = (.003)(198) + (.002)(98) – (.995)(2)
    = .594 + .196 – 1.990 = - 1.200 → -\$1.20

   b) The expectation of a person who purchases
      three tickets would be 3(–1.20) = –\$3.60.

   c) Expected value = Fair price – Cost
      –1.20 = Fair price – 2.00   \$.80 = Fair price

19. a) $E_{Cameron}$ = P(pic. card)(\$9) +
       P(not pic. card)(–\$3)

    $= \left(\dfrac{12}{52}\right)(9) - \left(\dfrac{40}{52}\right)(3) = \approx -\$0.23$

   b) $E_{Lindsey}$ = P(pic. card)(–\$9) + P(not pic.
      card)(\$3)

    $= \dfrac{-27}{13} + \dfrac{30}{13} = \dfrac{3}{13} \approx \$0.23$

   c) Cameron can expect to lose $(100)\left(\dfrac{3}{13}\right) \approx \$23.08$

20. E = P(sunny)(1000) + P(cloudy)(500) + P(rain)(100) = 0.4(1000) + 0.5(500) + 0.1(100) =
    400 + 250+10 = 660 people

21. a)

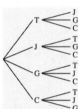

   b) Sample space:
      {TJ,TG,TC,JT,JG,JC,GT,GJ,GC,CT,CJ,CG}

   c) P(Gina is Pres. and Jake V.P.) = 1/12

22. a)

   b) Sample space:
      {H1,H2,H3,H4,T1,T2,T3,T4}

   c) P(heads and odd) = (1/2)(2/4) = 2/8 = ¼

   d) P(heads or odd) = (1/2)(2/4) + (1/2)(2/4)
      = 4/8 + 2/8 = 6/8 = 3/4

23. P(even and even) = (4/8)(4/8) = 16/64 = 1/4

24. P(outer is greater than 5 and inner is greater
    than 5)

    $= P(\text{outer is} > 5)\cdot P(\text{inner is} > 5) = \dfrac{3}{8}\cdot\dfrac{3}{8} = \dfrac{9}{64}$

25. P(outer odd and inner < 6)

    $= P(\text{outer odd}) \, P(\text{inner} < 6) = \dfrac{4}{8}\cdot\dfrac{5}{8} = \dfrac{1}{2}\cdot\dfrac{5}{8} = \dfrac{5}{16}$

26. P(outer is even or less than 6)
    = P(even) + P(< 6) – P(even and < 6)

    $= \dfrac{4}{8} + \dfrac{5}{8} - \dfrac{2}{8} = \dfrac{7}{8}$

27. P(inner even and not green) =

    $\dfrac{1}{2} + \dfrac{6}{8} - \dfrac{2}{8} = \dfrac{1}{2} + \dfrac{4}{8} = 1$

28. P(outer gold and inner not gold)

    $= \left(\dfrac{2}{8}\right)\left(\dfrac{6}{8}\right) = \left(\dfrac{1}{4}\right)\left(\dfrac{3}{4}\right) = \dfrac{3}{16}$

29. P(all 3 are GM) $= \dfrac{5}{12}\cdot\dfrac{4}{11}\cdot\dfrac{3}{10} = \dfrac{60}{1320} = \dfrac{1}{22}$

30. P(none are Kellogg's) =
    $\dfrac{8}{12}\cdot\dfrac{7}{11}\cdot\dfrac{6}{10} = \dfrac{336}{1320} = \dfrac{14}{55}$

31. P(at least one is Kellogg's) = 1 – P(none are
    Kellogg's)

    $= 1 - \dfrac{14}{55} = \dfrac{55}{55} - \dfrac{14}{55} = \dfrac{41}{55}$

32. P(GM, GM, Post)
    $= \dfrac{5}{12}\cdot\dfrac{4}{11}\cdot\dfrac{3}{10} = \dfrac{60}{1320} = \dfrac{1}{22}$

33. P(yellow) = 1/4

34. Odds against yellow   3:1
    Odds for yellow   1:3

35. $5 for red; $10 for yellow; $20 for green
    P(green) = ½; P(yellow) = ¼; P(red) = ¼
    EV = (1/4)(5) + (1/4)(10) + (1/2)(20) = $13.75

36. P(red, then green) = P(red)P(green)
    = (1/4)(1/2) = 1/8

37. P(not green) = 1/4 + 1/4 + 1/8 = 5/8

38. Odds in favor of green   3:5
    Odds against green   5:3

39. E = P(green)($10) + P(red)($5) +
    P(yellow)(–$20)
    = (3/8)(10) + (1/2)(5) – (1/8)(20)
    = (15/4) + (10/4) – (10/4) = 15/4 → $3.75

40. P(at least one red) = 1 – P(none are red)
    = 1 – (1/2)(1/2)(1/2) = 1-1/8 = 7/8

41. P(rated good) = 150/170 = 15/17.

42. P(good | dinner) = 85/95 = 17/19

43. P(poor | lunch) = 10/75 = 2/15

44. P(dinner | poor) = 10/20 = 1/2

45. P(right handed) = $\dfrac{230}{400} = \dfrac{23}{40}$

46. P(left brained | left handed) = $\dfrac{30}{170} = \dfrac{3}{17}$

47. P(right handed | no predominance) = $\dfrac{60}{80} = \dfrac{3}{4}$

48. P(right brained | left handed) = $\dfrac{120}{170} = \dfrac{12}{17}$

49. a) 4! = (4)(3)(2)(1) = 24
    b) E = (1/4)(10K) + (1/4)(5K) + (1/4)(2K)
        + (1/4)(1K) = (1/4)(18K) = $4,500.00

50. # of possible arrangements = $(_5C_2)(_3C_2)(_1C_1)$
    $= \left(\dfrac{5!}{3!2!}\right)\left(\dfrac{3!}{1!2!}\right)\left(\dfrac{1!}{1!}\right) = \dfrac{(5)(4)(3)}{(2)(1)} = 30$

51. $_{10}P_3 = \dfrac{10!}{7!} = (10)(9)(8) = 720$

52. $_9P_3 = \dfrac{9!}{6!} = \dfrac{(9)(8)(7)(6!)}{6!} = (9)(8)(7) = 504$

53. $_6C_3 = \dfrac{6!}{3!3!} = \dfrac{(6)(5)(4)}{(3)(2)(1)} = 20$

54. a) $_{15}C_{10} = \dfrac{15!}{5!10!} = \dfrac{(15)(14)(13)(12)(11)}{(5)(4)(3)(2)(1)} = 3003$

    b) number of arrangements = 10! = 3,628,800

55. a) P(match 5 numbers) = $\dfrac{1}{_{56}C_5}$

    $= \dfrac{1}{\dfrac{56!}{51!5!}} = \dfrac{51!5!}{56!} = \dfrac{1}{3,819,816}$

    b) P(Big game win) = P(match 5 #s and Big #)
        = P(match 5 #s) • P(match Big #)

    $= \left(\dfrac{1}{3,819,816}\right)\left(\dfrac{1}{46}\right) = \dfrac{1}{175,711,536}$

56. $(_{10}C_2)(_{12}C_5) =$

    $\left(\dfrac{10!}{8!2!}\right)\left(\dfrac{12!}{7!5!}\right)$

    $= \dfrac{(10)(9)(12)(11)(10)(9)(8)}{(2)(1)(5)(4)(3)(2)(1)}$

    $= 35,640$ possible committees

57. $(_8C_3)(_5C_2) =$

    $\left(\dfrac{8!}{5!3!}\right)\left(\dfrac{5!}{2!3!}\right) = \dfrac{(8)(7)(6)(5)(4)}{(3)(2)(1)(2)(1)} = 560$

58. P(two aces) = $\dfrac{_4C_2}{_{52}C_2} = \dfrac{\dfrac{4!}{2!2!}}{\dfrac{52!}{50!2!}}$

    $= \left(\dfrac{4!}{2!2!}\right)\left(\dfrac{50!2!}{52!}\right) = \dfrac{1}{221}$

59. P(all three are red) = $\left(\dfrac{5}{10}\right)\left(\dfrac{4}{9}\right)\left(\dfrac{3}{8}\right) = \dfrac{1}{12}$

60. P(1st 2 are red/3rd is blue) = $\left(\dfrac{5}{10}\right)\left(\dfrac{4}{9}\right)\left(\dfrac{2}{8}\right) = \dfrac{1}{18}$

61. P(1st red, 2nd white, 3rd blue)

$$= \left(\frac{5}{10}\right)\left(\frac{3}{9}\right)\left(\frac{2}{8}\right) = \frac{1}{24}$$

62. P(at least one red) = 1 – P(none are red)

$$= 1 - \left(\frac{5}{10}\right)\left(\frac{4}{9}\right)\left(\frac{3}{8}\right) = 1 - \frac{1}{12} = \frac{11}{12}$$

63. P(3 N&WRs) =

$$\frac{_5C_3}{_{14}C_3} = \frac{\frac{5!}{3!2!}}{\frac{14!}{3!11!}} = \frac{5!3!11!}{3!2!14!} = \frac{(5)(4)(3)}{(14)(13)(12)} = \frac{5}{182}$$

64. P(2 NWs & 1 Time) =

$$\frac{\left(_6C_2\right)\left(_3C_1\right)}{_{14}C_3} = \frac{\left(\frac{6!}{2!4!}\right)\left(\frac{3!}{1!2!}\right)}{\frac{14!}{3!11!}}$$

$$= \frac{(6)(5)(3)(3)(2)(1)}{(2)(1)(14)(13)(12)} = \frac{45}{364}$$

65. $\dfrac{_8C_3}{_{14}C_3} = \dfrac{\frac{8!}{3!5!}}{\frac{14!}{3!11!}} = \dfrac{8!3!11!}{3!5!14!}$

$$= \frac{(8)(7)(6)}{(14)(13)(12)} = \frac{336}{2184} = \frac{2}{13}$$

66. $1 - \dfrac{2}{13} = \dfrac{11}{13}$

67. a) $P(x) = {}_nC_x (0.6)^x (0.4)^{n-x}$

   b) $P(75) = {}_{100}C_{75}(0.6)^{75}(0.4)^{25}$

68. n = 5, x = 3, p = 1/5, q = 4/5

$$P(3) = {}_5C_3\left(\frac{1}{5}\right)^3\left(\frac{4}{5}\right)^2 = 10\cdot\left(\frac{1}{5}\right)^3\left(\frac{4}{5}\right)^2 =$$
0.0512

69. a) n = 5, p = 0.6, q = 0.4

$$P(0) = {}_5C_0 (0.6)^0 (0.4)^5$$

$$= (1)(1)(0.4)^5 = 0.01024$$

   b) P(at least 1) = 1 – P(0) = 1 – 0.01024 = 0.98976

Chapter Test

1. P(fishing for tuna) = $\dfrac{22}{40} = \dfrac{11}{20}$

2. $(P > 7) = \dfrac{2}{9} \approx 0.22$

3. P(odd) = $\dfrac{5}{9} \approx 0.55$

4. $P(\geq 4) = \dfrac{7}{9} \approx 0.78$

5. P(odd and > 4) = $\dfrac{3}{9} = \dfrac{1}{3} \approx 0.33$

6. P(both > 5) = $\dfrac{4}{9}\cdot\dfrac{3}{8} = \dfrac{12}{72} = \dfrac{1}{6}$

7. P(both even) = $\dfrac{4}{9}\cdot\dfrac{3}{8} = \dfrac{1\cdot1}{3\cdot2} = \dfrac{1}{6}$

8. P(1st odd, 2nd even) = $\dfrac{5}{9}\cdot\dfrac{4}{8} = \dfrac{5}{9}\cdot\dfrac{1}{2} = \dfrac{5}{18}$

9. P(neither > 6) = $\dfrac{6}{9}\cdot\dfrac{5}{8} = \dfrac{1\cdot5}{3\cdot4} = \dfrac{5}{12}$

10. P(red or picture)
    = P(red) + P(picture) – P(red and picture)
$$= \frac{26}{52} + \frac{12}{52} - \frac{6}{52} = \frac{32}{52} = \frac{8}{13}$$

11. 1 die   (6)(3) = 18

12.

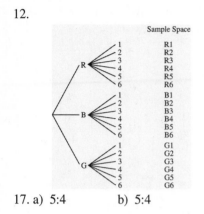

Sample Space

R — 1 R1
2 R2
3 R3
4 R4
5 R5
6 R6

B — 1 B1
2 B2
3 B3
4 B4
5 B5
6 B6

G — 1 G1
2 G2
3 G3
4 G4
5 G5
6 G6

17. a) 5:4   b) 5:4

19. E = P(club) ($8) + P(heart) ($4)
     + P(spade or diamond) (–$6)

$$= \left(\frac{1}{4}\right)(8) + \left(\frac{1}{4}\right)(4) + \left(\frac{2}{4}\right)(-6)$$

$$= \frac{8}{4} + \frac{4}{4} - \frac{12}{4} = \$0.00$$

20. d) $P(\text{GW Bridge} \mid \text{car}) = \dfrac{120}{214} = \dfrac{60}{107}$

22. $P(\text{neither is good}) = \dfrac{6}{20} \cdot \dfrac{5}{19} = \dfrac{3}{38}$

24. $_7C_3 = \dfrac{7!}{4!3!} = \dfrac{(7)(6)(5)}{(3)(2)(1)} = 35$

$$_5C_2 = \frac{5!}{3!2!} = \frac{(5)(4)}{(2)(1)} = 10$$

$$_{12}C_5 = \frac{12!}{7!5!} = \frac{(12)(11)(10)(9)(8)}{(5)(4)(3)(2)(1)} = 792$$

$$P(\text{3 red and 2 green}) = \frac{(35)(10)}{792} = \frac{350}{792} = \frac{175}{396}$$

13. $P(\text{green and 2}) = \dfrac{1}{18}$

14. $P(\text{red or 1}) = \dfrac{6}{18} + \dfrac{3}{18} - \dfrac{1}{18} = \dfrac{8}{18} = \dfrac{4}{9}$

15. $P(\text{not red or even}) = \dfrac{12}{18} + \dfrac{9}{18} - \dfrac{6}{18} = \dfrac{15}{18} = \dfrac{5}{6}$

16. Number of codes = (26)(9)(10)(26)(26)
     = 1,581,840

18. odds against Mark winning are 7:2 or

$$\frac{7}{2} = \frac{7/9}{2/9} = \frac{P(\text{not winning})}{P(\text{winning})}$$

Therefore, P(Mark wins) = 2/9

20. a) $P(\text{car}) = \dfrac{214}{456} = \dfrac{107}{228}$

   b) $P(\text{Golden Gate}) = \dfrac{230}{456} = \dfrac{115}{228}$

   c) $P(\text{SUV} \mid \text{Golden Gate}) = \dfrac{136}{230} = \dfrac{68}{115}$

21. $_6P_3 = \dfrac{6!}{(6-3)!} = \dfrac{6!}{3!} = 6 \cdot 5 \cdot 4 = 120$

23. $P(\geq 1 \text{ good}) = 1 - P(\text{neither -good}) =$
     $1 - \dfrac{3}{38} = \dfrac{35}{38}$

25. (0.3)(0.3)(0.3) = 0.027
     (0.3)(0.3)(0.3)(0.7)(0.7) = 0.01323

$$_5C_3 = \frac{5!}{3!2!} = \frac{(5)(4)}{(2)(1)} = 10$$

(10)(.01323) = 0.1323

CHAPTER THIRTEEN

STATISTICS

Exercise Set 13.1
1. **Statistics** is the art and science of gathering, analyzing, and making inferences (predictions) from numerical information obtained in an experiment.
3. Answers will vary.
5. Insurance companies, sports, airlines, stock market, medical profession
7. a) A **population** consists of all items or people of interest.   b) A **sample** is a subset of the population.
9. a) A **systematic sample** is a sample obtained by selecting every n^{th} item on a list or production line.
   b) Use a random number table to select the first item, then select every n^{th} item after that.
11. a) A **cluster sample** is a random selection of groups of units.
    b) Divide a geographic area into sections. Randomly select sections or clusters. Either each member of the selected cluster is included in the sample or a random sample of the members of each selected cluster is used.
13. a) An **unbiased sample** is one that is a small replica of the entire population with regard to income, education, gender, race, religion, political affiliation, age, etc.

15. Stratified sample
17. Cluster sample
19. Systematic sample
21. Convenience sample
23. Random sample
25. a) – c) Answers will vary.

27. President; four out of 42 U.S. presidents have been assassinated (Lincoln, Garfield, McKinley, Kennedy).

Exercise Set 13.2
1. Answers will vary.
3. Not all people who request a brochure will purchase a travel package.
5. Although the cookies are fat free, they still contain calories. Eating many of them may still cause you to gain weight.
7. The fact that Morgan's is the largest department store does not imply it is inexpensive.
9. People with asthma may move to Arizona because of its climate. Therefore, more people with asthma may live in Arizona.
11. The quality of a steak does not necessarily depend on the price of the steak.
13. There may be deep sections in the pond, so it may not be safe to go wading.
15. Half the students in a population are expected to be below average.

17. a)   b)

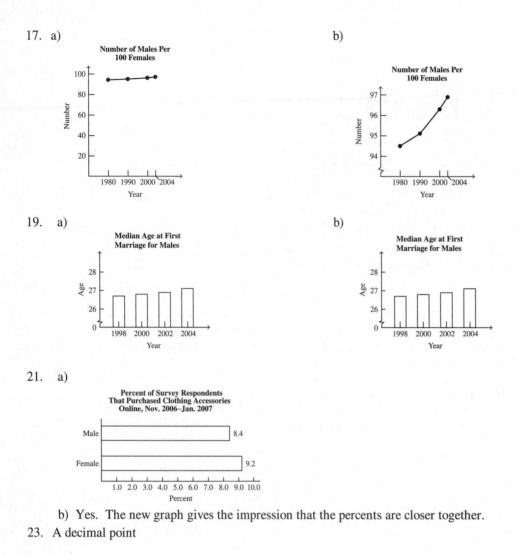

19. a)   b)

21. a)

b) Yes. The new graph gives the impression that the percents are closer together.

23. A decimal point

Exercise Set 13.3

1. A **frequency distribution** is a listing of observed values and the corresponding frequency of occurrence of each value.

3. a) 7   b) 16-22   c) 16   d) 22

5. The **modal class** is the class with the greatest frequency.

7. a) Number of observations = sum of frequencies = 20

   b) Width = $16 - 9 = 7$

   c) $\dfrac{16 + 22}{2} = \dfrac{38}{2} = 19$

   d) The modal class is the class with the greatest frequency. Thus, the modal class is 16 - 22.

   e) Since the class widths are 7, the next class would be 51 - 57.

9.

Number of Visits	Number of Students
0	3
1	8
2	3
3	5
4	2
5	7
6	2
7	3
8	4
9	1
10	2

11.

I.Q.	Number of Students
78 - 86	2
87 - 95	15
96 - 104	18
105 - 113	7
114 - 122	6
123 - 131	1
132 - 140	1

13.

I.Q.	Number of Students
80 - 90	8
91 - 101	22
102 - 112	11
113 - 123	7
124 - 134	1
135 - 145	1

15.

Placement test scores	No. of Students
472 - 492	9
493 - 513	9
514 - 534	5
535 - 555	2
556 - 576	3
577 - 597	2

17.

Placement test scores	No. of Students
472 - 487	4
488 - 503	9
504 - 519	7
520 - 535	3
536 - 551	2
552 - 567	2
568 - 583	2
584 - 599	1

19.

Circulation (10,000's)	Number of Magazines
157 - 306	34
307 - 456	9
457 - 606	2
607 - 756	1
757 - 906	1
907 - 1056	1

21.

Circulation (10,000's)	Number of Magazines
157 - 256	29
257 - 356	8
357 - 456	6
457 - 556	2
557 - 656	0
657 - 756	1
757 - 856	1
857 - 956	0
957 - 1056	1

23.

Population (millions)	Number of Cities
6.0 - 6.9	2
7.0 - 7.9	5
8.0 - 8.9	6
9.0 - 9.9	2
10.0 - 10.9	3
11.0 - 12.0	2

25.

Population (millions)	Number of Cities
5.5 - 6.6	2
6.6 - 7.6	4
7.7 - 8.7	6
8.8 - 9.9	3
9.9 - 10.9	3
11.0 – 12.0	2

27.

Percent	Number of States
5.6 -7.5	2
7.6 - 9.5	8
9.6 - 11.5	16
11.6 - 13.5	10
13.6 - 15.5	6
15.6 - 17.5	8

29.

Percent	Number of States
5.6 - 7.0	1
7.1 - 8.5	4
8.6 - 10.0	13
10.1 - 11.5	8
11.6 - 13.0	9
13.1 - 14.5	3
14.6 - 16.0	7
16.1 - 17.5	5

31. February, since it has the fewest number of days

Exercise Set 13.4

1. Answers will vary.

3. Answers will vary.

5. a) Answers will vary.

  b)

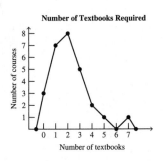

Number of Textbooks Required

7. a) Answers will vary.

  b)

Observed Values	Frequency
45	3
46	0
47	1
48	0
49	1
50	1
51	2

9. Tuition: $0.721(32,235) = \$23,241$

  Room: $0.151(32,235) = \$4867$

  Board: $0.111(32,235) = \$3578$

  Fees: $0.017(32,235) = \$548$

11. Breakfast: $\dfrac{46}{600} \approx 0.077 = 7.7\%$

  Dinner: $\dfrac{190}{600} \approx 0.317 = 31.7\%$

  Lunch: $\dfrac{293}{600} \approx 0.488 = 48.8\%$

  Snack: $\dfrac{71}{600} \approx 0.118 = 11.8\%$

Meals Eaten at a Fast Food Restaurant

13. a) and b)

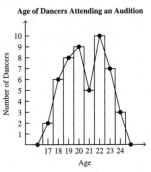

Age of Dancers Attending an Audition

15. a) and b)

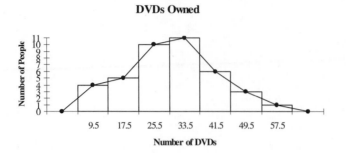

17. a) $2+4+8+6+4+3+1=28$
    b) 4
    c) 2
    d) $2(0)+1(4)+2(8)+6(3)+4(4)+3(5)+1(6)=75$

    e)

Number of TVs	Number of homes
0	2
1	4
2	8
3	6
4	4
5	3
6	1

19. a) 7 messages

    b) Adding the number of people who sent 6, 5, 4, or 3 messages gives: $4 + 7 + 3 + 2 = 16$ people

    c) The total number of people in the survey: $2 + 3 + 7 + 4 + 3 + 8 + 6 + 3 = 36$

    d)

Number of Messages	Number of People
3	2
4	3
5	7
6	4
7	3
8	8
9	6
10	3

    e)

Number of Text Messages Sent

21.

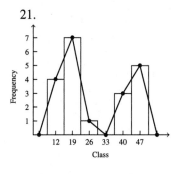

23. 1 | 2 represents 12

```
0 | 4 6 7 8
1 | 2 2 3 5 6 7 8 9
2 | 1 2 3 5 7
3 | 3 4
4 | 0
```

25. a)

Salaries (in $1000)	Number of Social Workers
27	1
28	7
29	4
30	3
31	2
32	3
33	3
34	2

b) and c)

Starting Salaries for 25 Different Social Workers

d) 2 | 8 represents 28

```
2 | 7 8 8 8 8 8 8 8 9 9 9 9
3 | 0 0 0 1 1 2 2 2 3 3 3 4 4
```

27. a)

Number of Performances	Number of Shows
1508 - 2548	33
2549 - 3589	7
3590 - 4630	4
4631 - 5671	1
5672 - 6712	3
6713 - 7753	1
7754 - 8794	1

b) and c)

Number of Performances for the 50 Longest Running Broadway Shows

29. a) – e) Answers will vary.

Exercise Set 13. 5

1. a) The **mean** is the balancing point of a set of data. It is the sum of the data divided by the number of pieces of data.

   b) The **median** is the value in the middle of a set of ranked data. To find the median, rank the data and select the value in the middle.

3. The median should be used when there are some values that differ greatly from the rest of the values in the set, for example, salaries.

5. The midrange should be used when the item being studied is constantly fluctuating, for example, daily temperature.

7. a) $\bar{x}$ b) μ

9. The second quartile is the median; the first and third quartiles are the medians of the lower and upper halves of the data

	mean	median	mode	midrange
11.	$\dfrac{117}{9} = 13$	12	12	$\dfrac{7+25}{2} = 16$
13.	$\dfrac{555}{7} \approx 79.3$	82	none	$\dfrac{52+100}{2} = 76$
15.	$\dfrac{64}{8} = 8$	$\dfrac{7+9}{2} = 8$	none	$\dfrac{1+15}{2} = 8$
17.	$\dfrac{118}{9} \approx 13.1$	11	1	$\dfrac{1+36}{2} = 18.5$
19.	$\dfrac{95}{8} \approx 11.9$	$\dfrac{12+13}{2} = 12.5$	13	$\dfrac{6+17}{2} = 11.5$
21.	$\dfrac{35}{10} = 3.5$ weeks	3 weeks	3 weeks	$\dfrac{1+7}{2} = 4$ weeks

	mean	median	mode	midrange
23. a)	$\dfrac{34}{7} \approx 4.9$	5	5	$\dfrac{1+11}{2} = 6$
b)	$\dfrac{37}{7} \approx 5.3$	5	5	$\dfrac{1+11}{2} = 6$
c)	Only the mean			
d)	$\dfrac{33}{7} \approx 4.7$	5	5	$\dfrac{1+10}{2} = 5.5$

The mean and the midrange

25. A 79 mean average on 10 quizzes gives a total of 790 points. An 80 mean average on 10 quizzes requires a total of 800 points. Thus, Jim missed a B by 10 points not 1 point.

27. a) Mean: $\dfrac{528.7}{10} \approx 52.9$ million

    c) Mode: none

    b) Median: $\dfrac{43.3+44.3}{2} = 43.8$ million

    d) Midrange: $\dfrac{37.6+85.9}{2} \approx 61.8$ million

29. a) Mean: $\dfrac{107.5}{10} \approx 10.8$ million

    c) Mode: none

    b) Median: $\dfrac{9.9+10.2}{2} \approx 10.1$ million

    d) Midrange: $\dfrac{7.8+15.1}{10} \approx 11.5$ million

31. Let x = the sum of his scores

$$\dfrac{x}{6} = 92$$

$$x = 92(6) = 552$$

33. One example is 72, 73, 74, 76, 77, 78.
Mean: $\dfrac{450}{6} = 75$, Median: $\dfrac{74+76}{2} = 75$, Midrange: $\dfrac{72+78}{2} = 75$

35. a) Yes
    b) Cannot be found since we do not know the middle two numbers in the ranked list
    c) Cannot be found without knowing all of the numbers
    d) Yes
points on his first four exams. Thus, he needs 400 - 307 = 93 or higher to get a B.

37. a) For a mean average of 60 on 7 exams, she must have a total of $60 \times 7 = 420$ points. Sheryl presently has 51+ 72 + 80 + 62 + 57 + 69 = 391 points. Thus, to pass the course, her last exam must be 420 − 391 = 29 or greater.
b) A C average requires a total of $70 \times 7 = 490$ points. Sheryl has 391. Therefore, she would need 490 − 391 = 99 or greater on her last exam.
c) For a mean average of 60 on 6 exams, she must have a total of $60 \times 6 = 360$ points. If the lowest score on an exam she has already taken is dropped, she will have a total of 72 + 80 + 62 + 57 + 69 = 340 points. Thus, to pass the course, her last exam must be 360 − 340 = 20 or greater.
d) For a mean average of 70 on 6 exams, she must have a total of $70 \times 6 = 420$ points. If the lowest score on an exam she has already taken is dropped, she will have a total of 340 points. Thus, to obtain a C, her last exam must be 420 − 340 = 80 or greater.

39. One example is 1, 2, 3, 3, 4, 5 changed to 1, 2, 3, 4, 4, 5.

    First set of data: Mean: $\dfrac{18}{6} = 3$, Median: $\dfrac{3+3}{2} = 3$, Mode: 3

    Second set of data: Mean: $\dfrac{19}{6} = 3.1\overline{6}$, Median: $\dfrac{3+4}{2} = 3.5$, Mode: 4

41. No, by changing only one piece of the six pieces of data you cannot alter both the median and the midrange.

43. The data must be arranged in either ascending or descending order.

45. He is taller than approximately 35% of all kindergarten children.

47. a) $Q_2 = $ Median $= \$25$

    b) $Q_1 = $ Median of the first 10 data values $= \$22$

    c) $Q_3 = $ Median of the last 10 data values $= \$34$

49. Second quartile, median

51. a) $530   b) $540   c) 25%   d) 25%   e) 17%   f) $100 \times \$550 = \$55,000$

53. a) Ruth: $\approx$ 0.290, 0.359, 0.301, 0.272, 0.315

    Mantle: $\approx$ 0.300, 0.365, 0.304, 0.275, 0.321

    b) Mantle's is greater in every case.

    c) Ruth: $\dfrac{593}{1878} \approx 0.316$; Mantle: $\dfrac{760}{2440} \approx 0.311$; Ruth's is greater.

    d) Answers will vary.

    e) Ruth: $\dfrac{1.537}{5} \approx 0.307$; Mantle: $\dfrac{1.565}{5} = 0.313$; Mantle's is greater.

    f) and g) Answers will vary.

55. $\Sigma xw = 84(0.40) + 94(0.60) = 33.6 + 56.4 = 90$

    $\Sigma w = 0.40 + 0.60 = 1.00$

    weighted average $= \dfrac{\Sigma xw}{\Sigma w} = \dfrac{90}{1.00} = 90$

57. a) – c) Answers will vary.

Exercise Set 13.6

1. To find the **range**, subtract the lowest value in the set of data from the highest value.

3. Answers will vary.

5. Zero since the mean is the same value as all of the data values. The spread about the mean is 0.

7. σ

9. In manufacturing or anywhere else where a minimum variability is desired

11. They would be the same since the spread of data about each mean is the same.

13. a) The grades will be centered about the same number since the mean, 75.2, is the same for both classes.

    b) The spread of the data about the mean is greater for the evening class since the standard deviation is greater for the evening class.

15. Range = 17 − 6 = 11

$$\overline{x} = \frac{55}{5} = 11$$

x	$x - \overline{x}$	$(x - \overline{x})^2$
11	0	0
9	−2	4
6	−5	25
12	1	1
17	6	36
	0	66

$$\frac{66}{4} = 16.5, s = \sqrt{16.5} \approx 4.06$$

17. Range = 136 − 130 = 6

$$\overline{x} = \frac{931}{7} = 133$$

x	$x - \overline{x}$	$(x - \overline{x})^2$
130	−3	9
131	−2	4
132	−1	1
133	0	0
134	1	1
135	2	4
136	3	9
	0	28

$$\frac{28}{6} \approx 4.67, s = \sqrt{4.67} \approx 2.16$$

19. Range = 15 − 4 = 11

$$\overline{x} = \frac{60}{6} = 10$$

x	$x - \overline{x}$	$(x - \overline{x})^2$
4	−6	36
8	−2	4
9	−1	1
11	1	1
13	3	9
15	5	25
	0	76

$$\frac{76}{5} = 15.2, \ s = \sqrt{15.2} \approx 3.90$$

21. Range = 12 − 7 = 5

$$\bar{x} = \frac{63}{7} = 9$$

x	$x - \bar{x}$	$(x - \bar{x})^2$
7	−2	4
9	0	0
7	−2	4
9	0	0
9	0	0
10	1	1
12	3	9
	0	18

$$\frac{18}{6} = 3, s = \sqrt{3} \approx 1.73$$

23. Range = 80 − 48 = $32

$$\bar{x} = \frac{660}{10} = \$66$$

x	$x - \bar{x}$	$(x - \bar{x})^2$
58	−8	64
58	−8	64
80	14	196
75	9	81
60	-6	36
75	9	81
78	12	144
48	−18	324
75	9	81
53	−13	169
	0	1240

$$\frac{1240}{9} \approx 137.78, s = \sqrt{137.78} \approx \$11.74$$

25. Range = 200 − 50 = $150

$$\bar{x} = \frac{1100}{10} = \$110$$

x	$x - \bar{x}$	$(x - \bar{x})^2$
50	−60	3600
120	10	100
130	20	400
60	−50	2500
55	−55	3025
75	−35	1225
200	90	8100
110	0	0
125	15	225
175	65	4225
	0	23,400

$$\frac{23,400}{9} = 2600, \; s = \sqrt{2600} \approx \$50.99$$

27. a) Range = 68 - 5 = $63

$$\overline{x} = \frac{204}{6} = \$34$$

x	$x - \overline{x}$	$(x - \overline{x})^2$
32	−2	4
60	26	676
14	−20	400
25	−9	81
5	−29	841
68	34	1156
	0	3158

$$\frac{3158}{5} = 631.6, s = \sqrt{631.6} \approx \$25.13$$

b) New data: 42, 70, 24, 35, 15, 78

The range and standard deviation will be the same. If each piece of data is increased by the same number, the range and standard deviation will remain the same.

c) Range = 78 - 15 = \$63

$$\overline{x} = \frac{264}{6} = \$44$$

x	$x - \overline{x}$	$(x - \overline{x})^2$
42	−2	4
70	26	676
24	−20	400
35	−9	81
15	−29	841
78	34	1156
	0	3158

$$\frac{3158}{5} = 631.6, s = \sqrt{631.6} \approx \$25.13$$ The answers remain the same.

29. a) - c) Answers will vary.

d) If each number in a distribution is multiplied by n, both the mean and standard deviation of the new distribution will be n times that of the original distribution.

e) The mean of the second set is $4 \times 5 = 20$, and the standard deviation of the second set is $2 \times 5 = 10$.

31. a) The standard deviation increases. There is a greater spread from the mean as they get older.

b) ≈ 133 lb

c) $\dfrac{175 - 90}{4} = 21.25 \approx 21$ lb

d) The mean weight is about 100 pounds and the normal range is about 60 to 140 pounds.

e) The mean height is about 62 inches and the normal range is about 53 to 68 inches.

f) 100% - 95% = 5%

33. a)

East	
Number of oil changes made	Number of days
15-20	2
21-26	2
27-32	5
33-38	4
39-44	7
45-50	1
51-56	1
57-62	2
63-68	1

West	
Number of oil changes made	Number of days
15-20	0
21-26	0
27-32	6
33-38	9
39-44	4
45-50	6
51-56	0
57-62	0
63-68	0

b)

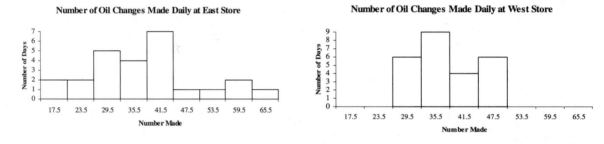

c) They appear to have about the same mean since they are both centered around 38.

d) The distribution for East is more spread out. Therefore, East has a greater standard deviation.

e) East: $\dfrac{950}{25} = 38$, West: $\dfrac{950}{25} = 38$

33. f)

East				West		
x	$x-\bar{x}$	$(x-\bar{x})^2$		x	$x-\bar{x}$	$(x-\bar{x})^2$
33	−5	25		38	0	0
30	−8	64		38	0	0
25	−13	169		37	−1	1
27	−11	121		36	−2	4
40	2	4		30	−8	64
44	6	36		45	7	49
49	11	121		28	−10	100
52	14	196		47	9	81
42	4	16		30	−8	64
59	21	441		46	8	64
19	−19	361		38	0	0
22	−16	256		39	1	1
57	19	361		40	2	4
67	29	841		34	−4	16
15	−23	529		31	−7	49
41	3	9		45	7	49
43	5	25		29	−9	81
27	−11	121		38	0	0
42	4	16		38	0	0
43	5	25		39	1	1
37	−1	1		37	−1	1
38	0	0		42	4	16
31	−7	49		46	8	64
32	-6	36		31	−7	49
35	−3	9		48	10	100
	0	3832			0	858

$\dfrac{3832}{24} \approx 159.67, \quad s = \sqrt{159.67} \approx 12.64$

$\dfrac{858}{24} = 35.75, \quad s = \sqrt{35.75} \approx 5.98$

35. 6, 6, 6, 6, 6

Exercise Set 13.7

1. A **rectangular distribution** is one where all the values have the same frequency.

3. A **bimodal distribution** is one where two nonadjacent values occur more frequently than any other values in a set of data.

5. A **distribution skewed to the left** is one that has "a tail" on its left.

7. A z-score measure how far, in terms of standard deviations, a given score is from the mean

9. a) Below the mean
   b) Above the mean

11. a) B b) C c) A

13. The distribution of outcomes from the roll of a die

15. J shaped right – consumer price index; J shaped left – value of the dollar

17. Normal

19. Skewed right

21. The mode is the lowest value, the median is greater than the mode, and the mean is greater than the median. The greatest frequency appears on the left side of the curve. Since the mode is the value with the greatest frequency, the mode would appear on the left side of the curve (where the lowest values are). Every value in the set of data is considered in determining the mean. The values on the far right of the curve would increase the value of the mean. Thus, the value of the mean would be farther to the right than the mode. The median would be between the mode and the mean.

23. Answers will vary.

25. In a normal distribution the mean, median, and the mode all have the same value.

27. 0.5000

29. (area to the left of 1) − (area to the left of −2)
$$= 0.8413 - 0.0228 = 0.8185$$

31. area to the right of 1.34
$$= 1 - (\text{area to the left of } 1.34)$$
$$= 1 - 0.9099 = 0.0901$$

33. area to the left of −1.78
0.0375

35. area between −1.32 and −1.64
$$0.0934 - 0.0505 = 0.0429$$

37. area to the left of −1.62
0.0536

39. $0.7611 = 76.11\%$

41. (area to the left of 2.24) − (area to the left of −1.34)
$$= 0.9875 - 0.0901 = 0.8974 = 89.74\%$$

43. area greater than −1.90
$$= \text{area less than } 1.90 = 0.9713 = 97.13\%$$

45. area less than $1.96 = 0.9750 = 97.50\%$

47. (area to the left of 2.14) − (area to the left of 0.72)
$$= 0.9838 - 0.7642 = 0.2196 = 21.96\%$$

49. a) Emily, Sarah, and Carol are taller than the mean because their z-scores are positive.
   b) Jason and Juan are at the mean because their z-scores are zero.
   c) Omar, Heather, and Kim are shorter than the mean because their z-scores are negative.

51. $0.5000 = 50\%$

53. $z_{23} = \dfrac{23-18}{4} = \dfrac{5}{4} = 1.25$

$$1.000 - 0.8944 = 0.1056 = 10.56\%$$

55. $z_{550} = \dfrac{550-500}{100} = \dfrac{50}{100} = 0.50$

area less than $0.5 = 0.6915 = 69.15\%$

57. $z_{550} = \dfrac{550-500}{100} = \dfrac{50}{100} = 0.50$

$z_{650} = \dfrac{650-500}{100} = \dfrac{150}{100} = 1.50$

area between 1.5 and 0.5
$$= 0.9332 - 0.6915 = 0.2417 = 24.17\%$$

59. $z_{525} = \dfrac{525-500}{100} = \dfrac{25}{100} = 0.25$

$z_{400} = \dfrac{400-500}{100} = \dfrac{-100}{100} = -1.00$

area between −1.00 and 0.25
$$= 0.5987 - 0.1587 = 0.4400 = 44.00\%$$

61. $z_{7.4} = \dfrac{7.4 - 7.6}{0.4} = \dfrac{-0.2}{0.4} = -0.50$

$z_{7.7} = \dfrac{7.7 - 7.6}{0.4} = \dfrac{0.1}{0.4} = 0.25$

$0.5987 - 0.3085 = 0.2902 = 29.02\%$

63. $z_{7.7} = \dfrac{7.7 - 7.6}{0.4} = \dfrac{0.1}{0.4} = 0.25$

$0.5987 = 59.87\%$

65. $0.5000 = 50.00\%$

67. $z_{56} = \dfrac{56 - 62}{5} = \dfrac{-6}{5} = -1.20$

$0.1151 = 11.51\%$

69. 11.51% of cars are traveling slower than 56 mph. (See Exercise 67.)

$(0.1151)(200) \approx 23$ cars

71. $z_{15.83} = \dfrac{15.83 - 16}{0.1} = -1.70$

$z_{16.32} = \dfrac{16.32 - 16}{0.1} = 3.2$

$0.9993 - 0.0446 = 0.9547 = 95.47\%$

73. $z_{15.83} = \dfrac{15.83 - 16}{0.1} = -1.70$

$1 - 0.9554 = 0.0446$

$(0.0446)(300,000) = 13,380$ boxes

75. $z_{3.1} = \dfrac{3.1 - 3.7}{1.2} = \dfrac{-0.6}{1.2} = -0.50$

area greater than $-0.50 =$ area less than 0.50

$0.6915 = 69.15\%$

77. $z_{6.7} = \dfrac{6.7 - 3.7}{1.2} = \dfrac{3}{1.2} = 2.5$

$1 - 0.9938 = 0.0062 = 0.62\%$

79. 69.15% of the children are older than 3.1 years. (See Exercise 75.)

$(0.6915)(120) \approx 83$ children

81. We need the percentage of customers with a weight loss of less than 5 lb.

$z_5 = \dfrac{5 - 6.7}{0.81} = \dfrac{-1.7}{0.81} = -2.10$

$1 - 0.9821 = 0.0179 = 1.79\%$

83. The standard deviation is too large. There is too much variation.

85. a) Katie: $z_{28,408} = \dfrac{28,408 - 23,200}{2170} = \dfrac{5208}{2170} = 2.4$

Stella: $z_{29,510} = \dfrac{29,510 - 25,600}{2300} = \dfrac{3910}{2300} = 1.7$

b) Katie. Her z-score is higher than Stella's z-score. This means her sales are further above the mean than Stella's sales.

87. Answers will vary.

89. Using Table 13.7, the answer is -1.18.

91. $\dfrac{0.77}{2} = 0.385$

Using the table in Section 13.7, an area of 0.385 has a z-score of 1.20.

$$z = \frac{x - \overline{x}}{s}$$

$$1.20 = \frac{14.4 - 12}{s}$$

$$1.20 = \frac{2.4}{s}$$

$$\frac{1.20s}{1.20} = \frac{2.4}{1.20}$$

$$s = 2$$

Exercise Set 13.8

1. The **correlation coefficient** measures the strength of the linear relationship between the quantities.

3. 1

5. 0

7. a) A positive correlation indicates that as one quantity increases, the other quantity increases.
    b) Answers will vary.

9. The **level of significance** is used to identify the cutoff between results attributed to chance and results attributed to an actual relationship between the two variables.

11. No correlation

13. Strong positive

15. Yes, $| \; 0.82 \; | > 0.684$

17. Yes, $| -0.73 \; | > 0.707$

19. No, $| -0.23 | < 0.254$

21. No, $| 0.75 | < 0.917$

Note: The answers in the remainder of this section may differ slightly from your answers, depending upon how your answers are rounded and which calculator you used.

23. a)

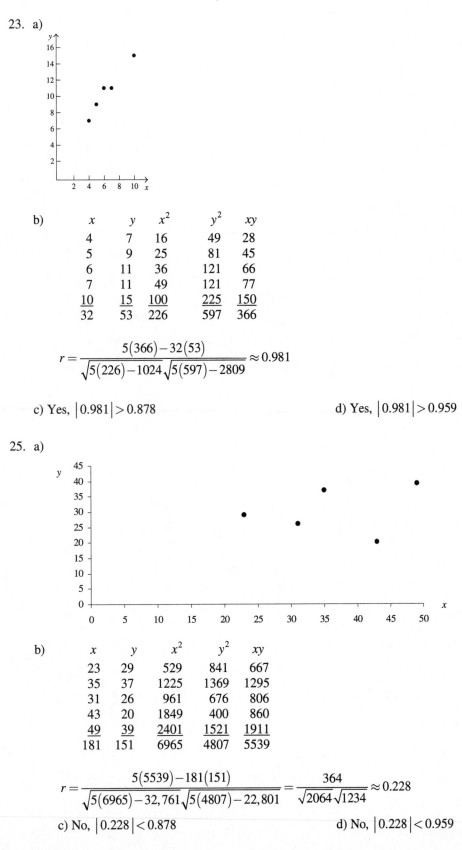

b)

x	y	x^2	y^2	xy
4	7	16	49	28
5	9	25	81	45
6	11	36	121	66
7	11	49	121	77
10	15	100	225	150
32	53	226	597	366

$$r = \frac{5(366) - 32(53)}{\sqrt{5(226) - 1024}\sqrt{5(597) - 2809}} \approx 0.981$$

c) Yes, $|0.981| > 0.878$

d) Yes, $|0.981| > 0.959$

25. a)

b)

x	y	x^2	y^2	xy
23	29	529	841	667
35	37	1225	1369	1295
31	26	961	676	806
43	20	1849	400	860
49	39	2401	1521	1911
181	151	6965	4807	5539

$$r = \frac{5(5539) - 181(151)}{\sqrt{5(6965) - 32,761}\sqrt{5(4807) - 22,801}} = \frac{364}{\sqrt{2064}\sqrt{1234}} \approx 0.228$$

c) No, $|0.228| < 0.878$

d) No, $|0.228| < 0.959$

27. a)

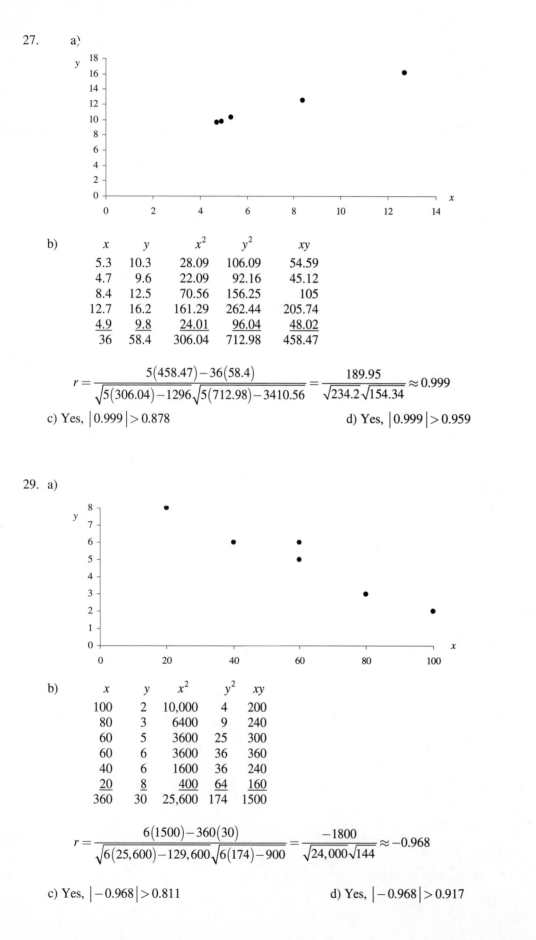

b)

x	y	x^2	y^2	xy
5.3	10.3	28.09	106.09	54.59
4.7	9.6	22.09	92.16	45.12
8.4	12.5	70.56	156.25	105
12.7	16.2	161.29	262.44	205.74
4.9	9.8	24.01	96.04	48.02
36	58.4	306.04	712.98	458.47

$$r = \frac{5(458.47) - 36(58.4)}{\sqrt{5(306.04) - 1296}\sqrt{5(712.98) - 3410.56}} = \frac{189.95}{\sqrt{234.2}\sqrt{154.34}} \approx 0.999$$

c) Yes, $|0.999| > 0.878$ d) Yes, $|0.999| > 0.959$

29. a)

b)

x	y	x^2	y^2	xy
100	2	10,000	4	200
80	3	6400	9	240
60	5	3600	25	300
60	6	3600	36	360
40	6	1600	36	240
20	8	400	64	160
360	30	25,600	174	1500

$$r = \frac{6(1500) - 360(30)}{\sqrt{6(25,600) - 129,600}\sqrt{6(174) - 900}} = \frac{-1800}{\sqrt{24,000}\sqrt{144}} \approx -0.968$$

c) Yes, $|-0.968| > 0.811$ d) Yes, $|-0.968| > 0.917$

31. From # 23: $m = \dfrac{5(366) - 32(53)}{5(226) - 1024} = \dfrac{67}{53} \approx 1.26$

$b = \dfrac{53 - \left(\dfrac{67}{53}\right)(32)}{5} \approx 2.51, \quad y = 1.26x + 2.51$

33. From # 25: $m = \dfrac{5(5539) - 181(151)}{5(6965) - 32{,}761} = \dfrac{364}{2064} \approx 0.18$

$b = \dfrac{151 - \dfrac{364}{2064}(181)}{5} \approx 23.82, \quad y = 0.18x + 23.82$

35. From # 27: $m = \dfrac{5(458.47) - 36(58.4)}{5(306.04) - 1296} = \dfrac{189.95}{234.2} \approx 0.81$

$b = \dfrac{58.4 - \dfrac{189.95}{234.2}(36)}{5} \approx 5.84, \quad y = 0.81x + 5.84$

37. From # 29: $m = \dfrac{6(1500) - 360(30)}{6(25{,}600) - 129{,}600} = \dfrac{-1800}{24{,}000} \approx -0.08$

$b = \dfrac{30 - \dfrac{-1800}{24{,}000}(360)}{6} \approx 9.50, \quad y = -0.08x + 9.50$

39.   a)

x	y	x^2	y^2	xy
50	40	2500	1600	2000
53	42	2809	1764	2226
60	45	3600	2025	2700
35	25	1225	625	875
43	34	1849	1156	1462
62	45	3844	2025	2790
303	231	15827	9195	12053

$r = \dfrac{6(12053) - (303)(231)}{\sqrt{6(15827) - 91809}\sqrt{6(9195) - 53361}} \approx 0.974$

b) Yes, $|\,0.974\,| > 0.811$

39. c) $m = \dfrac{6(12053) - (303)(231)}{(6)(15827) - 91809} = \dfrac{775}{1051} \approx 0.74$,

$b = \dfrac{231 - \dfrac{775}{1051}(303)}{6} \approx 1.26, \quad y = 0.74x + 1.26$

41. a)

x	y	x^2	y^2	xy
20	40	400	1600	800
40	45	1600	2025	1800
50	70	2500	4900	3500
60	76	3600	5776	4560
80	92	6400	8464	7360
100	95	10,000	9025	9500
350	418	24,500	31,790	27,520

$r = \dfrac{6(27{,}520) - 350(418)}{\sqrt{6(24{,}500) - 122{,}500}\sqrt{6(31{,}790) - 174{,}724}} = \dfrac{18{,}820}{\sqrt{24{,}500}\sqrt{16{,}016}} \approx 0.950$

b) Yes, $|0.950| > 0.917$

c) $m = \dfrac{6(27{,}520) - 350(418)}{6(24{,}500) - 122{,}500} = \dfrac{18{,}820}{24{,}500} \approx 0.77$, $b = \dfrac{418 - \dfrac{18{,}820}{24{,}500}(350)}{6} \approx 24.86$, $y = 0.77x + 24.86$

43. a)

x	y	x^2	y^2	xy
6.5	27	42.25	729	175.5
7	30	49	900	210
6	25	36	625	150
2	10	4	100	20
6.4	28	40.96	784	179.2
6	24	36	576	144
33.9	144	208.21	3714	878.7

$r = \dfrac{6(878.7) - (33.9)(144)}{\sqrt{6(208.21) - 1149.21}\sqrt{6(3714) - 20736}} = 0.99252 \approx 0.993$

b) Yes, $|\,0.993\,| > 0.811$

c) $m = \dfrac{6(878.7) - (33.9)(144)}{6(208.21) - 1149.21} = 3.90405 \approx 3.90$,

$b = \dfrac{144 - (3.90405)(33.9)}{6} \approx 1.94, \quad y = 3.90x + 1.94$

d) $y = 3.90(5) + 1.94 \approx 21.4$ kilocalories

45. a)

x	y	x^2	y^2	xy
27	23	729	529	621
31	22	961	484	682
35	20	1225	400	700
32	21	1024	441	672
30	24	900	576	720
30	22	900	484	660
185	132	5739	2914	4055

$$r = \frac{6(4055)-(185)(132)}{\sqrt{6(5739)-34225}\sqrt{6(2914)-17424}} \approx -0.804$$

b) No, $|-0.804| < 0.917$

c) $m = \dfrac{6(4055)-(185)(132)}{6(5739)-34225} = -0.43063 \approx -0.43$

$b = \dfrac{132-(-0.43063)(185)}{6} \approx 35.28, \quad y = -0.43x + 35.28$

Note, however, that since we have not found a significant correlation between x and y, the line of best fit may not be very useful for predicting y given x.

d) $y = -0.43(33) + 35.28 \approx 21.1$ mpg

47. a)

x	y	x^2	y^2	xy
1	80.0	1	6400.0	80.0
2	76.2	4	5806.4	152.4
3	68.7	9	4719.7	206.1
4	50.1	16	2510.0	200.4
5	30.2	25	912.0	151.0
6	20.8	36	432.6	124.8
21	326	91	20,780.7	914.7

$$r = \frac{6(914.7)-21(326)}{\sqrt{6(91)-441}\sqrt{6(20,780.7)-106,276}} = \frac{-1357.8}{\sqrt{105}\sqrt{18,408.2}} \approx -0.977$$

b) Yes, $|-0.977| > 0.917$

c) $m = \dfrac{6(914.7)-21(326)}{6(91)-441} = \dfrac{-1357.8}{105} \approx -12.93$, $\quad b = \dfrac{326-\dfrac{-1357.8}{105}(21)}{6} \approx 99.59, \quad y = -12.93x + 99.59$

d) $y = -12.93(4.5) + 99.59 \approx 41.4\%$

49. a) and b) Answers will vary.

   c)

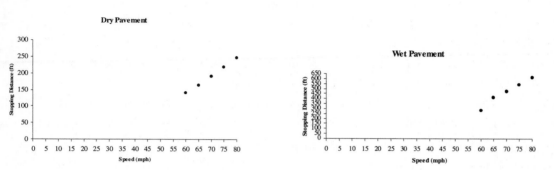

   d) The values in the last row of the calculation table are:

   $\Sigma x = 350$, $\Sigma y = 959$, $\Sigma x^2 = 24750$, $\Sigma y^2 = 191129$, $\Sigma xy = 68470$

   $$r = \frac{5(68470) - (350)(959)}{\sqrt{5(24750) - (350)^2}\sqrt{5(191129) - (959)^2}} \approx 0.999$$

   e) The values in the last row of the calculation table are:

   $\Sigma x = 350$, $\Sigma y = 2328$, $\Sigma x^2 = 24750$, $\Sigma y^2 = 1151074$, $\Sigma xy = 167015$

   $$r = \frac{5(167015) - (350)(2328)}{\sqrt{5(24750) - (350)^2}\sqrt{5(1151074) - (2328)^2}} \approx 0.990$$

   f) Answers will vary.

   g) $m = \dfrac{5(68470) - (350)(959)}{5(24750) - (350)^2} = 5.36$

   $b = \dfrac{959 - (5.36)(350)}{5} = -183.40$

   $y = 5.36x - 183.40$

   h) $m = \dfrac{5(167015) - (350)(2328)}{5(24750) - (350)^2} = 16.22$

   $b = \dfrac{2328 - (16.22)(350)}{5} = -669.80$

   $y = 16.22x - 699.80$

   i) Dry: $y = 5.36(77) - 183.40 \approx 229.3$ ft ft

   Wet: $y = 16.22(77) - 669.80 = 579.1$ ft ft

51. Answers will vary.

53. a) The values in the last row of the calculation table are:

   $\Sigma x = 12015$, $\Sigma y = 1097.4$, $\Sigma x^2 = 24060055$, $\Sigma y^2 = 201062.56$, $\Sigma xy = 2197621$

   $$r = \frac{6(2197621) - (12015)(1097.4)}{\sqrt{6(24060055) - (12015)^2}\sqrt{6(201062.56) - (1097.4)^2}} \approx 0.993$$

   b) Should be the same.

   c) The values in the last row of the calculation table are:

   $\Sigma x = 15$, $\Sigma y = 1097.4$, $\Sigma x^2 = 55$, $\Sigma y^2 = 201062.56$, $\Sigma xy = 2821$

   $$r = \frac{6(2821) - (15)(1097.4)}{\sqrt{6(55) - (15)^2}\sqrt{6(201062.56) - (1097.4)^2}} \approx 0.993$$

Review Exercises

1. a) A **population** consists of all items or people of interest.

    b) A **sample** is a subset of the population.

2. A **random sample** is one where every item in the population has the same chance of being selected.

3. The candy bars may have lots of calories, or fat, or sodium. Therefore, it may not be healthy to eat them.

4. Sales may not necessarily be a good indicator of profit. Expenses must also be considered.

5. a)

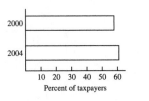

    b)

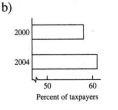

6. a)

Class	Frequency
35	1
36	3
37	6
38	2
39	3
40	0
41	4
42	1
43	3
44	1
45	1

    b) and c)

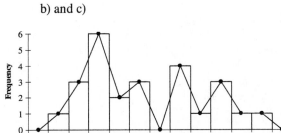

7. a)

High Temperature	Number of Cities
58 - 62	1
63 - 67	4
68 - 72	9
73 - 77	10
78 - 82	11
83 - 87	4
88 - 92	1

    b) and c)

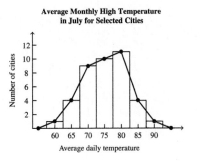

    d) 6 | 5 represents 65

```
5 | 8
6 | 3 6 6 7 8 8 9
7 | 0 1 1 1 2 2 3 3 3 4 5 5 5 6 6 7 9 9 9
8 | 0 0 0 0 1 2 2 2 3 4 4 7
9 | 1
```

8. $\overline{x} = \dfrac{480}{6} = 80$

9. $\dfrac{79+83}{2} = 81$

10. None

11. $\dfrac{65+93}{2} = 79$

12. $93 - 65 = 28$

13.

x	$x - \overline{x}$	$(x - \overline{x})^2$
65	−15	225
76	−4	16
79	−1	1
83	3	9
84	4	16
93	13	169
	0	520

$\dfrac{436}{5} = 87.2,\ s = \sqrt{87.2} \approx 9.34$

14. $\overline{x} = \dfrac{156}{12} = 13$

15. $\dfrac{12+14}{2} = 13$

16. 12 and 7

17. $\dfrac{4+23}{2} = 13.5$

18. $23 - 4 = 19$

19.

x	$x - \overline{x}$	$(x - \overline{x})^2$
4	-9	81
5	-8	64
7	-6	36
7	-6	36
12	-1	1
12	-1	1
14	1	1
15	2	4
17	4	16
19	6	36
21	8	64
23	10	100
	0	440

$\dfrac{440}{11} = 40,\ s = \sqrt{40} \approx 6.32$

20. $z_7 = \dfrac{7-9}{2} = \dfrac{-2}{2} = -1.00$

$z_{11} = \dfrac{11-9}{2} = \dfrac{2}{2} = 1.00$

$0.8413 - 0.1587 = 0.6826 = 68.26\%$

21. $z_5 = \dfrac{5-9}{2} = \dfrac{-4}{2} = -2.00$

$z_{13} = \dfrac{14-9}{2} = \dfrac{4}{2} = 2.00$

$0.9772 - 0.0228 = 0.9544 = 95.44\%$

22. $z_{12.2} = \dfrac{12.2-9}{2} = \dfrac{3.2}{2} = 1.6$

$0.9452 = 94.52\%$

23. Subtract the answer for Exercise 22 from 1:

$1 - 0.9452 = 0.0548 = 5.48\%$

24. $z_{7.8} = \dfrac{7.8-9}{2} = -\dfrac{1.2}{2} = -0.6$

$1-0.2743 = 0.7257 = 72.57\%$

25. $z_{20} = \dfrac{20-20}{5} = \dfrac{0}{5} = 0$

$z_{25} = \dfrac{25-20}{5} = \dfrac{5}{5} = 1.00$

$0.341 = 34.1\%$

26. $z_{18} = \dfrac{18-20}{5} = \dfrac{-2}{5} = -0.40$

$0.500-0.155 = 0.345 = 34.5\%$

27. $z_{22} = \dfrac{22-20}{5} = \dfrac{2}{5} = 0.40$

$z_{28} = \dfrac{28-20}{5} = \dfrac{8}{5} = 1.60$

$0.445-0.155 = 0.29 = 29.0\%$

28. $z_{30} = \dfrac{30-20}{5} = \dfrac{10}{5} = 2.00$

$0.500 - 0.477 = 0.023 = 2.3\%$

29. a)

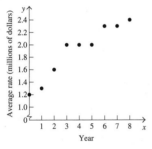

b) Yes; positive because generally as the year increases, the cost increases.

c) The values in the last row of the calculation table are:

$\Sigma x = 36,\ \Sigma y = 17.1,\ \Sigma x^2 = 204,\ \Sigma y^2 = 34.03,\ \Sigma xy = 77.6$

$r = \dfrac{9(77.6)-(36)(17.1)}{\sqrt{9(204)-(36)^2}\,\sqrt{9(34.03)-(17.1)^2}} \approx 0.957$.

d) Yes, $\left|\,0.957\,\right| > 0.666$

e) $m = \dfrac{9(77.6)-(36)(17.1)}{9(204)-(36)^2} = 0.1533 \approx 0.15$

$b = \dfrac{17.1-(0.1533)(36)}{9} \approx 1.29,\quad y = 0.15x + 1.29$

f) $y = 0.15(13)+1.29 \approx \$3.2$ million

30. a)

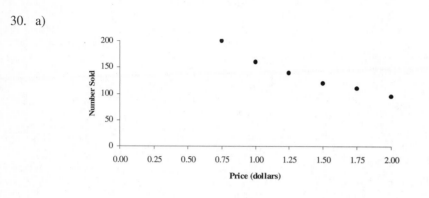

b)   Yes; negative because generally as the price increases, the number sold decreases.

c)

x	y	x^2	y^2	xy
0.75	200	0.5625	40,000	150
1.00	160	1	25,600	160
1.25	140	1.5625	19,600	175
1.50	120	2.25	14,400	180
1.75	110	3.0625	12,100	192.5
2.00	95	4	9025	190
8.25	825	12.4375	120,725	1047.5

$$r = \frac{6(1047.5) - 8.25(825)}{\sqrt{6(12.4375) - 68.0625}\sqrt{6(120,725) - 680,625}} = \frac{-521.25}{\sqrt{6.5625}\sqrt{43,725}} \approx -0.973$$

d) Yes, $\left|\ -0.973\ \right| > 0.811$

e) $m = \dfrac{6(1047.5) - 8.25(825)}{6(12.4375) - 68.0625} = \dfrac{-521.25}{6.5625} \approx -79.4$

$$b = \frac{825 - \dfrac{-521.25}{6.5625}(8.25)}{6} \approx 246.7, \quad y = -79.4x + 246.7$$

f) $y = -79.4(1.60) + 246.7 = 119.66 \approx 120$ sold

31. 180 lb

32. 185 lb

33. 25%

34. 25%

35. 14%

36. $(100)(192) = 19,200$ lb

37. $192 + (2)(23) = 238$ lb

38. $192 - (1.8)(23) = 150.6$ lb

39. $\bar{x} = \dfrac{150}{42} \approx 3.57$

40. 2

41. $\dfrac{3+3}{2} = 3$

42. $\dfrac{0+14}{2} = 7$

43. 14 - 0 = 14

44.

x	$x-\overline{x}$	$(x-\overline{x})^2$	x	$x-\overline{x}$	$(x-\overline{x})^2$	x	$x-\overline{x}$	$(x-\overline{x})^2$
0	−3.6	12.96	2	−1.6	2.56	4	0.4	0.16
0	−3.6	12.96	2	−1.6	2.56	5	1.4	1.96
0	−3.6	12.96	3	−0.6	0.36	5	1.4	1.96
0	−3.6	12.96	3	−0.6	0.36	5	1.4	1.96
0	−3.6	12.96	3	−0.6	0.36	6	2.4	5.76
0	−3.6	12.96	3	−0.6	0.36	6	2.4	5.76
1	−2.6	6.76	3	−0.6	0.36	6	2.4	5.76
1	−2.6	6.76	3	−0.6	0.36	6	2.4	5.76
2	−1.6	2.56	4	0.4	0.16	6	2.4	5.76
2	−1.6	2.56	4	0.4	0.16	7	3.4	11.56
2	−1.6	2.56	4	0.4	0.16	8	4.4	19.36
2	−1.6	2.56	4	0.4	0.16	10	6.4	40.96
2	−1.6	2.56	4	0.4	0.16	14	10.4	108.16
2	−1.6	2.56	4	0.4	0.16			332.32
2	−1.6	2.56						

$$\frac{332.32}{41} \approx 8.105, \quad s = \sqrt{8.105} \approx 2.85$$

45.

# of Children	# of Presidents
0 - 1	8
2 - 3	15
4 - 5	10
6 - 7	6
8 - 9	1
10 - 11	1
12 - 13	0
14 - 15	1

46. and 47.

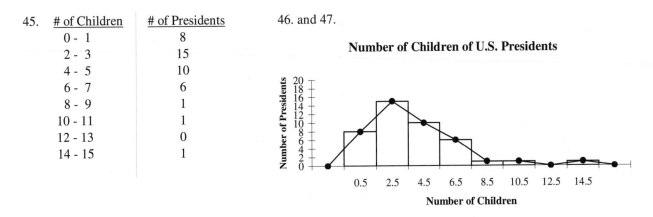

Number of Children of U.S. Presidents

48. No, it is skewed to the right.
49. No, some families have no children, more have one child, the greatest percent may have two children, fewer have three children, etc.
50. No, the number of children per family has decreased over the years.

Chapter Test

1. $\bar{x} = \dfrac{210}{5} = 42$

2. 43

3. 43

4. $\dfrac{27+52}{2} = 39.5$

5. $52 - 27 = 25$

6.
x	$x-\bar{x}$	$(x-\bar{x})^2$
27	−15	225
43	1	1
43	1	1
45	3	9
52	10	100
	0	336

$\dfrac{336}{4} = 84, \; s = \sqrt{84} \approx 9.17$

7.
Class	Frequency
25 - 30	7
31 - 36	5
37 - 42	1
43 - 48	7
49 - 54	5
55 - 60	3
61 - 66	2

8. and 9.

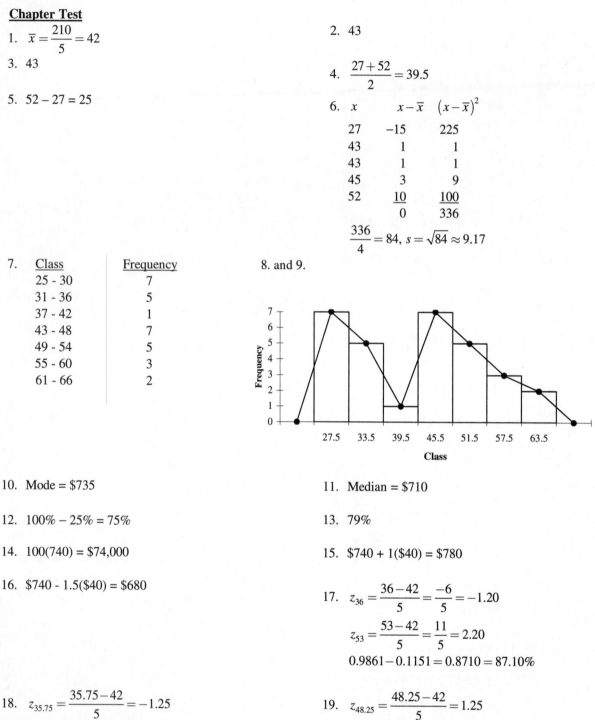

10. Mode = $735

11. Median = $710

12. $100\% - 25\% = 75\%$

13. 79%

14. $100(740) = \$74{,}000$

15. $\$740 + 1(\$40) = \$780$

16. $\$740 - 1.5(\$40) = \$680$

17. $z_{36} = \dfrac{36-42}{5} = \dfrac{-6}{5} = -1.20$

$z_{53} = \dfrac{53-42}{5} = \dfrac{11}{5} = 2.20$

$0.9861 - 0.1151 = 0.8710 = 87.10\%$

18. $z_{35.75} = \dfrac{35.75-42}{5} = -1.25$

$1 - 0.1056 = 0.8944 = 89.44\%$

19. $z_{48.25} = \dfrac{48.25-42}{5} = 1.25$

$1 - 0.8944 = 0.1056 = 10.56\%$

20. $z_{50} = \dfrac{50-42}{5} = 1.6$

$0.9452 = 94.52\%$

21. a)

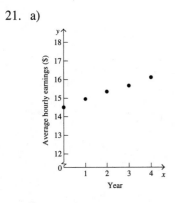

b)   Yes

c)   The values in the last row of the calculation table are:

$\Sigma x = 10, \ \Sigma y = 76.7, \ \Sigma x^2 = 30, \ \Sigma y^2 = 1178.11, \ \Sigma xy = 157.3$

$$r = \frac{5(157.3) - (10)(76.7)}{\sqrt{5(30) - (10)^2} \ \sqrt{5(1178.11) - (76.7)^2}} \approx 0.996$$

d) Yes, $\ |\ 0.996\ | > 0.878$

e) $m = \dfrac{5(157.3) - (10)(76.7)}{5(30) - (10)^2} = 0.39$

$b = \dfrac{76.7 - (0.39)(10)}{5} = 14.56, \quad y = 0.39x + 14.56$

f) $y = 0.39(29) + 14.56 = \$25.87$

CHAPTER FOURTEEN

GRAPH THEORY

1. A **graph** is a finite set of points, called **vertices**, that are connected with line segments, called **edges**.

3.

5. If the number of edges connected to the vertex is even, the vertex is **even**. If the number of edges connected to the vertex is odd, the vertex is **odd**.

7. a) A **path** is a sequence of adjacent vertices and the edges connecting them.
    b) A **circuit** is a path that begins and ends at the same vertex.
    c)

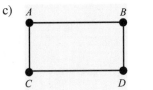

The path A, B, D, C is a path that is not a circuit.
The path A, B, D, C, A is a path that is also a circuit.

9.

A and B are odd.

11.

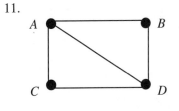

B and C are even. A and D are odd.

13.

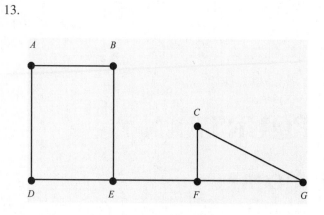

15. No. There is no edge connecting vertices *B* and *C*. Therefore, *A, B, C, D, E* is not a path.

17. No.

19. Yes. One example is *C, A, B, D, F, E, C, D*.

21.

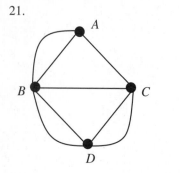

23.

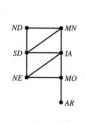

25.

27.

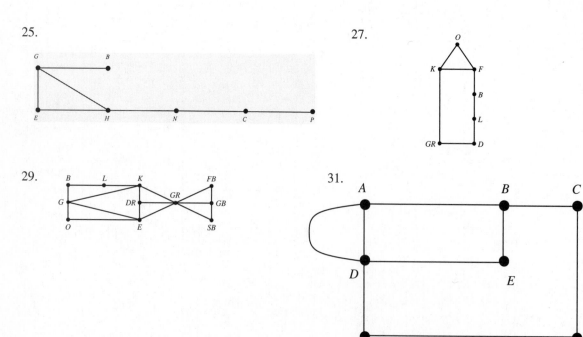

29.

31.

33. Disconnected. There is no path that connects
    A to C.
35. Connected
37. Edge AB
39. Edge EF
41.

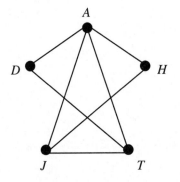

Other answers are possible.

43. It is impossible to have a graph with an odd number of odd vertices.
45. a) and b) Answers will vary.

Exercise Set 14.2

1. a) An **Euler path** is a path that must include each edge of a graph exactly one time.
    b) and c)

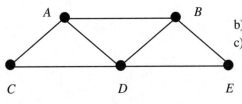

    b) The path A, B, E, D, C, A, D, B is an Euler path.
    c) The path A, B, E, D, C is a path that is not an Euler path.

3. a) Yes, according to Euler's Theorem.
    b) Yes, according to Euler's Theorem.
    c) No, according to Euler's Theorem.
5. If all of the vertices are even, the graph has an Euler circuit.
7. $A, B, D, B, C, D, E, A, C$; other answers are possible.
9. No. This graph has exactly two odd vertices. Each Euler path must begin with an odd vertex. B is an even vertex.
11. $A, B, A, C, B, E, C, D, A, D, E$; other answers are possible.
13. No. A graph with exactly two odd vertices has no Euler circuits.
15. $A, B, C, D, E, F, B, D, F, A$; other answers are possible.
17. $C, D, E, F, A, B, D, F, B, C$; other answers are possible.
19. $E, F, A, B, C, D, F, B, D, E$; other answers are possible.

21. a) Yes. There are zero odd vertices.

   b) Yes. There are zero odd vertices.

23. a) No. There are more than two odd vertices.

   b) No. There are more than zero odd vertices.

25. a) Yes. Each island would correspond to an odd vertex. According to item 2 of Euler's Theorem, a graph with exactly two odd vertices has at least one Euler path, but no Euler circuit.

   b) They could start on either island and finish at the other.

In Exercises 27-31, one graph is shown. Other graphs are possible.

27. a)

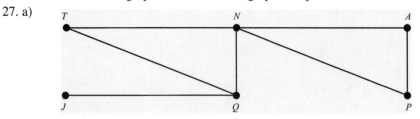

   b) Vertices *J* and *Q* are both odd. According to item 2 of Euler's Theorem, since there are exactly two odd vertices, at least one Euler path, but no Euler circuits exist.

   Yes; *J, Q, T, N, A, P, N, Q*

   c) No. (See part b) above.)

29. a)

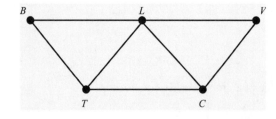

   b) Vertices *T* and *C* are both odd. According to item 2 of Euler's Theorem, since there are exactly two odd vertices, at least one Euler path, but no Euler circuits exist.

   Yes; *T, B, L, V, C, L, T, C*

   c) No. (See part b) above.)

31. a)

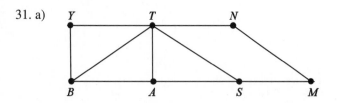

   b) Vertices *A, S, T,* and *B* are odd. According to item 3 of Euler's Theorem there is neither an Euler path nor an Euler circuit.

   c) No. (See part b) above.)

33. a) The graph representing the floor plan:

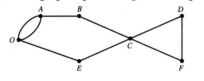

b) Yes
c) *O, A, B, C, D, F, C, E, O, A*

35. a) The graph representing the floor plan:

b) No; the graph has four odd vertices, and by Euler's theorem a graph with more than two odd vertices has neither and Euler path nor an Euler circuit.

37. a) Yes. The graph representing the map:

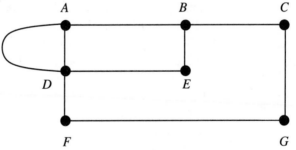

They are seeking an Euler path or an Euler circuit. Note that vertices *A* and *B* are both odd. According to item 2 of Euler's Theorem, since there are exactly two odd vertices, at least one Euler path, but no Euler circuits exist.

b) The residents would need to start at the intersection of Maple Cir., Walnut St., and Willow St. or at the intersection of Walnut St. and Oak St.

39. *B, E, D, A, B, C, A, D, C, E*; other answers are possible.
41. *H, I, F, C, B, D, G, H, E, D, A, B, E, F*; other answers are possible.
43. *A, E, B, F, C, G, D, K, G, J, F, I, E, H, A*; other answers are possible.
45. *A, C, D, G, H, F, C, F, E, B, A*; other answers are possible.
47. *A, B, C, E, B, D, E, F, I, E, H, D, G, H, I, J, F, C, A*; other answers are possible.
49. *OK, KS, CO, WY, UT, CO, NM, OK, CO*; other answers are possible.
51. *B, A, E, H, I, J, K, D, C, G, G, J, F, C, B, F, I, E, B*; other answers are possible.
53. *J, G, G, C, F, J, K, D, C, B, F, I, E, B, A, E, H, I, J*; other answers are possible.
55. a) No.

b) California, Nevada, and Louisiana (and others) have an odd number of states bordering them. Since a graph of the United States would have more than two odd vertices, no Euler path and no Euler circuit exist.

57. a)   b)   c)

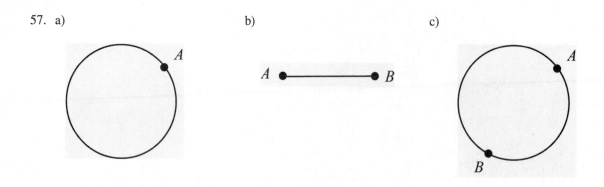

Exercise Set 14.3

1. a) A **Hamilton circuit** is a path that begins and ends with the same vertex and passes through all other vertices exactly one time.

   b) Both **Hamilton** and **Euler circuits** begin and end at the same vertex. A **Hamilton circuit** passes through all other *vertices* exactly once, while an **Euler circuit** passes through each *edge* exactly once.

3. a) A **weighted graph** is a graph with a number, or weight, assigned to each edge.

   b) A **complete graph** is a graph in which there is an edge between each pair of vertices.

   c) A **complete, weighted graph** is a graph in which there is an edge between each pair of vertices and each edge has a number, or weight, assigned to it.

5. a) The number of unique Hamilton circuits in a complete graph with n vertices is found by computing $(n-1)!$

   b) $n = 6; (n-1)! = (7-1)! = 6! = 720$

   c) $n = 10; (n-1)! = (10-1)! = 9! = 362,880$

7. To find the optimal solution using the **Brute Force method**, write down all possible Hamilton circuits and then compute the cost or distance associated with each Hamilton circuit. The one with the lowest cost or shortest distance is the optimal solution to the traveling salesman problem.

9. A, E, F, B, C, D, G and G, D, C, F, E, A, B; other answers are possible.

11. A, B, C, D, G, F, E, H and E, H, F, G, D, C, A, B; other answers are possible.

13. A, B, C, E, D, F, G, H and F, G, H, E, D, A, B, C; other answers are possible.

15. $A, B, C, F, I, H, E, G, D, A$ and $E, H, I, F, C, B, A, D, G, E$; other answers are possible.

17. $A, B, C, F, I, E, H, G, D, A$ and $A, E, B, C, F, I, H, G, D, A$; other answers are possible.

19.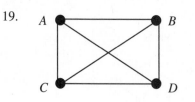

21. The number of unique Hamilton circuits within the complete graph with 7 vertices representing this situation is $(7-1)! = 6! = 6 \cdot 5 \cdot 4 \cdot 3 \cdot 2 \cdot 1 = 720$ ways

23. The number of unique Hamilton circuits within the complete graph with thirteen vertices representing this situation is $(13-1)! = 12! = 12 \cdot 11 \cdot 10 \cdot 9 \cdot 8 \cdot 7 \cdot 6 \cdot 5 \cdot 4 \cdot 3 \cdot 2 \cdot 1 = 479,001,600$ ways

In Exercises 25-31, other graphs are possible.

25. a)

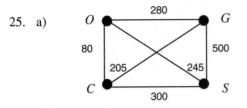

b)

Hamilton Circuit	First Leg/Distance	Second Leg/Distance	Third Leg/Distance	Fourth Leg/Distance	Total Distance
C, O, G, S, C	80	280	500	300	1160 miles
C, O, S, G, C	80	245	500	205	1030 miles
C, G, O, S, C	205	280	245	300	1030 miles
C, G, S, O, C	205	500	245	80	1030 miles
C, S, G, O, C	300	500	280	80	1160 miles
C, S, O, G, C	300	245	280	205	1030 miles

The shortest route is *C, O, S, G, C* or *C, G, O, S, C* or *C, G, S, O, C* or *C, S, O, G, C*

c) 1030 miles

27. a)

b)

Hamilton Circuit	First Leg/Distance	Second Leg/Distance	Third Leg/Distance	Fourth Leg/Distance	Total Distance
H, S, B, C, H	13	9	11	7	40 miles
H, S, C, B, H	13	18	11	5	47 miles
H, B, S, C, H	5	9	18	7	39 miles
H, B, C, S, H	5	11	18	13	47 miles
H, C, S, B, H	7	18	9	5	39 miles
H, C, B, S, H	7	11	9	13	40 miles

The shortest route is *H, C, S, B, H* or *H, B, S, C, H*

c) 39 miles

29. a)

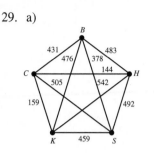

b) *S, B, C, H, K, S* for $1954
c) Answers will vary.

31. a)

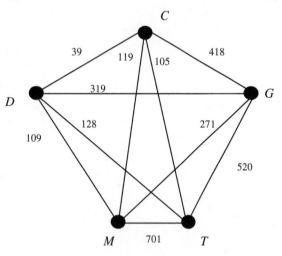

b) *C, D, M, G, T, C* for 39 + 109 + 271 + 520 + 105 = $1044
c) Answers will vary.

33. a) – d) Answers will vary.
35. *A, E, D, N, O, F, G, Q, P, T, M, L, C, B, J, K, S, R, I, H, A*; other answers are possible.

Exercise Set 14.4

1. A **tree** is a connected graph in which each edge is a bridge.
3. Yes, because removing the edge would create a disconnected graph.
5. A **minimum-cost spanning tree** is a spanning tree that has the lowest cost or shortest distance of all spanning trees for a given graph.
7.

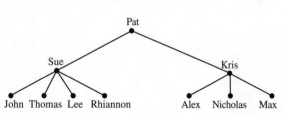

9.

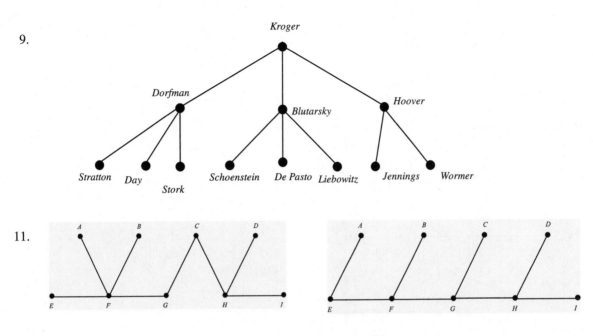

11.

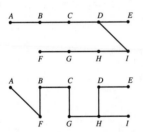

Other answers are possible.

13.

Other answers are possible.

15.

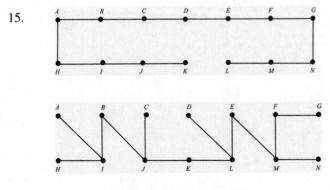

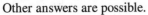

Other answers are possible.

17.

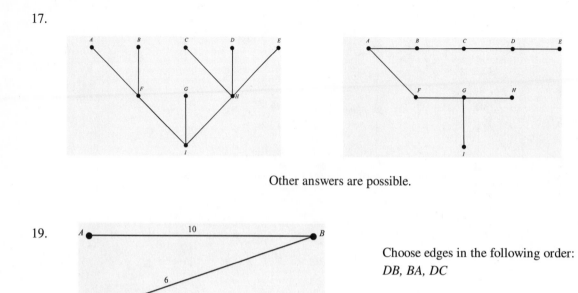

Other answers are possible.

19.

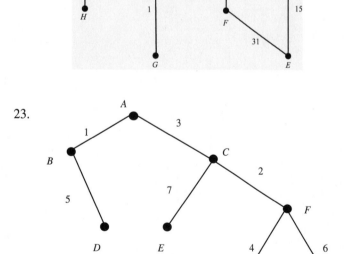

Choose edges in the following order:
DB, BA, DC

21.

Choose edges in the following order:
GB, BC, BA, AH, DE, CF, FE

23.

Choose edges in the following order:
AB, CF, AC, FG, BD, FH, EC

25.

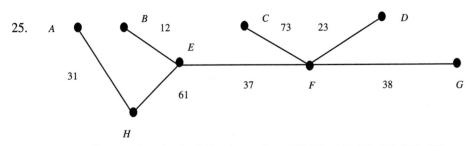

Choose edges in the following order: *BE*, FD, AH, EF, FG, HE, CF

27. a)

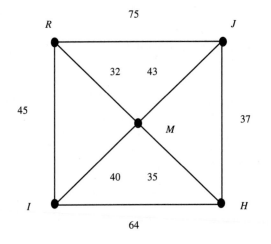

Other answers are possible.

b)

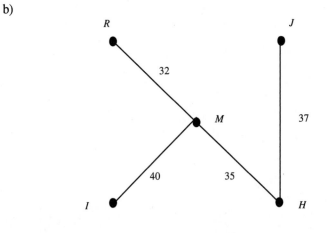

Choose edges in the following order:
RM, MH, JH, IM

c) $25(32 + 35 + 37 + 40) = 25(144) = \$3600$

29. a)

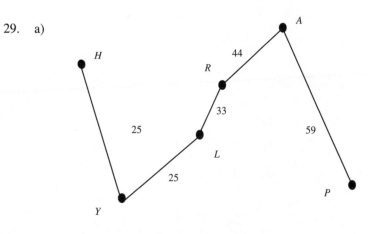

Choose edges in the following order: *HY, YL, LR, RA, AP*

b)   6800(25 + 25 + 33 + 44 + 59) = 6800(186) = $1,264,800

31. a)

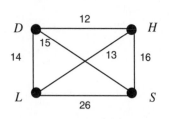

Other answers are possible.

b)

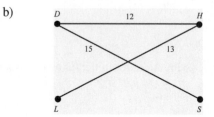

Choose edges in the following order:
DH, HL, DS

c) 3500 (12 + 13 + 15) = 3500 (40) = $140,000

33. a)

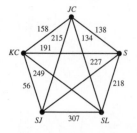

Other answers are possible.

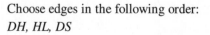

33. b)

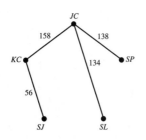

Choose edges in the following order: *KC-SJ, JC-SL, JC-SP, JC-KC*

c) 3700(56 + 134 + 138 + 158) = 3700(486) = $1,798,200

35. Answers will vary.
37. Answers will vary.

Review Exercises

1.

2.

3. *A, B, C, A, D, C, E, D*; other answers are possible.
4. No. The graph has two odd vertices, so a path that includes each edge exactly once must start at one odd vertex and end at the other; *B* is an even vertex.

5.

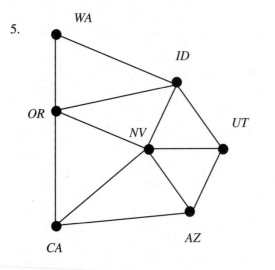

6.

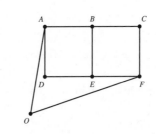

7. Connected

8. Disconnected. There is no path that connects *A* to *C*.

9. Edge *CD*

10. *D, A, B, C, E, H, G, F, D, B, E, G, D, E*; other answers are possible.

11. *E, D, B, E, G, D, A, B, C, E, H, G, F, D*; other answers are possible.

12. *B, C, A, D, F, E, C, D, E, B*; other answers are possible.

13. *E, F, D, E, C, D, A, C, B, E*; other answers are possible.

14. a)   No. The graph representing the map:

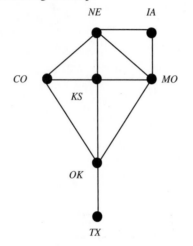

b) Vertices *CO* and *TX* are both odd. According to item 2 of Euler's Theorem, since there are exactly two odd vertices, at least one Euler path, but no Euler circuits exist.

Yes; *CO, NE, IA, MO, NE, KS, MO, OK, CO, KS, OK, TX*; other answers are possible.

c) No. (See part b) above.)

15. a)

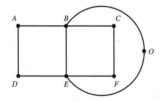

b) Yes; the graph has no odd vertices, so there is at least one Euler path, which is also an Euler circuit.

c) The person may start in any room or outside and will finish where he or she started.

16. a) Yes. The graph representing the map:

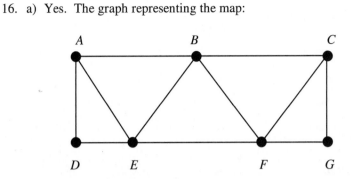

The officer is seeking an Euler path or an Euler circuit. Note that vertices *A* and *C* are both odd.
According to item 2 of Euler's Theorem, since there are exactly two odd vertices, at least one Euler
path but no Euler circuits exist.

   b) The officer would have to start at either the upper left-hand corner or the upper right-hand corner.
      If the officer started in the upper left-hand corner, he or she would finish in the upper right-hand corner,
      and vice versa.

17. *A, B, F, E, H, G, D, C, A, D, E, B*; other answers are possible.
18. *A, B, C, D, H, G, C, F, G, B, F, E, A*; other answers are possible.
19. *A, C, B, F, E, D, G* and *A, C, D, G, F, B, E*; other answers are possible.
20. *A, B, C, D, F, E, A* and *A, E, F, B, C, D, A*; other answers are possible.

21.

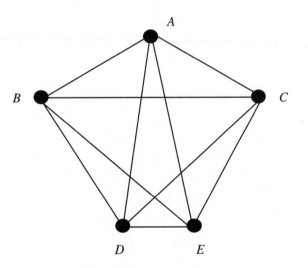

22. The number of unique Hamilton circuits within the complete graph with 5 vertices representing this
    situation is $(5-1)! = 4! = 4 \cdot 3 \cdot 2 \cdot 1 = 24$ ways

23. a)

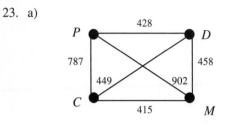

b)

Hamilton Circuit	First Leg/Cost	Second Leg/Cost	Third Leg/Cost	Fourth Leg/Cost	Total Cost
P, D, C, M, P	428	449	415	902	$2194
P, D, M, C, P	428	458	415	787	$2088
P, C, M, D, P	787	415	458	428	$2088
P, C, D, M, P	787	449	458	902	$2596
P, M, D, C, P	902	458	449	787	$2596
P, M, C, D, P	902	415	449	428	$2194

The least expensive route is *P, D, M, C, P* or *P, C, M, D, P*

c) $2088

24. a)

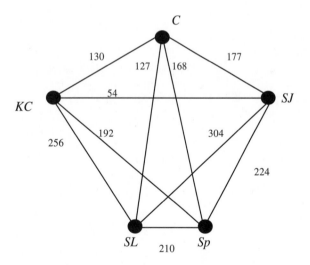

b) *SJ, KC, C, SL, Sp, SJ* traveling a total of 54 + 130 + 127 + 210 + 224 = 745 miles

c) *Sp, C, SL, KC, SJ, Sp* traveling a total of 168 + 127 + 256 + 54 + 224 = 829 miles

25.

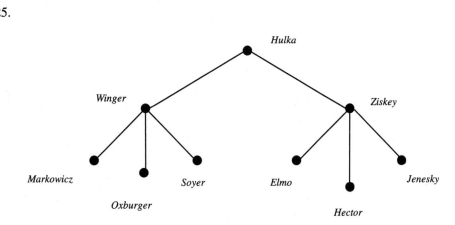

26.

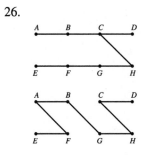

Other answers are possible

27.

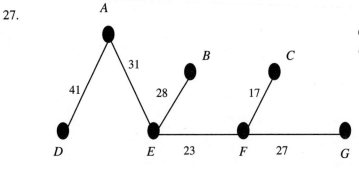

Choose edges in the following order:
CF, EF, FG, BE, AE, AD

28. a)

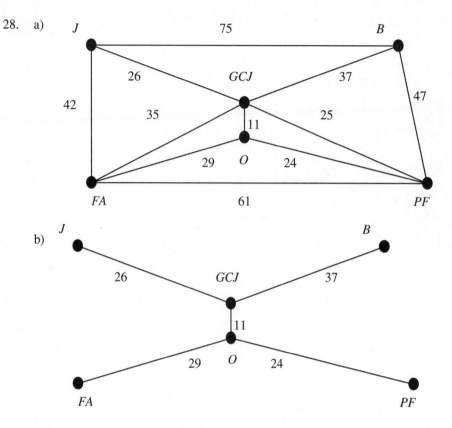

b)

Choose edges in the following order:
O GCJ, O PF, J GCJ, FA O, GCJ B

c) 2.50 (11 + 24 + 26 + 29 + 37) = 2.50 (127) = $317.50

Chapter Test

1.

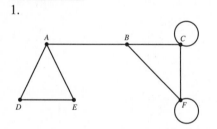

Edge *AB* is a bridge. There is a loop at vertex *G*.
Other answers are possible.

2.

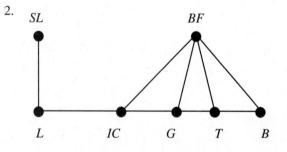

3. One example:

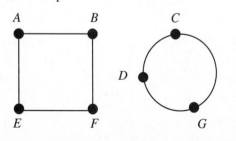

4. *A, B, C, F, H, G, E, F, B, E, D, A*;
   other answers are possible.

5. Yes. The person may start in room *A* and end in room *B* or vice versa.
6. *A, D, E, A, F, E, H, F, I, G, F, B, G, C, B, A*; other answers are possible.
7. *A, B, C, G, E, D, H, I, K, J, F, A*; other answers are possible.
8. The number of unique Hamilton circuits within the complete graph with 5 vertices representing

    this situation is $(5-1)! = 4! = 24$ ways

9. a)

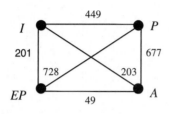

    b)

Hamilton Circuit	First Leg/Cost	Second Leg/Cost	Third Leg/Cost	Fourth Leg/Cost	Total Cost
I, P, EP, A, I	449	728	49	203	$1429
I, P, A, EP, I	449	677	49	201	$1376
I, A, P, EP, I	203	677	728	201	$1809
I, A, EP, P, I	203	49	728	449	$1429
I, EP, A, P, I	201	49	677	449	$1376
I, EP, P, A, I	201	728	677	203	$1809

    The least expensive route is *I, P, A, EP, I* or *I, EP, A, P, I* for $1376.

    c) *I, EP, A, P, I* for $1376

10.

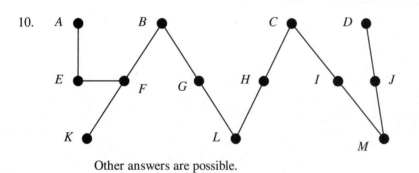

    Other answers are possible.

11.

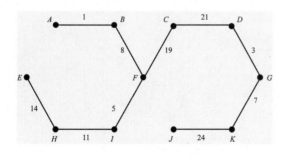

Choose edges in the following order:
AB, DG, FI, KG, BF, HI, EH, FC, CD, JK

12.   a)

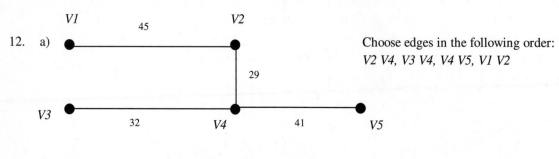

Choose edges in the following order:
V2 V4, V3 V4, V4 V5, V1 V2

b)   1.25 (29 + 32 + 41 + 45) = 1.25 (147) = $183.75

CHAPTER FIFTEEN

VOTING AND APPORTIONMENT

Exercise Set 15.1

1. When a candidate receives more than 50% of the votes.

3. Voters rank candidates from most favorable to least favorable. Each last place vote is awarded one point, each next to last place vote is awarded two points, each third from last place vote is awarded three points, etc. The candidate receiving the most points is the winner.

5. Voters rank the candidates. A series of comparisons in which each candidate is compared to each of the other candidates follows. If candidate A is preferred to candidate B, then A receives one point. If candidate B is preferred to candidate A, then B receives one point. If the candidates tie, each receives ½ point. The candidate receiving the most points is declared the winner.

7. A preference table summarizes the results of an election.

9. By ranking their choices, voters are able to provide more information with the Borda count method.

11. a) Jeter is the winner; he received the most votes using the plurality method.

   b) No. $\dfrac{265128}{192827+210361+265128} = \dfrac{265128}{668316} \approx 0.40$ is not a majority.   Majority is $> 334,158$ votes.

13.

Number of votes	3	1	2	2	1
First	H	G	S	S	G
Second	G	H	H	G	S
Third	S	S	G	H	H

15. $10 + 5 + 4 + 2 = 21$ employees

17. Wrench wins with the most votes (10).

19. A majority out of 21 votes is 11 or more votes.
   First choice votes: (W) 10, (H) 9, (F) 2
   None receives a majority, thus F with the least votes is eliminated.
   Second round: (W) 10, (H) 5+4+2 = 11
   Water heater wins with a majority of 11 votes.

21. Y: 4 1st place votes = $(4)(3) = 12$
       2 2nd place votes = $(2)(2) = 4$
       3 3rd place votes = $(3)(1) = 3$
    R: 2 1st place votes = $(2)(3) = 6$
       4 2nd place votes = $(4)(2) = 8$
       3 3rd place votes = $(3)(1) = 3$
    S: 3 1st place votes = $(3)(3) = 9$
       3 2nd place votes = $(3)(2) = 6$
       3 3rd place votes = $(3)(1) = 3$
    Y = 19 points; R = 17 points; S = 18 points
    Yellowstone wins with 19 points.

23. Y vs. R: Y = $3+2+1 = 6$ R = $2+1 = 3$
    Y gets 1 pt.
    Y vs. S: Y = $3+1 = 4$ S = $2+2+1 = 5$
    S gets 1 pt.
    R vs. S: R = $3+2 = 5$ S = $2+1+1 = 4$
    R gets 1 pt.
    All get 1 point, which indicates no winner.

25. Votes: (S) $8+3+2 = 13$ (L) $6+3 = 9$
    (H) $4+3+2 = 9$ (T) 1
    San Antonio wins with the most votes.

27. A majority out of 32 votes is 17 or more votes.
    First choice votes: (S) 13, (L) 9, (H) 9, (T) 1
    None receives a majority, thus T with the least
    votes is eliminated.
    Second round: (S) 13, (L) 9, (H) 10
    No majority, thus eliminate L.
    Third round: (S) 16, (H) 16
    Since S and H tied, there is no winner.

29. W: 5 1st place votes = $(5)(3) = 15$
       4 2nd place votes = $(4)(2) = 8$
       3 3rd place votes = $(3)(1) = 3$
    D: 1 1st place votes = $(1)(3) = 3$
       7 2nd place votes = $(7)(2) = 14$
       4 3rd place votes = $(4)(1) = 4$
    J: 6 1st place votes = $(6)(3) = 18$
       1 2nd place votes = $(1)(2) = 2$
       5 3rd place votes = $(5)(1) = 5$
    W = 26 points; D = 21 points; J = 25 points
    Williams wins with 26 points.

31. W vs. D: W = $5+4 = 9$ D = $1+2 = 3$
              W gets 1 pt.
    W vs. J: W = 5   J = $1+4+2 = 7$
              J gets 1 pt.
    D vs. J: D = $5+1 = 6$ J = $4+2 = 6$
              D and J get 0.5 pt.
    W = 1 pt.   D = 0.5 pt.   J = 1.5 pts.
    Johnson wins with 1.5 points.

33. A majority out of 12 votes is 7 or more votes.
    Most last place votes: (W) 3, (D) 4, (J) 5
    Thus J with the most last place votes is eliminated.
    Second round using the most last place votes:
    (W) $1+2 = 3$, (D) $5+4 = 9$
    Williams wins with the least last place votes.

35. L: 5 1st place votes = $(5)(3) = 15$
       6 3rd place votes = $(6)(1) = 6$
    E: 2 1st place votes = $(2)(3) = 6$
       9 2nd place votes = $(9)(2) = 18$
    O: 4 1st place votes = $(4)(3) = 12$
       2 2nd place votes = $(2)(2) = 4$
       5 3rd place votes = $(5)(1) = 5$
    L = 21 points; E = 24 points; O = 21 points
    Erie Road wins with 24 points.

37. L vs. E: L = 5   E = 2+4 = 6   E gets 1 pt.
     L vs. O: L = 5   O = 2+4 = 6   O gets 1 pt.
     E vs. O: E = 5+2 = 6   O = 4   E gets 1 pt.
     Erie Road wins with 2 points.

39. a) Votes: (B): 112, (L): 2, (D): 1
       Becker wins with the most votes.
   b) B: 12 1st place votes = (12)(4) = 48
          5 2nd place votes = (3)(3) = 9
      L: 3 1st place votes = (2)(4) = 8
          6 2nd place votes = (8)(3) = 24
          6 3rd place votes = (5)(2) = 10
      M: 9 3rd place votes = (10)(2) = 20
          6 4th place votes = (5)(1) = 5
      D: 1 1st place vote = (1)(4) = 4
          4 2nd place votes = (4)(3) = 12
          10 4th place votes = (10)(1) = 10
      B = 57 points; L = 42 points; M = 25 points,
      D = 26 points   B wins with 57 points.

39. c) A majority out of 15 votes is 8 or more votes.
       First choice votes: (B) 12,   (L) 2
                            (M) 0,   (D) = 2
       Because B already has a majority, B wins.
    d) B vs. L: B = 8+4+1 = 13   L = 2
       B gets 1 pt.
       B vs. M: B = 8+4+2+1 = 15   B gets 1 pt.
       B vs. D: B = 8+4+2 = 14   D = 1
       B gets 1 pt.
       L vs. M: L = 8+4+2+1 = 15   L gets 1 pt.
       L vs. D: L = 8+2 = 10   D = 4+1 = 5
       L = gets 1 pt.
       M vs. D: S = 8+2 = 10   D = 4+1 = 5
       M gets 1 pt.
       B wins with 3 points.

41. a) G: 8 1st place votes = (8)(4) = 32
           3 2nd place vote = (3)(3) = 9
           4 3rd place votes = (4)(2) = 8
           14 4th place votes = (14)(1) = 14
       I: 3 1st place votes = (3)(4) = 12
           7 2nd place vote = (7)(3) = 21
           19 3rd place votes = (19)(2) = 38
       P: 14 1st place votes = (14)(4) = 56
           8 2nd place votes = (8)(3) = 24
           3 3rd place votes = (3)(2) = 6
           4 4th place votes = (4)(1) = 4
       Z: 4 1st place votes = (4)(4) = 16
           11 2nd place vote = (11)(3) = 33
           3 3rd place votes = (3)(2) = 6
           11 4th place votes = (11)(1) = 11
       G = 63 points; I = 71 points; P = 91 points;
       Z = 66 points
       P wins with 91 points.

41. b) Votes: (G): 8, (I): 3, (P): 14, (Z): 4
       P wins with the most votes.
    c) A majority out of 29 votes is 15 or more votes.
       First choice votes: (G) 8,   (I) 3
                            (P) 14,   (Z) = 4
       None receives a majority, thus I with the least
       votes is eliminated.
       Second round: (G) 11, (P) 14, (Z) 4
       No majority, thus eliminate Z.
       Third round: (G) 15, (P) 14
       G wins with 15 votes.
    d) G vs. P: G = 15   P = 14   G gets 1 pt.
       G vs. Z: G= 11   Z = 18   Z gets 1 pt.
       G vs. I: G = 8   I = 21   I gets 1 pt.
       P vs. Z: P = 25   Z = 4   P gets 1 pt.
       P vs. I: P = 22   I = 7   P gets 1 pt.
       Z vs. I: Z = 15   I = 14   Z gets 1 pt.
       P and Z tie with 2 points.

43. a) If there were only two columns then only two of the candidates were the first choice of the voters. If each of the 15 voters cast a ballot, then one of the voters must have received a majority of votes because 15 cannot be split evenly.
    b) An odd number cannot be divided evenly so one of the two first choice candidates must receive more than half of the votes.

47. a) Each voter casts 4+3+2+1 = 10 votes.
        (15)(10) = 150 votes
    b) 150 – (35+40+25) = 150 – 100 = 50 votes
    c) Yes. Candidate D has more votes than each of the other 3 candidates.

45. a) C: $4 + 1 + 1 = 6$ R: $4 + 4 + 3 = 11$
        W: $3 + 3 + 2 + 2 + 1 + 1 = 12$
        T: $4 + 3 + 2 + 2 = 11$
        The Warriors finished 1st, the Rams and the Tigers tied for 2nd, and the Comets were 4th.
    b) C: $5 + 0 = 5$ R: $5 + 5 + 3 = 13$
        W: $3 + 3 + 1 + 1 + 0 + 0 = 8$
        T: $5 + 3 + 1 \ 1 = 10$
        Rams - 1st, Tigers - 2nd, Warriors - 3rd, and Comets - 4th.

49. Answers will vary.

Exercise Set 15.2

1. If a candidate receives a majority of first place votes, then that candidate should be declared the winner.

3. If a candidate is favored when compared individually with every other candidate, then that candidate should be declared the winner.

5. A candidate that is preferred to all others will win each pairwise comparison and be selected with the pairwise comparison method.

7. If a candidate receives a majority of first place votes, then that candidate should be declared the winner. Plurality counts only the 1st place votes.

9. The plurality method yields Tampa is the winner with a majority of 10 1st place votes. However, if the Borda count method is used:
    Tampa $(10)(3) + (3)(2) + (6)(1) = 30 + 6 + 6 = 42$
    Portland $(6+3)(3) + (10)(2) = 27 + 20 = 47$
    SA $(6)(2) + (10+3)(1) = 12 + 13 = 19$
    The winner is Portland using the Borda count method, thus violating the majority criterion.

11. Total votes = $3+2+1+1 = 7$ Candidates A is the candidate of choice with a majority of 5 votes.
    A: 5 1st place votes = $(5)(4) = 20$
        4 4th place votes = $(4)(1) = 4$
    B: 4 1st place vote = $(4)(4) = 16$
        5 2nd place vote = $(5)(3) = 15$
    S: 3 2nd place vote = $(3)(3) = 9$
        5 3rd place vote = $(5)(2) = 10$
        1 4th place vote = $(1)(1) = 1$
    Y: 1 2nd place vote = $(1)(3) = 3$
        4 3rd place votes = $(4)(2) = 8$
        4 4th place votes = $(4)(1) = 4$
    A gets 24 pts., B gets 31 pts., S gets 20 pts., and Y gets 15 pts. Candidate B wins by the Borda count method.

13. P: 4 1st place votes = (4)(3) = 12
       2 2nd place votes = (2)(2) = 4
       3 3rd place votes = (3)(1) = 3
    L: 3 1st place vote = (3)(3) = 9
       5 2nd place vote = (5)(2) = 10
       1 3rd place vote = (1)(1) = 1
    S: 2 1st place votes = (2)(3) = 6
       2 2nd place vote = (2)(2) = 4
       5 3rd place vote = (5)(1) = 5
    P = 19 votes;   L = 20 votes;   S = 15 votes
    P vs. L: P = 4+1 = 5   L = 4   P gets 1 pt.
    P vs. S: P = 4+1 = 5   S = 4   P gets 1 pt.
    L vs. S: L = 4+1+2 = 7   S = 2   L gets 1 pt.
    Because Parking wins its head-to-head comparisons and the Lounge Areas win by Borda count method, the head-to-head criterion is not satisfied.

15. A majority out of 21 votes is 11 or more votes.
    Plurality with elimination:
    First choice votes: A = 6, B = 8, C = 7
    Eliminate A. B = 8, C = 13; C wins

    Pairwise comparison:
    A vs. B: A = 7+6 = 13   B = 8   A gets 1 pt.
    A vs. C: A = 8+6 = 14   C = 7   A gets 1 pt.
    B vs. C: B = 8   C = 7+6=13   C gets 1 pt.

    No, because C wins by plurality with elimination but A wins using head-to-head comparison.

17. Votes: A: 12,   B: 4,   C: 9; thus, A wins.
    If B drops out, we get the following:
    Votes: A: 12,   C: 9 + 4 = 13, thus C would win.
    The irrelevant alternatives criterion is not satisfied.

19. A receives 53 points, B receives 56 points, and C receives 53 points. Thus, B wins using the Borda count method. If C drops out, we get the following: B receives 46 points, and C receives 35 points. Thus, B still wins. The irrelevant alternatives criterion is satisfied.

21. A majority out of 32 voters is 17 or more votes.
    Votes: A: 8 + 3 = 11, B: 9, C: 12; none has a majority, thus eliminate B.
    Votes: A:8 + 3 = 11,   C: 9 +12 = 21, thus C wins. If the three voters who voted for A,C,B change to C,A,B, the new set of votes becomes:
    Votes: A: 8,   B: 9,   C: 15; none has a majority, thus eliminate A.
    Votes: B: 17, C:15, B wins.
    Thus, the monotonicity criterion is not satisfied.

23. A vs. B: A = 13   B = 13   A gets .5, B gets .5.
    A vs. C: A = 13   C = 13   A gets .5, C gets .5.
    A vs. D: A = 13   D = 13   A gets .5, D gets .5.
    B vs. C: B = 13   C = 13   B gets .5, C gets .5.
    B vs. D: B = 13   D = 13   B gets .5, D gets .5.
    C vs. D: C = 12   D = 14   D gets 1 pt.
    D wins with 2 pts.
    After the change in votes:
    A vs. B: A = 8   B = 16   B gets 1 pt.
    A vs. C: A = 13   C = 13   A gets .5, C gets .5.
    A vs. D: A = 13   D = 13   A gets .5, D gets .5.
    B vs. C: B = 18   C = 6   B gets 1 pt.
    B vs. D: B = 13   D = 13   B gets .5, D gets .5.
    C vs. D: C = 15   D = 11   C gets 1 pt.
    B wins with 2.5 pts.
    The monotonicity criterion is not satisfied.

25. A receives 2 points, B receives 3 point, C receives 2 points, D receives 1 point, and E receives 2 pts. B wins by pairwise comparison.
    After A, C and E drop out, the new set of votes is B: 2   D: 3, thus D wins. The irrelevant alternatives criterion is not satisfied.

27. Total votes = 7   A wins with a majority of 4 votes.
    A: 4 1st place votes = (4)(3) = 12
       3 3rd place votes = (3)(1) = 3
    B: 2 1st place vote = (2)(3) = 6
       5 2nd place vote = (5)(2) = 10
    C: 1 1st place votes = (1)(3) = 3
       2 2nd place votes = (2)(2) = 4
       4 3rd place vote = (4)(1) = 4
    A = 15 points;   B = 16 points;   C = 11 points
    B wins with 16 points. No. The majority criterion is not satisfied.

29. a) Washington, with 12 out of 23 votes
    b) Again, Washington, since it has a majority.
    c) N = (20)(3) + (3)(1) = 63
       W = (12)(4) + (3)(3) + (8)(1) = 65
       P = (8)(4) + (15)(2) = 62
       B = (3)(4) + (8)(2) + (12)(1) = 40
       Washington wins.
    d) Again, Washington, since it has a majority.

e) N vs. W: N = 8   W = 15, W gets 1 pt.
   N vs. P: N= 12   P = 11, N gets 1 pt.
   N vs. B: N= 20   B = 3, N gets 1 pt.
   W vs. P: W = 15   P = 8, W gets 1 pt.
   W vs. B: W= 15   B = 8, W gets 1 pt.
   P vs. B: P = 20   B = 3, P gets 1 pt.
   Washington wins with 3 pts.
   f) None of them.

31. a) A majority out of 82 votes is 42 or more votes.
    First choice votes: (A) 28, (L) 30, (W) 24
    None receives a majority, thus D with the least votes is eliminated.
    Second round: (A) 52, (L) 30
    Thus, Jennifer Aniston is selected..
    b) No majority on the 1st vote; C is eliminated with the fewest votes.
    Second round: (A) 38, (W) 44
    Denzel Washington is chosen.
    c) Yes.

33. A candidate who holds a plurality will only gain strength and hold and even larger lead if more favorable votes are added.

37. AWV

39. AWV

Exercise Set 15.3

1. If we divide the total population by the number of items to be apportioned we obtain a number called the standard divisor.
3. The standard quota rounded down to the nearest whole number.
5. An apportionment should always be either the upper quota or the lower quota.
7. Jefferson's method, Webster's method, Adams's method
9. a) Webster's method   b) Adams's method   c) Jefferson's method

11. a) $\dfrac{8000000}{160} = 50,000$ = standard divisor

   b)

State	A	B	C	D	Total
Population	1,345,000	2,855,000	982,000	2,818,000	8,000,000
Standard Quota	26.9	57.1	19.64	56.36	

13. a) and b) Modified divisor: 49,300

State	A	B	C	D	Total
Population	1,345,000	2,855,000	982,000	2,818,000	8,000,000
Modified Quota	27.31	57.97	19.939	57.218	
Jefferson's Apportionment (round down)	27	57	19	57	160

15. a) and b) Modified divisor: 50,700

State	A	B	C	D	Total
Population	1,345,000	2,855,000	982,000	2,818,000	8,000,000
Modified Quota	26.529	56.312	19.369	55.582	
Adams's Apportionment (round up)	27	57	20	56	160

17. a) and b) Standard divisor: 50,000

State	A	B	C	D	Total
Population	1,345,000	2,855,000	982,000	2,818,000	8,000,000
Standard Quota	26.9	57.1	19.64	56.36	
Webster's Apportionment (standard rounding)	27	57	20	56	160

19. a) Standard divisor = $\dfrac{\text{total}}{25} = \dfrac{675}{25} = 27$

   b)

Hotel	A	B	C	Total
Amount	306	214	155	675
Standard Quota	11.33	7.93	5.74	

21. a) and b)

Hotel	A	B	C	Total
Amount	306	214	155	675
Modified Quota	11.86	8.29	6.01	
Jefferson's Apportionment (rounded down)	11	8	6	25

23. a) and b)

Hotel	A	B	C	Total
Amount	306	214	155	675
Modified Quota	10.55	7.38	5.34	
Adams's Apportionment (rounded up)	11	8	6	25

25. a) and b)

Store	A	B	C	Total
Amount	306	214	155	675
Standard Quota	11.33	7.93	5.74	
Webster's Apportionment (standard rounding)	11	8	6	25

27. a) A standard divisor $= \dfrac{\text{total}}{50} = \dfrac{550}{50} = 11$

b) and c)

Resort	A	B	C	D	Total
Rooms	86	102	130	232	550
Standard Quota	7.82	9.27	11.82	21.09	
Lower Quota	7	9	11	21	48
Hamilton's Method	8	9	12	21	50

29. Using a divisor of 11.4:

Resort	A	B	C	D	Total
Rooms	86	102	130	232	550
Modified Quota	7.54	8.95	11.40	20.35	
Adams's Method	8	9	12	21	50

31. a) Standard divisor $= \dfrac{\text{total}}{250} = \dfrac{13000}{250} = 52$

b) and c)

School	LA	Sci.	Eng.	Bus.	Hum	Total
Enrollment	1746	7095	2131	937	1091	13000
Standard Quota	33.58	136.44	40.98	18.02	20.98	
Lower Quota	33	136	40	18	20	247
Hamilton's Apportionment	34	136	41	18	21	250

33. A divisor of 51.5 was used.

School	LA	Sci.	Eng.	Bus.	Hum	Total
Enrollment	1746	7095	2131	937	1091	13000
Modified Quota	33.90	137.77	41.38	18.19	21.18	
Jefferson's Apportionment (round down)	33	137	41	18	21	250

35. a) A standard divisor = $\dfrac{\text{total}}{150} = \dfrac{13500}{150} = 90$

b) and c)

Dealership	A	B	C	D	Total
Annual Sales	4800	3608	2990	2102	13500
Standard Quota	53.33	40.09	33.22	23.36	150.00
Hamilton's Apportionment	53	40	33	24	150

37. A divisor of 89.5 was used.

Dealership	A	B	C	D	Total
Annual Sales	4800	3608	2990	2102	13500
Modified Quota	53.63	40.31	33.41	23.47	
Webster's Apportionment	54	40	33	23	150

39. a) Standard divisor = $\dfrac{75,000}{100} = 750$

b) and c)

Route	A	B	C	D	E	F	Total
Passengers	9070	15,275	12,810	5720	25,250	6875	75,000
Standard Quota	12.09	20.37	17.08	7.63	33.67	9.17	
Lower Quota	12	20	17	7	33	9	98
Hamilton's Method	12	20	17	8	34	9	100

41. The divisor 765 was used.

Route	A	B	C	D	E	F	Total
Passengers	9070	15,275	12,810	5720	25,250	6875	75,000
Modified Quota	11.86	19.97	16.75	7.48	33.01	8.99	
Adams's Method	12	20	17	8	34	9	100

43 a) Standard divisor = $\dfrac{\text{total}}{200} = \dfrac{2400}{200} = 12$

b) and c)

Shift	A	B	C	D	Total
Room calls	751	980	503	166	2400
Standard Quota	62.58	81.67	41.92	13.83	
Lower Quota	62	81	41	13	197
Hamilton's Method	62	82	42	14	200

45. The divisor 11.9 was used.

Shift	A	B	C	D	Total
Room calls	751	980	503	166	2400
Modified Quota	63.11	82.35	42.27	13.95	
Jefferson's Apportionment	63	82	42	13	200
(round down)					

47. Standard divisor = $\dfrac{3615920}{105} = 34437.33$

   a) Hamilton's Apportionment: 7, 2, 2, 2, 8, 14, 4, 5, 10, 10, 13, 2, 6, 2, 18
   b) Jefferson's Apportionment: 7, 1, 2, 2, 8, 14, 4, 5, 10, 10, 13, 2, 6, 2, 19
   c) States that Benefited: Virginia   States Disadvantaged: Delaware

49. One possible answer is A; 734, B: 367, C: 432, D: 491, E: 519, F: 388

Exercise Set 15.4

1. The Alabama paradox occurs when an increase in the total # of items results in a loss of items for a group.
3. The population paradox occurs when group A loses items to group B, although group A's population grew at a higher rate than group B's.
5. Adams's, Webster's

7. New divisor = $\dfrac{1080}{61} = 17.71$

Office	A	B	C	D	E	Total
Employees	246	201	196	211	226	1080
Standard Quota	13.98	11.35	11.07	11.91	12.76	
Lower Quota	13	11	11	11	12	58
Hamilton's Apportionment	14	11	11	12	13	61

No. No office suffers a loss so the Alabama paradox does not occur.

9. a) Standard divisor = $\dfrac{900}{30} = 30$

State	A	B	C	Total
Population	161	250	489	900
Standard Quota	5.37	8.33	16.30	
Hamilton's Apportionment	6	8	16	30

   b) New divisor = $\dfrac{900}{31} = 29.03$

State	A	B	C	Total
Population	161	250	489	900
Standard Quota	5.55	8.61	16.84	
Hamilton's Apportionment	5	9	17	31

   Yes, state A loses 1 seat and states B and C each gain 1 seat.

11. a) Standard divisor = $\dfrac{25000}{200} = 125$

City	A	B	C	Total
Population	8130	4030	12,840	25,000
Standard Quota	65.04	32.24	102.72	
Hamilton's Apportionment	65	32	103	200

b) New divisor = $\dfrac{25125}{200} = 125.625$

City	A	B	C	Total
New Population	8150	4030	12,945	25,125
Standard Quota	64.88	32.08	103.04	
Hamilton's Apportionment	65	32	103	200

No. None of the Cities loses a bonus.

13. a) Standard divisor = $\dfrac{5400}{54} = 100$

Division	A	B	C	D	E	Total
Population	733	1538	933	1133	1063	5400
Standard Quota	7.33	15.38	9.33	11.33	10.63	
Lower Quota	7	15	9	11	10	52
Hamilton's Apportionment	7	16	9	11	11	54

b) New divisor = $\dfrac{5454}{54} = 101$

Division	A	B	C	D	E	Total
Population	733	1539	933	1133	1116	
Standard Quota	7.26	15.24	9.24	11.22	11.05	
Lower Quota	7	15	9	11	11	53
Hamilton's Apportionment	8	15	9	11	11	54

Yes. Division B loses an internship to Division A even though the population of division B grew faster than the population of division A.

15. a) Standard divisor $= \dfrac{4800}{48} = 100$

Tech. Data	A	B	Total
Employees	844	3956	4800
Standard Quota	8.44	39.56	
Lower Quota	8	39	47
Hamilton's Apportionment	8	40	48

b) New divisor $= \dfrac{5524}{55} = 100.44$

Tech. Data	A	B	C	Total
Employees	844	3956	724	5524
Standard Quota	8.40	39.39	7.21	
Lower Quota	8	39	7	54
Hamilton's Apportionment	9	39	7	55

Yes. Group B loses a manager.

17. a) Standard divisor $= \dfrac{990000}{66} = 15,000$

State	A	B	C	Total
Population	68970	253770	667260	990000
Standard Quota	4.60	16.92	44.48	
Hamilton's Apportionment	5	17	44	66

b) New divisor $= \dfrac{1075800}{71} = 15,152.11$

State	A	B	C	D	Total
Population	68970	253770	667260	85800	1075800
Standard Quota	4.55	16.75	44.04	5.66	
Hamilton's Apportionment	4	17	44	6	71

Yes. State A loses a seat.

Review Exercises

1. a) Robert Rivera wins with the most votes (15).

   b) A majority out of 34 voters is 18 or more votes. Robert Rivera does not have a majority.

2. a) Michelle MacDougal wins with the most votes (231).

   b) Yes. A majority out of 413 voters is 207 or more votes.

3.

# of votes	3	2	1	3	1
First	B	A	D	C	D
Second	A	C	C	B	A
Third	C	D	A	A	B
Fourth	D	B	B	D	C

4.

# of votes	2	2	2	1
First	C	B	A	C
Second	A	A	C	B
Third	B	C	B	A

5. Number of votes = 8 + 5 + 4 + 2 + 1 = 20

6. DVD player wins with a plurality of 8 votes.

7. I: 59 points,   D: 48 points,   C: 53 points,
   N: 40 points. iPod wins with 59 points.

8. A majority out 20 voters is 11 or more votes.
   Votes: I: 4, D: 8, C:7, N: 1.
   None has a majority, thus eliminate N.
   Votes: I: 4, D: 9, C: 7
   None has a majority, thus eliminate I
   Votes: D: 9, D: 11
   The digital camera wins.

9. The camera and iPod tie with 2 points each.
   D and N each have 1 point.

10. Votes: Votes: I: 4, D: 8, C:7, N: 1   None has a
    majority, thus eliminate D with most last place
    votes. Votes: I: 12, C: 7, N: 1; the iPod wins.

11. 38+30+25+7+10 = 110 students voted

12. Volleyball wins with a plurality of 40 votes.

13. S: 223 pts., V: 215 pts., B: 222 pts.
    Soccer wins.

14. A majority out of 110 voters is 56 or more votes.
    Votes: S: 38, V: 40, B: 32; None has a majority,
    thus eliminate B. Votes: S: 45, V: 65
    Volleyball wins.

15. S: 1 pt.,   V: 1 pt.,   B: 1 pt.   A 3-way tie

16. Votes: S: 38,   V: 40,   B: 32   None has a
    majority, thus eliminate V with the most last place
    votes.   Votes: S: 68, B: 42.   Soccer wins.

17. a) Votes: A: 161+134 = 295,   F: 45,   M: 12,
       P: 0   AARP wins.
    b) Yes. A majority out of 372 voters is 187 or more
       votes. AARP receives a majority.
    c) A: 1387 pts.,   F: 740 pts.,   M: 741 pts.,
       P: 852 pts.   AARP wins.
    d) 187 or more votes is needed for a majority.
       Votes: A: 295,   F: 45,   M: 12,   P: 0
       AARP wins.
    e) A: 3 pts.,   F: 1 pt.,   M: 1 pt.,   P: 1 pt.
       AARP wins.

18. Votes: C: 25, S: 80, D: 45, L: 50
    a) A majority out of 200 voters is 101 or more
       votes. None of the cities has a majority.
    b) Seattle, with a plurality of 80 votes.
    c) C: 480, S: 440, D: 490, L: 590; Las Vegas wins.
    d) C: 25, S: 80, D: 45, L: 50; no city has a majority
       so eliminate C.
       S: 80, D: 45, L: 75; no city has a majority so
       eliminate D.
       S: 80, L: 120; Las Vegas wins.
    e) Las Vegas wins with 3 points; Dallas has 2 and
       Chicago 1.

19. a) A majority out of 16 voters is 9 or more votes.
       Votes: (EB): 4+3+ = 7,   (FW): 1+1 = 2,
       (G): 0,   (WB): 6+1 = 7   None has a majority,
       thus eliminate G.   Votes: (EB): 4+3 = 7,
       (FW): 1+1 = 2,   (WB): 6 + 1 = 7   None has a
       majority, thus eliminate FW
       Votes: (EB): 4+3+1 = 8,   (WB): 6+1+1 = 8.
       Thus, EB and WB tie.
    b) Use the Borda count method to break the tie.
       (EB) = 46 points,   (WB) = 50 points;
       World Book wins.

19. c) (EB) vs. (WB): EB: 4+3+1 = 8 points,
       (WB): 6+1+1 = 8 points.
       EB and WB tie again.

20. A: 44 pts.,   B: 53 pts,   C: 46 pts.,   D: 27 pts.
    Using the Borda count, candidate B wins.
    However, B only has 2 first place votes, thus the
    majority criterion is not satisfied.

21. A wins all its head-to-head comparisons but B
    wins using the Borda count method.
    The head-to-head criterion is not satisfied.

22. a) A majority out of 42 voters is 21 or more votes.
Votes: A: 12,   B: 10+6 = 16,   C: 14
None has the majority, thus eliminate A.
Votes: B :10+6 = 16,   C: 14+12 = 26   C wins.

b) The new preference table is

Number of votes	10	14	6	12
First	B	C	C	A
Second	A	B	B	C
Third	C	A	A	B

Votes: A: 12,   B: 10,   C: 20;   None has a majority, thus eliminate B.
Votes: A: 22,   C: 20   A wins.   When the order is changed A wins.   Therefore, the monotonicity criterion is not satisfied.

22. c) If B drops out the new table is

Number of votes	10	14	6	12
First	A	C	C	A
Second	C	A	A	C

Votes: A: 10+12 = 22,   C: 14+6 = 20   A wins.
Since C won the first election and then after B dropped out A won, the irrelevant criterion is not satisfied.

23. a) B to O: 59 to 55, B to N: 58 to 56, B to H: 58 to 56, so Ball Park wins the head-to-head comparison.

b) Oscar Mayer with a plurality of 34.

c) B; 289, O: 237, N: 307, H: 307; a tie between Nathan's and Hebrew National.

d) In the first round N is eliminated, then B is eliminated, and Hebrew National wins with 80 votes.

e) Ball Park wins with 3 points.

f) Plurality, Borda count, and plurality with elimination all violate the head-to-head criterion.

24. a) Yes. Fleetwood Mac is favored when compared to each of the other bands.

b) Votes: R: 15,   B: 34,   J: 9+4 = 13, F: 25   Boston wins.

c) R: 217 points,   B: 198 points,   J: 206 points, F: 249 points   Fleetwood Mac wins.

d) A majority out of 87 voters is 44 or more votes.
Votes: R: 15,   B: 34,   J: 13,   F:25
None has a majority, thus eliminate J.
Votes: R: 15+9+4 = 28,   B: 34,   F: 25
None has a majority, thus eliminate F.
Votes: R: 28+25 = 53,   B: 34   REO wins.

e) R = 2 pts., B = 0 pts., J = 1 pt., F = 3 pts.
Thus, Fleetwood Mac wins.

f) Plurality and plurality w/elimination methods

25. The Borda count method

27. Pairwise comparison and Borda count methods

26. Plurality and plurality w/elimination methods

28. Standard divisor = $\dfrac{6000}{10} = 600$

Region	A	B	C	Total
Number of Houses	2592	1428	1980	6000
Standard Quota	4.32	2.38	3.30	
Lower Quota	4	2	3	9
Hamilton's Apportionment	4	3	3	10

29. Using the modified divisor 500.

Region	A	B	C	Total
Number of Houses	2592	1428	1980	6000
Modified Quota	5.18	2.86	3.96	
Jefferson's Apportionment (rounded down)	5	2	3	10

30. Using the modified divisor 700.

Region	A	B	C	Total
Number of Houses	2592	1428	1980	6000
Modified Quota	3.70	2.04	2.83	
Adams's Apportionment (rounded up)	4	3	3	10

31. Using the modified divisor 575.

Region	A	B	C	Total
Number of Houses	2592	1428	1980	6000
Modified Quota	4.51	2.48	3.4	
Webster's Apportionment (normal rounding)	5	2	3	10

32. Yes. Hamilton's Apportionment becomes 5, 2, 4.   Region B loses one truck.

33. Standard divisor $= \dfrac{870}{29} = 30$

Course	A	B	C	Total
Number of Students	371	279	220	870
Standard Quota	12.37	9.30	7.33	
Lower Quota	12	9	7	28
Hamilton's Apportionment	13	9	7	29

34. Use the modified divisor 28.

Course	A	B	C	Total
Number of Students	371	279	220	870
Modified Quota	13.25	9.96	7.86	
Jefferson's Apportionment (round down)	13	9	7	29

35. Use the modified divisor 31.2

Course	A	B	C	Total
Number of Students	371	279	220	870
Modified Quota	11.89	8.94	7.05	
Adams's Apportionment (round up)	12	9	8	29

36. Use the modified divisor 29.5

Course	A	B	C	Total
Number of Students	371	279	220	870
Modified Quota	12.58	9.46	7.46	
Webster's Apportionment (standard rounding)	13	9	7	29

37. The new divisor is $\dfrac{877}{29} = 30.24$

Course	A	B	C	Total
Number of Students	376	279	222	698
Standard Quota	12.75	9.46	7.53	
Lower Quota	12	9	7	28
Hamilton's Apportionment	13	9	7	29

No. The apportionment remains the same.

38. The Standard divisor $= \dfrac{55000}{55} = 1000$

State	A	B	Total
Population	4862	50138	55,000
Standard Quota	4.86	50.14	
Hamilton's Apportionment	5	50	55

39. The apportionment is 4, 51.

40. The apportionment is 5, 50.

41. The apportionment is 5, 50.

42. The new divisor is $\dfrac{60940}{60} = 1015.67$

State	A	B	C	Total
Population	4862	50138	5940	60940
Standard Quota	4.79	49.36	5.85	
Hamilton's Apportionment	5	49	6	60

Yes. State B loses a seat.

Chapter Test

1. $7 + 6 + 6 + 5 = 24$ members voted.

2. No candidate has a majority of ≥ 13 votes.

3. Pizza wins with a plurality of 11 votes.

4. T: 49, P: 53, B: 42; pizza wins.

5. B is eliminated and T then has a plurality of 13; tacos wins.

6. T: 1.5 pts., P 1 pt., B 0.5 pt.; Tacos wins.

7. a) Votes: H: $26+14 = 40$, I: 29, L: 30, S: 43
    Thus, the salamander wins.

  b) (H) 1st $(40)(4) = 160$
        2nd $(59)(3) = 177$
        3rd $(0)(2) = 0$
        4th $(43)(1) = 43$ H receives 380 points.
    (I) 1st $(29)(4) = 116$
        2nd $(40)(3) = 120$
        3rd $(73)(2) = 146$
        4th $(0)(1) = 0$ I receives 382 points
    (L) 1st $(30)(4) = 120$
        2nd $(43)(3) = 129$
        3rd $(43)(2) = 86$
        4th $(26)(1) = 26$ L receives 361 points

7. b) (S) 1st $(43)(4) = 172$
        2nd $(0)(3) = 0$
        3rd $(26)(2) = 52$
        4th $(73)(1) = 73$ S receives 297 points.
    The iguana (I) wins with the most points.

  c) A majority out of 142 voters is 72 or more votes.
    Votes: H: 40, I: 29, L: 30, S: 43; None has a majority, thus eliminate I. Votes: H: 69, L: 30, S: 43 None has a majority, thus eliminate L. Votes: H: 99, S: 43
    The hamster wins.

  d) H vs. I: I gets 1 pt. H vs. L: L gets 1 pt.
    H vs. S: H gets 1 pt. I vs. L: L gets 1 pt.
    I vs. S: I gets 1 pt. L vs. S: L gets 1 pt.
    The lemming wins with 3 points.

8. Plurality: Votes: W: 86, X: 52+28 = 80, Y: 60, Z: 58 W wins.

Borda count: W gets 594 points, X gets 760 points, Y gets 722 points, Z gets 764 points Z wins

Plurality with elimination: A majority out of 284 voters is 143 or more votes.

Votes: W: 86, X: 80, Y: 60, Z: 58

None has a majority, thus eliminate Z.

Votes: W: 86, X: 80+58 = 138, Y: 60

None has a majority, thus eliminate Y.

Votes: W: 86, X: 138+60 = 198 X wins.

8. Head-to-Head: When Y is compared to each of the others, Y is favored. Thus Y wins the head-to-head comparison.

Plurality, Borda count and Plurality with elimination each violate the head-to-head criterion. The pairwise method never violates the head-to-head criterion.

9. A majority out of 35 voters is 18 or more votes. El Capitan (E) has a majority.

However, the mule deer (M) wins using the Borda count method with 115 points. Thus the majority criterion is violated.

10. a) The standard divisor = $\dfrac{33000}{30} = 1100$

State	A	B	C	Total
Population	6933	9533	16534	33,000
Standard Quota	6.30	8.67	15.03	
Hamilton's Apportionment	6	9	15	30

b) The divisor 1040 was used.

State	A	B	C	Total
Population	6933	9533	16534	33,000
Modified Quota	6.67	9.17	15.90	
Jefferson's Apportionment (round down)	6	9	15	30

c) The divisor used was 1160

State	A	B	C	Total
Population	6933	9533	16534	33,000
Standard Quota	5.98	8.22	14.25	
Adams's Apportionment	6	9	15	30

d) The divisor 1100 was used.

State	A	B	C	Total
Population	6933	9533	16534	33,000
Modified Quota	6.30	8.67	15.03	
Webster's Apportionment	6	9	15	30

10. e) The new divisor 1064.52

State	A	B	C	Total
Population	6933	9533	16534	33,000
Standard Quota	6.51	8.96	15.53	
Hamilton's Apportionment	6	9	16	31

The Alabama paradox does not occur, since none of the states loses a seat.

f) The divisor $= \dfrac{33826}{30} = 1127.53$

State	A	B	C	Total
Population	7072	9724	17030	33,826
Standard Quota	5.98	8.62	15.10	
Hamilton's Apportionment	6	9	15	30

The Alabama paradox does not occur, since none of the states loses a seat.

g) The new divisor is $\dfrac{38100}{36} = 1058.33$

State	A	B	C	D	Total
Population	6933	9533	16534	5100	38100
Standard Quota	6.55	9.01	15.62	4.82	
Hamilton's Apportionment	6	9	16	5	36

The new states paradox does not occur, since none of the existing states loses a seat.

APPENDIX

GRAPH THEORY

1. A **vertex** is a designated point.
3. To determine whether a vertex is odd or even, count the number of edges attached to the vertex.
   If the number of edges is odd, the vertex is **odd**. If the number of edges is even, the vertex is **even**.

5. 5 vertices, 7 edges
7. 7 vertices, 11 edges
9. Each graph has the same number of edges from the corresponding vertices.
11. Odd vertices: *C, D*
    Even vertices: *A, B*
13. Yes. The figure has exactly two odd vertices, namely *C* and *D*. Therefore, the figure is traversable. You may start at *C* and end at *D*, or start at *D* and end at *C*.
15. Yes. The figure has no odd vertices. Therefore, the figure is traversable. You may start at any point and end where you started.
17. No. The figure has four odd vertices, namely *A, B, E,* and *F*. There are more than two odd vertices. Therefore, the figure is not traversable.
19. Yes. The figure has exactly two odd vertices, namely *A* and *C*. Therefore, the figure is traversable. You may start at *A* and end at *C*, or start at *C* and end at *A*.
21. a) 0 rooms have an odd number of doors.
    5 rooms have an even number of doors.
    b) Yes because the figure would have no odd vertices.
    c) Start in any room and end where you began. For example: *A* to *D* to *B* to *C* to *E* to *A*.
23. a) 2 rooms have an odd number of doors.
    4 rooms have an even number of doors.
    b) Yes because the figure would have exactly two odd vertices.
    c) Start at *B* and end at *F*, or start at *F* and end at *B*.
    For example: *B* to *C* to *F* to *E* to *D* to *A* to *B* to *E* to *F*
25. a) 4 rooms have an odd number of doors.
    1 room has an even number of doors.
    b) No because the figure would have more than two odd vertices.
27. a) 3 rooms have an odd number of doors.
    2 rooms have an even number of doors.
    b) No because the figure would have more than two odd vertices.
29. The door must be placed in room *D*. Adding a door to any other room would create two rooms with an odd number of vertices. You would then be unable to enter the building through the door marked "enter" and exit through the new door without going through a door at least twice.

31. Yes because the figure would have exactly two odd vertices. Begin at either the island on the left or on the right and end at the other island.

33.

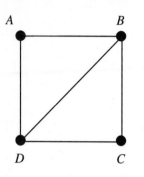

35. a) Kentucky, Virginia, North Carolina, Georgia, Alabama, Mississippi, Arkansas, Missouri
     b) Illinois, Arkansas, Tennessee

37. a) 4
     b) 4
     c) 11

39.

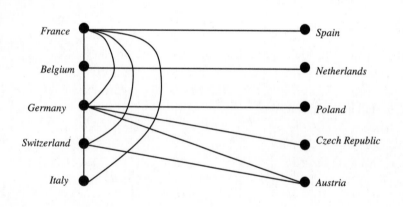

41. a) Yes, the graph has exactly two odd vertices, namely *C* and *G*.
     b) *C, A, B, E, F, D, G, C*